THE OBJECT OF LABOR

Martha Lampland

THE OBJECT OF LABOR

Commodification in Socialist Hungary

THE UNIVERSITY OF CHICAGO PRESS
Chicago & London

Martha Lampland is associate professor of sociology at the University of California, San Diego.

The University of Chicago Press, Chicago 60637
The University of Chicago Press, Ltd., London
© 1995 by The University of Chicago
All rights reserved. Published 1995
Printed in the United States of America
04 03 02 01 00 99 98 97 96 95   1 2 3 4 5
ISBN: 0-226-46829-1 (cloth)
        0-226-46830-5 (paper)

Library of Congress Cataloging-in-Publication Data

Lampland, Martha.
   The object of labor : commodification in socialist Hungary / Martha Lampland.
      p.   cm.
   Includes bibliographical references (p.   ) and index.
   1. Agricultural laborers—Hungary. 2. Farmers—Hungary.
3. Work ethic—Hungary. 4. Hungary—Rural conditions.
5. Collectivization of agriculture—Hungary. I. Title.
   HD 1536.H9L36   1995
   331.7'63'09439—dc20   95-11554
                                                              CIP

Maps on pp. 30–31 by Ellen O'Brien.
Photograph on cover and title page: Wheat harvest in the 1940s, courtesy Néprajzi Múzeum, Budapest.

∞ The paper used in this publication meets the minimum requirements of the American National Standard for Information Sciences—Permanence of Paper for Printed Library Materials, ANSI Z39.48-1984.

For my parents

Elizabeth Hedback Lampland and Donald Oscar Qualee Lampland

and the people of Sárosd

CONTENTS

	List of Tables	*ix*
	List of Figures	*xi*
	Linguistic Conventions	*xii*
	Acknowledgments	*xiii*
1	Introduction	1
2	Struggling for Possession	35
3	Severing Ties	109
4	Reforming Notions	167
5	Planning as Science, Planning as Art	233
6	The Space to Work	273
7	Conclusion	335
	Bibliography	*361*
	Index	*385*
	Gallery follows p. 232	

LIST OF TABLES

2.1 Social distribution and religious affiliation of families *38*
2.2 Distribution of landed properties in 1935, nationwide *39*
2.3 Manorial and peasant holdings in 1930, in Sárosd *41*
2.4 Distribution of lands in cultivation in 1930, in Sárosd *41*
2.5 Distribution of crops in acreage in 1930, in Sárosd *42*
2.6 Animals in 1930, in Sárosd *43*
2.7 Machinery in use in 1930, in Sárosd *50*
2.8 Craftsmen, merchants, officials, and professionals in 1927, in Sárosd *55*
2.9 Employment in three villages in 1927 *65*
2.10 Trend in the number of migrant workers, by county *68*
2.11 Personnel employed at manors surrounding Sárosd in 1930 *74*
2.12 Salary contracts at three manors in 1930, Sárosd *75*
2.13 Land reform in 1924 *93*
3.1 Number and size of landed properties appropriated in the course of the land reform (1945–1947) *120*
3.2 Grantees in the land reform (1945–1947) *121*
3.3 Distribution of new landowners, nationwide (1945–1947) *122*
3.4 Size and number of properties in Sárosd in 1949 *137*
3.5 Expansion of state property during the three-year plan *138*
3.6 Delivery quotas in Hungary for 1946–1948 *142*
3.7 Proportion of produce and animal products in relation to land tax, nationwide *143*

List of Tables

3.8 Increase in the number of cooperative farms, their membership, and their acreage (1948–1953) *145*

3.9 Distribution of plow land (1949–1953) *146*

3.10 Social background of Progress Cooperative Farm membership *148*

3.11 New members of Progress Cooperative Farm *148*

3.12 Number of families in Hungarian cooperatives *149*

3.13 Income of cooperative members compared to that of individual farmers, nationwide (1951–1955) *150*

3.14 Real income of the major population groups (1950–1956) *151*

3.15 Average monthly wages (1949, 1955) *151*

3.16 Drop in membership and acreage of cooperative farms (June 30, 1953–December 31, 1954) *158*

4.1 Changes in wages (1956–1957) *168*

4.2 Distribution of members according to the size of arable land contributed to the cooperative farm *177*

4.3 Number and size of cooperative and private farms between December 1956 and December 1958 *177*

4.4 Investments in agriculture according to financial-technical compositions and social sector (1960–1970) *202*

4.5 Distribution of tractors in Hungary, by social sector *202*

4.6 Percentage of mechanization of major agricultural tasks (1960–1970) *203*

4.7 Changes in numbers and proportions in work groups, nationwide *206*

4.8 Distribution of 1972 sample according to type of work group and length of membership in the cooperative *208*

4.9 Annual earnings in May 1st Cooperative, 1976 (forints) *209*

4.10 Original family social status of interviewees *210*

4.11 Agricultural tasks and their assigned work unit values *220*

6.1 Ancillary enterprises within agricultural cooperatives: number of units engaged in nonagricultural activity *284*

6.2 Distribution of households pursuing agricultural small-scale production *306*

6.3 Time-budget of Hungarians in 1977 and 1986 *307*

LIST OF FIGURES

1.1 Map of Hungary *30*
1.2 The village of Sárosd and surrounding manors *31*
4.1 Organizational structure of Harmony Farm (1960s) *227*
4.2 Organizational structure of Harmony Farm (1980s) *228*

LINGUISTIC CONVENTIONS

In the text, the plural for Hungarian words has been formed by using the English convention of adding an s, and possession by 's. Proper names, in accord with Hungarian usage, have been written with family name first and first name second (e.g., Bartók Béla or Jászi Oszkár).

Communists have changed the name of their party twice since 1945, first when absorbing their partners in the coalition front in 1948, and then following the 1956 revolution. Between 1945 and 1948, the Communists were members of the Hungarian Communist Party (Magyar Kommunista Párt). In the period between 1948 and 1956, the party was known as the Hungarian Workers' Party (Magyar Dolgozók Pártja). In the days when the 1956 revolution was being crushed, the party changed its name once again: to the Hungarian Socialist Worker's Party (Magyar Szocialista Munkáspárt). To convey the continuity between these different organizations, I have employed the term "Communist party" to refer to all three of these groups.

All translations which appear in the text, unless otherwise indicated, are my own.

ACKNOWLEDGMENTS

The object of my labor is this book. But to call this book my own would obscure the social world in which it was made. Friends and family, villagers here and gone, colleagues close to home and strangers far away have contributed substantially, giving me countless hours of their time, numerous ideas and many arguments, much needed money, and precious moral lessons.

I have been fortunate to have been provided generous monetary and institutional support throughout the various phases of researching and writing this book. I wish to thank the Dissertation Research Program of the International Research and Exchanges Board and the U.S. Department of Education's Fulbright-Hays Doctoral Dissertation Fellowship for their support of my fieldwork in Hungary. I am grateful that the Committee on Eastern Europe of the American Council of Learned Societies gave me a grant while I wrote the dissertation on which this book is based. I was the happy recipient of a Mellon Postdoctoral Fellowship at the Center for Russian and East European Studies at the University of Michigan, Ann Arbor. Over the last seven years, the Department of Sociology at the University of California, San Diego, has consistently supported my work, providing a comfortable home for an unrepentant anthropologist.

The Ethnographic Research Institute of the Hungarian Academy of Sciences was my research base while I conducted fieldwork; colleagues and staff assisted me in all manner of tasks. Sárkány Mihály was particularly helpful, guiding me early in my stay and helping to select a village site. Lukács László and his colleagues at the István Király Museum in Székesfehérvár offered an important local base during my stay in Sárosd. In fact, it was Laci's idea that I choose Sárosd for my field site, a choice I have never regretted. I have benefitted enormously from the attention and assistance given by Farkas Gábor and his colleagues at the Fejér County Archives.

Special thanks go to the Egyetértés Termelöszövetkezet of Sárosd, the cooperative farm where I conducted my research. From my very first visit, the membership—from president to night watchman—showed interest and enthusiasm for my project. I was welcomed into countless conversations and many homes. My questions were tolerated with patience and answered with care. People demanded to ask a few questions of their own, sparking debates and heated discussions. I also preyed upon others within the village of Sárosd who were not employed or associated with the cooperative farm. They were equally generous. I wish to thank the Gyimesi Károly family and the Jankovits István family in particular for permitting me to live in their homes; their hospitality is well remembered. I would also like to thank the following families for their generosity and kindness: the Bóda János family, the Imre Ferenc family, and the ifj. Jankovits István family. I cannot measure the contribution offered by Szarka Ferenc and his family to my well-being, and to my understanding of the community. Their home became my home, and my school, in the village.

I have been nurtured, supported, and sustained in this long effort by many friends in the academic village. Early on in this project I was encouraged and educated by discussions with Barney Cohn, Fernando Coronil, John Kelly, Gail Kligman, Nancy Munn, Rafael Sanchez, and Julie Skurski. Benedek Katalin, Biró Judit, Jávor Kata, and Szegedy-Maszák Mihály helped me find valuable sources and answer vexing questions. Susan Gal and Kit Woolard steered me away from some of my more foolish notions. Batta Imre, Halustyik Anna, Judit Hersko, and Tamás Anna kept my heart in place. For intellectual inspiration and persistent disagreements, I would like to thank Rick Biernacki, Michael Burawoy, Joanna Goven, Jeff Haydu, Éva Huseby-Darvas, Juhász Pál, Maria Teresa Koreck, Magyar Bálint, Ruth Mandel, Carole Nagengast, Jud Newborn, Ákos Róna-Tas, Iván Szelényi, Anna Szemere, and Christena Turner. I am especially grateful to John Comaroff for his abiding support and encouragement.

In the final stages of writing, valuable assistance was again at hand. The comments of several anonymous reviewers improved the manuscript. Trisha Stewart typed complex tables, a thankless task. Ellen O'Brien drew two beautiful maps. I am indebted to Kathy Mooney's skill and intelligence for an excellent index. The beautiful touches adorning the book bear the traces of Julia Robling-Griest's fine hand and keen eye. Finally, Carol Saller did an

Acknowledgments

excellent job of editing the manuscript. I appreciate her careful and considered reading of the text; the book is far better for it.

Special thanks are due Fern Barioni and Sally Rumbaugh for their infinite patience and generosity. I would also like to thank a large group of friends from San Diego, who shall remain anonymous. Their care and compassion has sustained me beyond measure.

I have the good fortune of belonging to a large and loving family. I wish to acknowledge in particular the support of my sisters Sarah Lampland and Susan Woodward. Finally, I wish to thank my mother who, among so many other things, taught me to write, and my father, who has always encouraged me to live up to the highest standards.

ONE

Introduction

The subject at hand is the object of labor. The moral purpose of village life. The privileged site of socialist transformation. The tangible property of a commodified world. The value of labor and its various manifestations form the substance of this inquiry. My object is to understand historical process, taking work as the location of this investigation. How are ideologies of work and value lived in particular relations of production and property, in specific social configurations of class and community; and moreover, how do these complex relations change over time? To this end, I have studied the collectivization of agriculture in one Hungarian village. I have sought to understand how patterns of work changed, how social communities reconstituted themselves, how forms of thought altered in the wake of collectivization. I was particularly curious about how social change occurs when it is mandated by national policy and thrust upon communities against their will. In the following account, I describe how economic and political practices have altered, how notions of labor, money, and time have been transformed in village life. To discover the specific role collectivization played in this process, I have chronicled the complex changes in economic and political practices—transformations in social action and cultural belief—which have taken place in rural Hungary since World War I. My argument, in brief, is that the process of commodifying labor has been fully realized under socialism in conditions thought to be inimical to capitalist development generally, and to commodification in particular.

While working in the village during the early 1980s, I was struck by the rampant individualism and utilitarianism of everyday life. Far from the proclaimed goals of socialist ideologues, villagers were preoccupied with achieving personal advantage, and consistently interpreted the actions of their fellow villagers in this light. Horatio Alger could easily have been teaching at the local school or preaching in the local pulpit, for individual initiative and responsibility were the preferred explanations for success or failure in the community. This bootstrap mentality would seem to have been alien to socialist principles of community action and social accountability. In contrast to the party's view that all realms of society should be subordinated to socialist ideology, villagers drew a stark distinction between public and private life in speech and in deed. Politics were dismissed as irrelevant and vacuous, while economic forces were understood to determine nearly every aspect of people's lives. This held true despite the party's control of all public institutions—economic and governmental—and despite the party's call for politically committed behavior on the part of Hungarian citizens. These entreaties, as all else the party demanded, fell on deaf ears.

All of these qualities—heightened individualism, utilitarianism, economism, a stark distinction between public and private spheres—were reminiscent of capitalism. Could it be possible that socialism had not influenced villagers at all? This hardly seems a plausible claim. Indeed, many villagers professed a strong belief in the goals of socialism—full employment, health care, maternity leave, old-age pensions—although they were quite critical of the way in which socialism was practiced by the Hungarian Communist party. Many of those employed in the agrarian sector had come to see the advantages of cooperative production, even though they had been thoroughly against collectivization when it was finally forced on them in the 1960s. The advantages villagers mentioned included the welfare benefits listed above which everyone associated with socialism. Some mentioned economies of scale for grain production, and most everyone lauded the elimination of drudgery and toil which they associated with private agricultural production prior to mechanization. But a crucial factor in villagers' support of cooperative production was the leeway given farm members to slough off on the job. It was almost impossible to force cooperative employees to exert themselves on the job, a freedom that they used to divert their energies into working hard and long in the private, second economy. Indeed, one need only consider the degree to which opposition to public socialist institutions was a central component of everyday life, visible most glaringly in the thriving second economy, to see that socialism had wrought profound effects. One possible explanation, then, for the sturdy capitalist attitudes and consistent capitalist practices of villagers was the continuation of prewar capitalist ideologies of work and morality, which had survived unscathed in private economic practices and the home lives of rural residents.

In light of the ethnographic literature on prewar rural communities, however, this claim was highly questionable. A survey of the ethnographic literature on Hungarian village life would reveal a consensus among scholars on the entrenched traditionalism of agrarian society. Even though migrant labor and day labor were common practices in the 1930s, exchange of labor was portrayed in the ethnographic literature as occurring primarily through extended kinship and neighborhood networks. Despite the fact that many peasants sold their produce for money at local markets, natural goods and products were depicted as the dominant means of payment. The terms "semifeudal" or "protofeudal" were frequently applied to these forms of exchanges, and to social life more generally (see, for example, Jávor 1978; Petánovics 1987). Other ethnographers were more attentive to the vast changes that had occurred in village life since the abolition of serfdom in 1848, speaking explicitly about the rise of capitalism during this period (Bell 1984; Fél and Hofer 1969; Sárkány 1978; Vasary 1987). They chronicled vast changes in marriage practices, inheritance patterns, and family size; they discussed shifting practices of crop rotation, animal husbandry, and tool use, and the rise of migrant labor in rural communities across the nation. All agreed that village communities in the interwar period bore little resemblance to those perched on the edge of feudalism's demise a mere seven decades earlier. However, none of these observers noted the substantial shift in the villagers' conceptual world. Most notably, the reconceptualization of labor as a tangible quality of personhood goes unremarked in these works. Although villagers may have been drowning in the inexorable tide of penury, these writers would claim that they continued to embrace an absolute commitment to an ideology of land possession and self-sufficient farming (Bell 1984; Sárkány 1978; Vasary 1987). Meanwhile, slight notice is taken of alternative conceptions of identity, such as the integrity of owning one's labor, which structured local hierarchies of privilege and morality. Thus, while these authors discuss the extensive changes in agricultural production or shifts in the dimensions of social community, they do not consider the conceptual or ideological transformations that accompanied these complex and quite far-reaching changes.

The representation of village life as "traditional" prior to socialism was also premised on the notion that rural settlements were isolated from other communities and national agencies, most notably the state. It would be easy to assume, listening to Hungarians in the 1980s or reading much of the literature on Communist party rule in Hungary and Eastern Europe, that the state did not play an active role in village life prior to 1945. Perhaps more remarkably, the Marxist-Leninist states of Eastern Europe have long been regarded as unique social experiments. However, the process of state-sponsored social change is neither new nor unusual. Quite well known are the modernizing projects of European states in the nineteenth century, pursued either at home in Europe or in colonial contexts abroad.[1] Explaining the radical restructuring of social forms which was envisioned in these imperial designs and enacted with quite perplexing consequences occupies the pages of much modern history and social science. However, what was unique in the social programs of many socialist states was the integral role of the collectivization of agricultural properties in the national project of industrial modernization.[2]

Although collectivization was an innovation unique to socialist policies, other sorts of state policies—forced deliveries, strict labor policies, and grand projects restructuring the rural landscape—had existed prior to the socialist period, some starting as early as the mid-eighteenth century. The modernizing project of the Hungarian state in the nineteenth century entailed the regular and quite intrusive intervention of state authorities in local relations of politics and production. Policies of regressive taxation, strict labor contracts for agrarian workers, and punitive practices waged against labor unrest had already been fully institutionalized prior to World War I.

The view that the state did not intervene in rural life has been reinforced by ethnographic studies. The most significant political body mentioned in these works is the local list of prominent taxpayers (*virilisták*) (Bell 1984; Sárkány 1978; Vasary 1987). But we know full well that during the interwar period, for example, the state maintained a close watch over the activities of labor recruiters, fearing agitation among the poor by radical leftist elements. The Ministry of the Interior regularly reviewed the by-laws of organizations to ensure that local groups not abrogate national laws concerning rights of association. These policies continued the venerable nineteenth-century tradition of state supervision of local

---

[1] For examples of recent approaches to these questions in the non-European context, see Cockcroft 1983; Comaroff and Comaroff 1991; Cooper 1992, 1993; Cooper and Stoler 1989; Guha and Spivak 1988; Ludden 1992; Sanderson 1984; Trouillot 1989.

[2] Among European socialist states, Poland and Yugoslavia did not collectivize agricultural properties; all other nations in the region did.

morality and politics. Yet the state also introduced new policies during the twentieth century. These include the introduction of paramilitary organizations (such as the *Levente*), organized by the centralized authorities in local communities. Paramilitary organizations were a significant force for instilling conservative nationalist and often irredentist sentiments in villagers during the interwar period. Another significant innovation was the policy of forced taxation in foodstuffs. Generally associated with Stalinist practices, forced taxation had already been introduced during the Second World War, at the height of the war economy.

Clearly, neither proposition—that socialism had no substantial impact on social life, or that Hungarian rural life had been untouched by capitalist practices prior to the war—is tenable. The question to answer nonetheless remains: how had socialist institutions and practices altered the character of work and attitudes toward work in village life? How could it be that the imposition of collective property relations by the state and repeated party campaigns promoting socialist ethics had produced such atomized, individualist, utilitarian attitudes toward work, property, money, and morality? To answer the question about the socialist state's impact on the world of local labor, it was necessary to know more about social life during the interwar period. Was it, in fact, a world of semifeudal peasants who were impervious to capitalist markets in labor and objects? Or in contrast to the vision offered of a community timelessly traditional, was rural life excruciatingly capitalistic in its character? If capitalism had developed in the countryside, what were its distinctive features, and how did they differ from the individualism and economism of the socialist period?

I will argue that the transition to capitalism in Hungary, though well underway by the 1930s, was significantly furthered by policies of the socialist state. The full commodification of labor, the final blossoming of commodity fetishism, was achieved by the policies and practices of Hungarian socialism. This seems a curious conclusion. After all, socialist economies were planned, deliberately eliminating, or at least restricting, market activities within the nation. Some would argue that markets—in labor or in commodities—were so limited by socialist policies that they cannot be compared to the presumably robust markets of capitalist societies. And so, the logic goes, commodification could not occur in such circumstances. I, on the other hand, would point to the increase in the possibilities of selling one's labor or the products of one's labor in later socialism as crucial for understanding many of the social practices and cultural attitudes toward work and morality I wish to chronicle in village life.

The growth of markets in labor and the strengthening of consumption in the 1970s and 1980s, however, seem to me an insufficient explanation for the substantial shift in attitudes toward work and economic practices discernible in village communities. Indeed, I wish to argue that the assumption that commodification is a process bred and fostered solely through markets is problematic. To wit, the complex history of commodification in other regions of Europe entailed far more than the simple growth of markets in labor and commodities. The transition to capitalism also entailed the development of quite complex organizational bodies, such as factories, commercial enterprises, and state bureaucracies serving the burgeoning economic sector. So, too, did these shifts affect the composition of home life, redefining the economic, political, and sexual activities of families and friends in the now private sphere. Substantial changes in everyday practice accompanied the introduction of these new forms, changes that we can identify in Hungary during the socialist period. Some scholars may assume that these complex institutional, organizational innovations accompany, or perhaps even precede, the growth of formally capitalist markets in labor (see Kriedte, Medick, and Schlumbohm 1981). But in many discussions, markets stand alone, occupying an undisputed position in the debate. My purpose here is to unpack this often conflated whole, to make the wide range of institutions and practices more visible, and to emphasize their relative independence from markets per se.[3]

Other aspects of socialist economic practice besides markets thus warrant attention, specifically the broad range of practices begun with industrialization and the establishment of collectivized agricultural production. Collectivization and industrialization during the socialist period substantially altered the work experience of most adult Hungarians, who were now regularly employed in formal, and often complex enterprises. The structural features of state employment included: the elaboration of hierarchical organizations of modern management, the setting of temporal boundaries to the workday, the tying of wages to tasks or to hourly segments throughout the workday, the acquiring of welfare benefits

---

[3] The complex history of markets—both as an empirical event and in social science theorizing—has come under renewed scrutiny. Some recent work on the character of markets has attempted to examine the culturally embedded character of market practices (Haskell and Teichgraeber III 1993), while others have focused on the concept of the market itself in social analysis (Dilley 1992).

through one's work contract, and the introduction of trade unions (though in the absence of effective representation of worker interests within the enterprise). Granted, not all of these practices were first introduced by the socialist state. Some were already in place, certainly in the rudimentary industrial sector and to some extent in agricultural work, before World War II. Daily monetary wages had been common for agricultural day laborers, but wages had not been calibrated by the hour. Prior to the war, Hungarian villagers worked from sunrise to sunset, seeing the task through to completion. Manorial workers and migrant laborers had stipulations in their contracts concerning when and for how long they could rest through the day, but their breaks were judged by the position of the sun, not by a clock or timepiece. Nor had wages been calculated according to levels of productivity, as would become the case with work units or piece rates in socialism. Sharecropping contracts stipulated a set proportion of one's crop (e.g., one-third to one-half) to be turned over to the land's owner; natural conditions determined production levels, not a norm setter committed to the abstract calculation of necessary returns. Workers at manorial estates had bosses watching over them, but not the hordes of middle and upper management who would come to inhabit socialist industrial complexes, and later, cooperative farm offices. Accompanying the institutional transformation under socialism were important changes in the economic and political boundaries of the family: the delineation of public and private work, and a new politics intervening in the constitution and character of family lives.

These innovations in social life deserve greater attention, to displace the unfortunate preoccupation with the relative freedom of markets in socialism as an indication of the degree or possibility of commodification, or as the singular feature shared by capitalist and socialist economies. In short, I firmly believe that much of socialist experience has been structured by practices that bear a striking resemblance to those of capitalist political economy. This accounts for the strange consequence of commodification under socialism.

Commodification of Labor

What do I mean by commodification of labor, or commodity fetishism? Marx developed these terms in his classic study of English capitalism (1967, 1973; see also Elson 1979; Kay 1979; Postone 1993). Marx's crucial insight into the fetishism of commodities was to suggest that in capitalism, a curious displacement occurred, in which social relations came to be seen as objects, and objects were accorded the qualities of persons. In capitalism, Marx argued, the complex social relations that produce objects for consumption were obscured by the elaborate activities that surrounded the exchange of those objects. Markets were described by scholars and workers alike as active players in the economic domain, producing effects and responding to historical events, even though, in Marx's eyes, markets were clearly a series of actions taken by human beings. Workers, on the other hand, who were immediately responsible for producing the objects circulated by the market, were referred to in the same terms as the products they made: as objects to be moved around, as an additional tool in the capitalist's machine shop. Marx found this representation of the working class both personally offensive and analytically skewed. To offer an alternative perspective on capitalist society, Marx engaged in an analysis of factory production and market relations to reveal the magical qualities of capitalist fetishization.[4]

Commodities may be defined simply as objects that may be sold at the market. In Marx's perspective, however, a strange series of consequences results from selling labor at the market as one would wheat or clothing. This historical shift, in which labor is conceptualized as another object for sale at the local town square—indeed, in which labor itself acquires an objective quality akin to other objects at the town square—was central to the revolutionary transformation of capitalist society, breeding new forms of social subjectivity, intentionality, and radical reformulations of space and time (Lukács 1971). The rise of commodity fetishism is manifested in the development of labor as an abstract category of social value and a concrete physical property of social actors.

> What renders the fabric of that underlying social structure [of capitalism] so peculiar, for Marx, is that it is constituted by labor, by the historically specific quality of labor in capitalism. Hence, the social relations specific to, and

[4] Needless to say, a century of debate has swirled around questions of value, labor, and prices in Marx's theory, most notably the labor theory of value and the so-called transformation problem. These debates have engaged economists who are eager to understand, confirm, or refute their application to capitalism (e.g., Böhm-Bawerk 1949; Mandel and Freeman 1984; Meek 1956, 1976; Morishima 1974; Morishima and Catephores 1976; Samuelson 1971; Sraffa 1960), as well as socialist theorists and planners who are concerned about implementing an alternative framework to price theory in capitalism (see Lewin 1974; Nove 1983). However, one need not participate in the convoluted arguments surrounding Marx's theories of prices or value to agree with his representation of the pervasive misplaced concreteness in the social world of capitalism.

characteristic of, capitalism exist only in the medium of labor. Since labor is an activity that necessarily objectifies itself in products, commodity-determined labor's function as a socially mediating activity is inextricably intertwined with the act of objectification: commodity-producing labor, in the process of objectifying itself as concrete labor in particular use values, also objectifies itself as abstract labor in social relations. (Postone 1993, 153)

This process of objectifying labor is captured succinctly in the Hungarian notion of *dolog*, which in the span of one hundred years shifts in connotation from defining the *process* of activity, such as harvest labor, to referring to activity *as an object*.[5] And as an object, one's labor could be alienated from one's person to be sold to others. This transformation is intimately tied to changes in the assessment and calculation of value, of effort and time in production, and of changes in the constitution of pride and of honor in rural communities (Lampland, n.d.).[6] In other words, the history of commodifying labor in Hungarian rural communities illustrates in very explicit form how labor becomes both a tangible and quite finite object, and at the same time, the most abstract form of social value and generalized category of cultural estimation and esteem.

Thus commodity fetishism entails far more than the appearance of wage labor per se. We know of cases of wage labor in Eastern Europe from as early as the sixteenth and eighteenth centuries (Kula 1976; Pach 1961), though these societies were definitely feudal in character.[7] Many of the concepts and practices of com-

---

[5] *Dolog* is a very complex term, and has a variety of meanings, including "work, task, obligation, affair, life circumstances, fate, conceptual or imaginary object, tangible object, belongings, tool" (Benkö 1967, 1: 657). In the ensuing discussion, I wish to emphasize the historical shift in usage discernible in agrarian contexts: once referring to the activity of working, *dolog* eventually refers both to activity and object simultaneously.

[6] Biernacki's analysis of the woolen textile industry in Germany and Britain reveals comparable notions of labor and time as object, as a substantial quality. "British employers and workers supposed that labor could be literally embodied in a product that served as a vessel to transfer it—as if human activity could become a substance. . . . The German specification of labor's commodity form supposed that the employer could purchase time itself, . . . that one [could] purchase the disposition over workers' time as a 'thing'" (1995, 384–385).

[7] It is often assumed that monetary payment replaces payment in kind, and that such a shift indicates the transition to capitalism. Interestingly, in Hungary at the end of the sixteenth century, impoverished peasants were paid in money, complementing the labor of serfs on manorial estates (Pach 1961, 4). Later manor owners reverted entirely to employing servile labor. Hence there was no simple progression from in-kind payment to monetary instruments. The presumably 'natural' progression from servile

modity fetishism—new categories of time, altered use of money, shifting practices of labor, the provocative manipulation of wealth—are discernible in the early modern period. But it is the interrelated constellation of these ideas and actions which issues in the eventual social world defined by Marx as commodity fetishism. Capitalism is, indeed, a sum greater than its constituent parts.[8] For clocks, money, markets, or even explicitly formulated categories of economic wealth and value long predate the ascendance of commodity fetishism in Europe. In the fourteenth to fifteenth centuries in Italy we find new units of time correlated with daily tasks, forged in battles over the length of the workday by urban artisans and developed in the writings of new humanist thinkers (Le Goff 1980). The dynamics of price movements, markets, and money in late feudalism have fascinated modern scholars, and prompted quite intense debates (Abel 1980; Day 1987; Le Roy Ladurie and Goy 1982; Postan 1973). Economic thinkers of the fifteenth to eighteenth centuries developed quite extensive theories of the source of value, its production, and its circulation. Indeed, the writings of the British, and particularly Scottish, school, form the cornerstone of modern political economy (Tribe 1978). A crucial development in a shift toward a concept of labor as commodity is demonstrated in Biernacki's careful analysis of formal political economy writings in Britain and Germany, where he identifies the rise of a notion in which "labor becomes the metric measure of value."[9] This lengthy process of scholarly debate and political struggle culminates in quite radical concepts of economic value and revolutionary practices of production. Commodity fetishism is the end result.

---

labor to wage labor in Eastern Europe is complicated by the history of the second serfdom, when many changes that had paralleled the demise of feudalism in Western Europe were reversed, subjugating serfs to much more onerous feudal duties than in the preceding two or three centuries (Ashton and Philpin 1985; Blum 1957, 1978; Brenner 1976; see also Lampland, n.d.).

[8] Marx was not alone in drawing attention to the systematicity of capitalist modernity, or to the need to analyze social forms as a unique totality. Durkheim's famous call to study society sui generis echoes these sentiments. While it is often read as a general exhortation to resist reductionism in social analysis, Durkheim's position should rather be interpreted as a commentary on the systemic properties of capitalism, which alerts us once more to the intimate relation between theories of society that were bred in the nineteenth century and the particular character of the societies under investigation here.

[9] Biernacki's recent comparison of wool factories in England and Germany is an exemplar of the type of close analysis of the various components of capitalist industrial production which illuminates the cultural project of commodification.

Introduction

The breaking up of the day into discrete segments, the manifestation of value in the objects of sale (be they products or labor), the measuring and calculation of time and activity all signal a shift in social activity and social thought. But these significant developments in European economic history do not in and of themselves form the fully constituted world of commodity fetishism. How may we distinguish these discrete social projects from commodity fetishism? I define commodification in the classic sense as the conflation of labor's objectification in particular acts of production with its more general status as the source and arbiter of value: that is, I would emphasize the strange parallel of labor's acquiring a concrete, physical character while it takes on the general or dominant role in structuring social action and creating cultural value (Marx 1967; Lukács 1971).

Another way of stating my position is to say that labor becomes a category of economic life, social action, and cultural speculation only with the arrival of capitalism. This appears to many to be an absurd claim, since we are well aware of the role of production, of labor, of work in societies across the globe that are far from capitalist in character. I am not suggesting that people do not work (in the most general sense) outside capitalism. Rather, following Marx's analysis (and reinforced in my analysis by Postone's recent reinterpretation of Marx), I am arguing that labor is a category unique to capitalism. In capitalism, labor is the touchstone of social life: it is a material property of human actors, bearing physical, nearly tangible qualities. It is also the touchstone, the foundation, of subjectivity and morality.[10] Labor does not occupy such a position outside capitalism.

> The historically specific "essence" of capitalism appears in the immanent analysis as a physiological, ontological essence, a form that is common to all societies: "labor." The category of abstract labor presented by Marx is thus an initial determination of what he explicates with his no-

---

[10] "In capitalist society alone could a concept of labor serve as the organizing principle for a multiplicity of humble practices. Where labor has not been subsumed under the commodity form, labor may be recognized as the source of material sustenance, but it does not take on the social function of structuring the relation of person to person through the exchange of abstract labor time. Definitions of labor may not surface at all in kin-based or precapitalist societies as a principle for structuring social relations; should they arise, they remain subordinate to other categories coordinating social reproduction. Only in capitalist society is labor both a form of understanding and the integrative principle that regulates social relations in society as a whole; only there does it bridge lived experience and the invisible functioning of a system" (Biernacki 1995, 208).

tion of the fetish: because the underlying relations of capitalism are mediated by labor, hence are objectified, they appear not to be historically specific and social but transhistorically valid and ontologically grounded forms. The appearance of labor's mediational character in capitalism as physiological labor is the fundamental core of the fetish of capitalism. (Postone 1993, 170)

The theoretical power of Marx's formulation of fetishism is clear. It also provides an astonishingly accurate description of Hungarian agrarian history, where the literally concrete form of labor is so striking. Moreover, this tangible quality arises in Hungarian social thought precisely as the character of work comes to be the central, dominant means of identifying social value within the community, value understood here not only as wealth, but also as honor and respect.

In the final analysis, it is important to recognize the fundamental shift in the character of social relationality which arises with capitalism. The character of domination changes, deflecting our comprehension of the structural features of the economy and obscuring the politics of inequality. "In Marx's analysis, social domination in capitalism does not, on its most fundamental level, consist in the domination of people by other people, but in the domination of people by abstract social structures that people themselves constitute" (Postone 1993, 30). This dispersal of social responsibility onto a system accounts for another aspect of fetishization, that is, the naturalization of the social world (J. Comaroff and J. L. Comaroff 1991; Taussig 1987). The categories of capitalism represent themselves as transhistorical, natural, given. The inchoate quality of oppression that deprives people of a sense of their central role in production reinforces social inequalities. The dynamics of class politics are much more difficult to apprehend, while other differences—such as race, gender, and sexuality—are defined as biologically fixed and so appear unassailable on social terms.[11]

The impersonal domination of capitalism, in which obscure market forces prevail, is replaced in socialism by the highly personalized character of party rule. The impersonal market is eliminated, replaced by the inspired leadership of the party, whose omnipotent knowledge makes planning the national economy possi-

---

[11] Laqueur's analysis of the rise of biological explanations for social hierarchy within the natural sciences in Europe during the eighteenth century (1990) and Jordanova's work on science and medicine (1989) provide important evidence for this shift in social thinking.

ble. This accounts, as some people have argued, for the "transparency of class" (Burawoy 1985, 195) and so for the heightened class consciousness of socialist workers. Yet the party's representation of its almighty abilities falls short of an accurate rendering of social forces that structure action and belief. No matter how personal are the bonds of Communist party rule, the lives of Hungarians in the late socialist period are nonetheless dominated by the abstract qualities of utility, time, and money. Strangely enough, villagers who laud the goals of socialism find the entire project contrary to human nature, and the same people who complain about the injustices of markets and capitalism consider them a necessary evil. Such convoluted logics of morality, justice, and necessity speak volumes about the consequences and contradictions of a history of socialism in which commodity fetishism has taken hold. It is for this reason that we must attend more directly to the qualities of naturalization, objectification, and systematicity that constitute the particular dynamics of social relations and social thought in a commodified world, be it socialist or capitalist.

Commodity fetishism thus connotes a radical and quite grand restructuring of the social universe. And precisely because of the momentous character of this transformation, the transition often takes generations to achieve. The argument put forward in this book rests on the assumption that capitalism does not spring, like Athena, fully developed from its historical sire, feudalism. On the contrary, as an extended historical project, capitalism develops in subtle and complex ways over time.[12] However, the challenge is to capture this process historically: to provide a clear picture of changes in social form and social thought which contribute to a broad transformation eventually to be defined as capitalism. It is this complex process I wish to present: the history of the rise of commodity fetishism since World War I.[13]

The history of Hungarian agricultural labor over the last 150 years provides an exemplary illustration of the long process of commodifying labor, a process often dated to the abolition of serfdom in 1848. In the first two decades following the elimination of servile duties, few changes in agricultural production could be dis-

---

[12] My model for this argument is Sider's splendid book, *Culture and Class in Anthropology and History* (1986), in which he offers a subtle analysis of the shift from merchant capitalism to family fishing in Newfoundland.

[13] The transition from feudalism to capitalism during the eighteenth and nineteenth centuries, the earlier period in which the objectification of labor begins, is the subject of my forthcoming book, *Transforming Objects: The Transition from Feudalism to Capitalism in Hungary during the Nineteenth Century*.

cerned. Peasants devoted little time to farming, and shunned any overture from aristocrats to work on their large estates. But by the early decades of the twentieth century, a transformation of agricultural practices and agrarian ideologies was underway. Across the nation, people were working from sunrise to sunset, intensifying agricultural production as much as was humanly possible. Labor had become the yardstick used to measure the quality of work and the integrity of workers. People's actions were determined in increasingly significant ways by the intensity of their labor and their moral commitment to diligence. Not incidentally, this notion of labor as a property of action and a moral indication of character had become quite widespread at a time when impoverished peasants identified their economic and political interests as separate from those of manor owners and wealthier peasants, contributing to the quite substantial agrarian socialist movement of the 1890s. This process signaled an important shift away from a feudal universe, in which relations of service defined social action and moral character. One might say, then, that capitalism had begun to develop by the turn of the century in Hungary.

At the beginning of the present narrative in the 1920s, villagers were quite familiar with wage labor, either having contracted for daily wages at the manor or peasant farms, or having hired farmworkers themselves to help out the family in the harvest. They did not think of their labor primarily in monetary terms, however, nor did they understand work to be bounded by the fine divisions of clock-time. Nonetheless, labor was seen to be a property of their persons, which more and more had to be sold to others to make a living. Diligence was a central means by which moral character and personal worth were evaluated. This combination of a moral vision of diligence with the objectification of the activity of labor suggests that capitalism had taken hold, though the particular character of social relations was far from the world of objectification, naturalization, and systematicity—structured around the axes of labor, time, and money—which would come into being during socialism.

In the course of shifting relations of production and exchange after the war, labor came to be equated with time and money. Activity was carved into temporal segments and assigned a monetary value. New notions of intentionality were crafted. People considered the utility of their actions and reasoned through a calculus of meaning and value. These altered forms of intentionality and purpose arose as conceptual analogues to the equivalencies of the once disparate social categories of time, labor, space, and money. Arrows of utility—relations among relations—had come to be seen as the sole motive of behavior and so its ultimate explanation.[14] In light of this brief historical description, I will designate the interwar period as capitalist, though I will also distinguish it from the fully realized commodity fetishism that comes into being during the late socialist period. Simply put, the "labor property" world of the interwar period is to be distinguished from the "relational equation" of commodity fetishism discernible in 1980s Hungary. This simple description illustrates the historical argument that is at the heart of this analysis: capitalism—as social project and cultural form—changes over time.

In the last few decades, social history has flourished in debates concerning the history of European capitalism. A number of provocative works have appeared which marry theoretical rigor with careful empirical analysis. These studies are fiercely committed to theoretical generalization; nonetheless, they attend closely to the diversity of cultural and regional experience, illustrating curious peculiarities within European history (see, for example, Ashton and Philpin 1985; Blackbourn and Eley 1984; Corrigan and Sayer 1985; Hobsbawm 1964; Sewell 1980; Thompson 1963; Wood 1991). Such studies are complemented by the wealth of studies examining the rise of capitalism and class politics in non-European contexts (e.g., Cooper 1980; Garon 1987; Gordon 1985; Guha 1983; Katznelson 1981; Mallon 1983; Marks and Rathbone 1982; Roseberry 1983; Sider 1986; Stoler 1985; van Onselen 1982). Studies of the dynamics of gender, sexuality, and race as crucial aspects of the culture of capitalism within and beyond Europe have further enlightened our understanding of the quite diverse and varied character of capitalist history (Chatterjee 1989; Fredrickson 1981; Mosse 1985; Packard 1989; Stansell 1982; Stoler 1989; Tilly and Scott 1978). These significant contributions to social theory and social history lead us to the conclusion that capitalism as a complex historical form is a cultural as well as socioeconomic project. The idiosyncrasies of capitalism in Europe bred eccentricities too, the most notable of which was socialism. The history of socialism—as theory and as prac-

---

[14] The phrase "relations among relations" is a nod to Saussurean, and later structuralist, theories of meaning within the field of anthropology. It is mentioned here in order to draw the parallel that can be found between anthropological theories of meaning and concepts of utility and value which inhabit the social world of Hungarian peasants. I have discussed the interesting similarities between the construction of value in capitalist commodification and theories of meaning in anthropology—highlighting the strongly European history of social theory and commodification—in an earlier article (1991).

tice—is a direct product of the cultural dilemmas and socioeconomic contradictions of European capitalism. In Hungary, as elsewhere, socialism confronted, and eventually transformed, the social ills and cultural fortunes it inherited from the past.

The Peculiarities of Socialism

I consider the commodification of labor to be the salient feature of Hungarian village economy in the late socialist period. This raises a series of questions about the relation between capitalism and socialism, a topic that has long been the subject of serious debate among social scientists. From Trotsky's argument on state capitalism to Djilas's class analysis, from modernization theorists to the new institutionalists, social analysts have long argued over the degree of similarity and difference between the policies, politics, and production of socialist and capitalist states. I wish to situate my work within this larger debate, to clarify the sometimes stark, often subtle differences that distinguish my approach from that of several schools of thought or prominent theorists in this field.

My overriding concern throughout this project has been to understand the dynamics of social history. Three key assumptions form the bedrock of my investigation. The first is that social change takes time. Change happens incrementally; everyday life has consequences. Accordingly, social process must be analyzed carefully, to understand how change actually occurs over time. The second assumption is that social history is necessarily cultural history. Social communities are distinguished by their histories, subtly and profoundly. Formulated somewhat differently, economics and politics are always cultural forms, and must be analyzed as such.

The third assumption I hold is that society is the product of social actors. Social community takes form through the laborious projects of everyday experience. How people understand their world and hence create their communities are central components of a rigorous political economy. Hence studies of socialism which neglect the conceptual world of socialist citizens will necessarily fail in understanding why socialism takes the form it does in various places and at various times. So, too, the tendency to see socialism as a superficial and failed attempt to alter the basic contours of society seriously misconstrues the manner in which social actors build their world. Hegemonic forms, though often quite resilient, are nonetheless historical products. Thus, the ongoing struggles over morality and meaning in socialism have irreparably altered social thought and cultural practice. Allow me to elaborate these thoughts by drawing a road map through, between, and around the theorists and schools who have dominated recent debates over socialist societies.

If capitalist attitudes have been bred by socialist policies and practices, may Hungarian socialism be most accurately described as state capitalism? I think not. Arguments in this vein (following Trotsky's seminal piece on Bolshevik abuses of power) assume that Marxist-Leninist parties continue to run factory enterprises and other state institutions as had their bourgeois predecessors (Bettelheim 1976, 1978; Mandel 1978; Sweezy and Bettelheim 1971; Trotsky 1937).[15] This assumption seriously misrepresents the character of industrial and agricultural production in socialist societies. I have emphasized what I see to be similarities between capitalist firms and socialist enterprises, including formal organizational structures with managerial hierarchies, wage packages, work shifts, and limited influence of the rank and file over managerial decision-making. In itself, such a constellation of factors could be easily subsumed under classic modernization theories, which indeed have characterized some approaches to socialist political economy (see, for example, Johnson 1970 and a critique of this approach in Stark and Nee 1989).[16] As Stark and Nee point out in their critique of modernization approaches to socialist societies, this sort of analysis bears an eerie resemblance to reductionist models of Marxism which privilege technological imperatives as the driving force of historical change (1989, 6). The unfortunate consequence of such approaches is that they obscure the social process of economic activity, ignoring important differences in economic decision-making, such as planning, which have a profound effect on the organizational practices of enterprises. Furthermore, modernization theories, bound to a particular trajectory of historical necessity, are profoundly ahistorical. The dynamic of technological growth is self-contained, impervious to local cultures of

---

[15] Corrigan, Ramsay, and Sayer's critique of Bolshevism is a good example of this approach. "[T]he Bolsheviks continued to adhere to that impoverished notion of what production is, and what is entailed in developing it, which we met in the Second International's 'theory of productive forces,' and . . . this in turn eventuated in the separation of politics from production, and thus an effective fracturing of the spiral of productive and political liberation fundamental to a socialist programme . . . In short, and unsurprisingly, to foster capitalist forms of productive activity eventuates in the reproduction of various defining relations of the bourgeois Stateform that is their condition and consequence." (1978, 39, 46)

[16] Another variation on this theme is convergence theory, which posits an increasing similarity between capitalist and socialist regimes over time (Tinbergen 1961; see also Ellman 1980).

economic history and politics. Such analyses do not look to the particular histories of societies prior to the onset of modernization which may account for differences. Moreover, they are not attuned to the local processes of implementing socialism. Any divergence from the classic path to modernity is seen to be an exception, or a curious mix of modern and traditional elements.

The concept of neotraditionalism was introduced by Jowitt (1983) to illuminate the singularity of Soviet politics and has been further developed by Walder in his study of Chinese industry (1986). The emphasis in this approach is the political character of organizations in Soviet or socialist economies. Jowitt argued that Leninism represents an alternative to capitalist modernity, forging features of both modern and traditional societies into "a novel form of charismatic organization" (1983, 277). Jowitt's approach is focused explicitly on the particular character of Leninist institutions, and the quandary of a revolutionary party's routinization in solving the tasks of the nation's political economy. Walder's analysis is situated more directly within industry, studying (from afar) the authority relations in factory settings in order to speculate on the character of wider political relations within society. Joining Jowitt in rejecting convergence theories, Walder argues for a particular route for socialist economies, identifying "an evolution toward a historically new system of institutionalized clientelism" (1986, 8). In both cases, I find refreshing the attention to how party organizations are lived in practice. However, the tendency to contrast the formal rules employed in capitalist bureaucracies with the strongly personalized, informal practices of socialist bureaucracies is unfortunate. This opposition is exaggerated, in my opinion. It employs an idealized view of capitalism, ignoring the significant role of informal practices in capitalist bureaucracies. Moreover, neotraditionalism attempts to explain socialist forms of interpersonalism as a mixture of tradition and modernity. This construction of socialist politics and economics suffers from categories of tradition and modernity which are essentially ahistorical, flattening out the complex history of socialism and what precedes it in ways inimical to serious historical analysis. This cripples not only our understanding of the longue durée of social transformation; it also impedes a clear understanding of recent socialist experience.

The approach to the study of socialist economy and politics which Stark and Nee have termed the new institutionalism (1989) has been an important innovation in the study of socialist economies, offsetting the unfortunate homogenizing views of modernization theory by clarifying the *particular* dynamics of socialist industrial production. This approach has been especially fruitful in the study of Hungarian socialism (Fazekas 1984; Galasi 1982; Haraszti 1978; Héthy and Makó 1974, 1978; Kertesi and Sziráczki 1985; Makó 1985; Nee and Stark 1989; Simonyi 1983; Stark 1986, 1989). Studies of internal labor markets and the politics of the shop floor have illuminated important qualities of industrial management and worker negotiation in socialism, tying these dynamics to phenomena—such as plan fulfillment, shortage, and storming—which are distinctive features of socialist political economy. The complex dynamics of industrial communities are also significantly influenced by the second economy, which although not a distinctive feature of socialism, has an important history in the character of socialist economics.[17] While the attempt by new institutionalists to offer an analysis of the particularity of socialist institutions may share similarities with those promoting neotraditionalism, the new institutionalists differ by coming to these questions from the perspective of society at large. Rather than study the party and state elites, who occupy a privileged role in neotraditionalism, these analysts start from the shop floor, wander about in backyards and makeshift shops, and visit short-lived efforts and long-term communities forged in work and in resistance to party and bureaucracy (Stark and Nee 1989, 30).[18]

The particular histories of socialism derive as centrally from the ideological platforms of Marxist-Leninists as from the institutional structures that arise from their manifestos. A commitment to planning the national economy, promoted to eliminate the pernicious effects of capitalist market economies, gives birth to new dynamics within the economy at large (Kornai 1992), and has a significant impact on the organization of labor and the consciousness of workers within socialist enterprises (Burawoy and Lukács 1992). The party was committed to the growth of the economy, waging an all-out effort toward rapid industrialization and modernization. Certain sectors of the economy were privileged in this quest, lead-

[17] The new institutionalism has not been employed as commonly in the study of agrarian enterprises (Nee's work in China is an exception: 1989, 1991a). Ironically, however, many of the innovative strategies of enticing labor and maintaining productivity in 1980s Hungary studied by those associated with new institutionalism were techniques drawn from reforms in cooperative farming.

[18] Lisa Rofel's analysis of factory discipline and spatial reorganization in China provides an important model for studying the *process* of socialism, in that she remains attentive always to the historical and cultural specificity of these forms. I consider her study to be a provocative approach to Chinese socialist practice, but I am also sympathetic to her clear articulation of the principle of the always local quality of modernity (1992).

ing to sectoral distortions that had adverse consequences for agriculture, commerce, and the service sector. As a general principle, the character of growth promoted by the party was a simple one: brute increase over time. Enterprises were evaluated in terms of quantitative gross output, with little attention paid to the costs of production (Swain 1992, 71). Jowitt refers to this conception as "an 'arithmetical' conception of the economy [which] is consistent with a traditional ontology that stresses the discrete and physical quality of social reality" (1983, 280). This conception of socialist economy as a natural system—that is, a direct representation of social reality—was opposed quite self-consciously to a notion of capitalism as artifice, as a world driven by the cunning and convoluted categories of capital, markets, and profit.[19] In the course of constructing an alternative world, Marxist-Leninists devised a system of economic decision-making with its own curious dimensions. Planning became its own end, putting a strange twist on the entire economic process. The national imperative to fulfill the plan introduced practices quite different than appear to be common in capitalist firms, with consequences for the everyday lives of workers and managers. For the transformation of worker attitude and behavior in Hungary, I would argue, does not rest solely on new institutional contexts. It also depends upon the economic policies and political visions that motivated the character of socialist industrial production and collective agriculture.

My argument about the commodification of labor is offered to explain village society in the 1980s. However, my approach differs from Szelényi's work on agricultural entrepreneurs (1988). In elaborating a theory of interrupted embourgeoisement, Szelényi emphasized the tenacity of cultural notions of entrepreneurship among a small group of once-protocapitalists, who found opportunities in the growing second economy to return to capitalistic activities, reaching back to the prewar capitalist era to resuscitate attitudes and practices. I argue, on the other hand, that the combined efforts of socialist economists, officials, managers, and workers have given birth to a much more widely shared universe of capitalist attitudes and practices. The set of practices and beliefs I identify as similar to capitalism—individualism, utilitarianism, economism—is shared by far more than the small subset of

---

[19] A means by which the Comunist party legitimated its economic activity was to represent its practices as natural and timeless, even as the particular character of these activities was consistently changing over time. This is a familiar hegemonic move: social projects being represented as historical imperatives, political necessities, and scientific truths.

entrepreneurs Szelényi is intent upon describing. In contrast to Szelényi's image of a stable account of cultural capital which can be drawn upon in altered social circumstances, I see capitalist attitudes to have arisen in the complex struggles throughout Hungarian society over policies, politics, and production during the socialist period.

The difficulties of shortage and of living with the irrationalities of planning in socialism are well documented. Though the confusing experiences of planning and plan bargaining, of internal labor markets, and of second-economy participation have been exhaustively analyzed in Hungarian sociology, they have not been studied in relation to shifting cultural attitudes (two prominent exceptions being Kemény 1982 and Rév 1987). Hence the relation between work and belief, between personal convictions and institutional arrangements—at both the factory and in the home—have not been explored. In fact, public and private activities have been treated in many studies of Hungarian socialism as entirely separate spheres: the public domain maligned, private space hallowed. My argument deliberately contests this representation. The politics of resistance in socialism distanced the cultures of public and private life, although they shared much in common. I contend that the economism of planning has contributed to the character of private economic activity. So too, the elaborate (nay, baroque) strategies employed by citizens (see Kenedi 1981) in exploiting, circumventing, and avoiding socialist policies leads, I argue, to the strange economism of late socialism in Hungary. While the alternative politics of late socialism, as well as the economic policies of the party, thrived on the separation of public and private domains, it is incumbent upon social observers to acknowledge the similarities between economic behaviors and utilitarian logic characterizing *both* the public and private spheres. Although social actors regularly distinguished these two forms in everyday practice and common sense, the economism of socialism became the common idiom of both public and private worlds.

The differences between the aims of socialist and capitalist production are related not only to the process of planning and the economic goals of Marxist-Leninists. They are also related to the political structures of socialist society. Konrád and Szelényi's now famous argument on the role of intellectuals in socialism argues that socialist societies may be best described as "rational redistributive," awarding intellectuals a preeminent role in deciding how surplus is to be extracted, where it is to be invested, and by whom it is to be consumed (1979; see also Szelényi 1982). The control of surplus by elites has also been explained in the work of Fehér, Heller, and Márkus (1983) as a project designed to expand the power of the party, an approach anticipated in Djilas's early treatise on the privileged political class under socialism (1957). These studies, complementary in important ways, provide significant insights into the manner in which power is allocated and privilege appropriated by elites, within and beyond the party. While these particular dynamics differ from the structure of power and class known in capitalist societies, they are not experienced as radically different by workers on the shop floor or in cooperative farms. Workers themselves noted no radical change in the character of exploitation between socialism and capitalism, although there were moments when some still hoped for change. The bitter quality of workers' criticism of socialism derives from the fact that in socialist rhetoric, workers are lauded as the controllers of their destiny. This has given rise to a number of jokes comparing capitalism and socialism, the most famous being: "What's the difference between capitalism and socialism? Capitalism is the exploitation of man by man, and in socialism, it is the other way around."[20]

To comprehend the particular history of commodifying labor in socialist Hungary requires a close reading of the process of change. It demands a serious analysis of the cultures of communities and the politics of everyday life. I have emphasized the similarity between capitalist and socialist industrial development to argue for the influence such innovations have had in transforming the consciousness and practice of workers, of changing the attitudes and behaviors of former peasants and proletarians during the socialist period. But I do not wish to neglect the significant differences Hungarian socialism has introduced in industrial and agricultural production. The changes chronicled herein include the proletarianization of Hungarian citizens, depriving them of agricultural, industrial, and commercial properties to render them workers for the state. Their initiation into new roles within the economy has en-

---

[20] The question of elite formation since the transition has been the site of much controversy (Nee 1991b; Staniszkis 1991; Stark 1990; Szelényi and Szelényi 1990; see Róna-Tas 1994 for a thorough overview of the debate). A recent study undertaken of elite formation in Eastern Europe and Russia, directed by Iván Szelényi and Donald Treiman, will answer many questions raised by theorists and social observers since the transition. Needless to say, villagers did not expect much change in the composition of elites, and in any case, they could be secure in the knowledge that they would be ignored. Now the joke being circulated about the return of capitalism is the following: "The one thing worse than being exploited in capitalism is not being exploited in capitalism."

tailed new behaviors within the workplace, and with time, new behaviors in markets. I will recount how these changes have had profound effects, most visible in diverging patterns of work between young and old. In the book to follow, I will touch upon politics, institutions, structures of production, and forms of family to weave together the various strands of this argument, emphasizing the particularity of Hungarian socialist experience, while arguing for the general phenomenon of commodification.

The Story Told

This book is composed of seven chapters. I begin in chapter 2 with a description of agricultural work and relations among villagers and agrarian workers in the interwar period. Achieving the ideal of rural life—the ability to master one's own affairs—was a near impossibility for most villagers. They struggled continuously for possession of their land, labor, and dignity. Chapter 3 covers the period between 1945 and 1956. I discuss the establishment of the socialist command economy, the initial phase of collectivization, and villagers' resistance to socialist intrusion in their lives. This chapter ends with the revolution of 1956, concluding with a discussion of the rural manifestations of the revolution. Chapter 4 chronicles the period from late 1956 to the early 1980s. I describe the particular constellation of economic and political forces which developed in socialist agricultural production during this period: the final phase of collectivization, mechanization of production, economic reform, the second economy, and the rise of technocracy.

The final two chapters consider the character of economics and politics during the early 1980s. Chapter 5 discusses planning, both as an exercise in the scientism of politics, and as an everyday practice in village life. I argue that the rampant economism and utilitarianism of the early 1980s is related in part to the practice of planning, despite the consistent attempt of Hungarians (be they villagers or urban social scientists) to distance the cultures of private and public life. In chapter 6, I look at the spatial distribution of work, illustrating the differences and similarities between collective production and private farming. Curious similarities and significant differences characterize the practices and attitudes of young and old in the village, demonstrating the slow yet significant historical shift at issue in the analysis.

In the book's conclusion, I comment in general terms on village affairs since the transition in 1989. I offer a few speculative comments on the transition and on events to come. My musings are intended to caution against some assumptions being made about the return of capitalism to Hungary, assumptions I consider unfounded. Finally, I offer some general remarks on the historical consequences of state-sponsored change and raise questions about European social theory and the history of commodification.

Site Selection and Methods

Hungary was an auspicious choice for studying socialism and the fortunes of agrarian workers following collectivization. Hungary's agricultural sector was thriving in the early 1980s. The policies of reform which had permitted the growth of the second economy in agriculture were widely discussed, and even brought socialist colleagues from as far as China to investigate. Many observers commented on the successful compromise achieved between collective production and private work in Hungary, a success that reverberated across the entire national economy. Thus agriculture was an ideal subject for studying the complex relations of public and private work which characterized Hungarian socialism in the 1980s. Though within a few short years, innovative organizational forms and private entrepreneurship would become much more common in other sectors of the economy (Róna-Tas 1990; Stark 1986, 1989), at the time of my fieldwork, the second economy in agriculture was by far the longest lived and most stable form of nonsocialist production in Hungary.

A number of ethnographic studies of socialist peasants have been conducted in Hungary. In addition to important studies conducted by sociologists and economists (Donáth 1976, 1977; Juhász 1973, 1976, 1979, 1983, 1984, n.d.; Juhász and Magyar 1984; Magyar 1986, 1988; Márkus 1973; Orbán 1972; Rév 1987; Swain 1985, 1992; Szelényi 1988; Zavada 1984, 1986), anthropologists and ethnographers have also conducted a series of valuable studies of recent peasant life (Bell 1984; Bodrogi 1978; Fél and Hofer 1969; Hann 1980; Hollos and Maday 1983; Huseby-Darvas 1984; Mátyus and Tausz 1984; Petánovics 1987; Sozan 1985; Vasary 1987). Studies of socialist peasants and collectivization in other regions of Eastern Europe and the Soviet Union have also been quite valuable (Hann 1985; Humphrey 1983; Kideckel 1976, 1977, 1982, 1993; Kligman 1988; Nagengast 1991; Verdery 1983; Wädekin 1982).

Though for years the language proved a daunting barrier to many who wished to study there, interest in the particular road Hungary was taking among socialist economies proved too strong for many Western scholars to resist, resulting in a flurry of studies in the 1980s (Burawoy and Lukács 1992; Comisso and Marer 1986; Goven 1993; Hare, Radice, and Swain 1981; Jenkins, n.d.; Marrese 1985; Stark 1980, 1986, 1989). Considering the attention given to the Hungarian "model," many would argue that the path Hungary took toward reform was unique among socialist states, and so may influence my argument about the connections between socialism and capitalism. I would argue, on the contrary, that among European socialist states, the difference in the role of markets and money, of factories and farms, was only one of degree. Indeed, other scholars have identified similar issues in socialist development in the region (Bauman 1974; Corrigan, Ramsay, and Sayer 1978; Nagengast 1991). Although I see many parallels with other European socialist states, in the final analysis my argument pertains only to Hungary.

The question at hand is how the process of collectivization influenced the structure of agricultural work and attitudes toward work within a Hungarian village community. I assumed when first posing this question that patterns of work, property relations, and attitudes toward work were intertwined in complex ways. Certainly it had been noted in the ethnographic and sociological literature on Hungarian rural communities prior to 1948 that those who owned land worked differently from those who worked at manorial estates, who owned little more than the clothes on their backs. My primary consideration in choosing a village for study, then, was to find a community whose members had different histories in agricultural production prior to collectivization. In short, I wished to find a community where there had been both landowning peasants and landless workers, preferably families who had lived and worked at manorial estates. This would afford me the opportunity to examine the impact of collectivization on very different social groups within the limited bounds of a village community, groups who had different histories of property ownership and labor experience.

There were many in Hungary's agrarian community prior to the war who did not own land, and whose experience of agricultural labor differed from that of subsistence peasants. Nearly three-fourths of the agrarian labor force had been landless or nearly so prior to the land reform in 1945 and the beginnings of collectivization in 1949. They included day laborers, migrant workers, and manorial workers. I was particularly interested in manorial workers. Prewar manorial production shared many of the attributes of early collective farms: a large organizational unit, with a complex and diverse production profile including crop production, animal husbandry, and in some cases, small industries such as distilleries or sugar refineries. Work was done in large groups, divided up among the various branches of the manorial estate. This contrasted starkly with the family mode of peasant production, in which land was owned by the family and worked by its members, with the possible addition of day laborers or yearly servants. Because of the stark difference in these two communities, I was interested in comparing landowning, subsistence peasants with manorial workers. I was curious to learn whether former manorial workers, having long worked at large farms, would experience collectivization differently from landowning peasants, perhaps finding the change less radical than peasant families. Some in the literature on collectivization (Donáth 1977; Orbán 1972) had pointed out that former manorial workers were less reluctant to join collective farms than their landowning neighbors, lending credence to my supposition.

My intent, then, was to compare the process of collectivization among two potentially different communities of agrarian workers. How would they greet collectivization, how would they participate, and how might they respond differently to the state's imposition of collective ownership and production? Would I find very different attitudes toward work among these groups even after twenty years of cooperative farming? But there was an additional reason to look at former manorial workers as the comparison group with subsistence peasants. This had to do with the party's representation of the process of collectivization. Not surprisingly, the party shared many of the assumptions I have listed above about the connections between property ownership, the social relations of production, and ideological attitudes. Indeed, collectivization was initiated in part to alter attitudes the socialist state considered to be inimical to socialist development; the other reason was clearly to facilitate industrial advancement. Collectivization was heralded in all official forums as the first and most important step toward modernized, large-scale production, a necessary advance over the limitations of peasant family production. From these sources, one could easily have believed that nearly all those working in agricultural production had been subsistence peasants. Admittedly the evils and backwardness of manorial production were often cited, if only to contrast these latifundia with the radically new and modernized cooperatives and state farms. But the continuity or parallels between the experiences of those working at manorial estates and collective farms went unacknowledged in official forums. Thus an additional aspect of the comparison I wished to make centered on the similarities between manorial estates and cooperative farms, similarities that would explain the different attitudes former manorial workers had toward collectivization and cooperative work.

This disregard for families living on manorial estates or other impoverished families in the agrarian community prior to the war was also fairly common in the ethnographic literature. Peasant studies focused primarily on landowning peasants, seen to be the embodiment of nationalist traditions and rural culture. During the 1930s, a few monographs were written about poorer communities by the so-called sociographic school to combat the romanticized representation of rural life prevalent in the media, nationalist writings, and political propaganda. However, with the exception of Illyés's 1936 sociological essay à la memoir (*People of the Puszta*), most of the work of the sociographs still focused on impoverished peasants and migrant workers (Darvas 1943; Kovács 1937; Szabó 1937). During the socialist period, especially in the 1950s, historians and ethnographers attempted to rectify this picture, providing studies of the agrarian proletariat, both in village communities and as activists in political struggles (Katona 1958; Merey 1956; Sándor 1955; Sándor, Szabad, and Vörös 1952). In the final analysis, however, manorial workers were poorly represented in studies of agrarian communities overall in Hungary (a notable exception being Mátyus and Tausz 1984). The neglect of manorial communities in the literature also drew me to study the lives of former manorial workers.

I was looking for a village community inhabited by both former manorial workers and former landowning, subsistence peasants. Many villages in Hungary could have easily filled the bill. I had additional concerns, howver, which narrowed my focus. I was not interested in working in a community with a history of ethnic or religious divisions. Many communities in Hungary were torn in two by the animosities born of religious intolerance or ethnic hatreds. This would only complicate the clear class divisions I hoped to examine within the agrarian community. An additional consideration was the quality or character of the cooperative farm. I did not wish to study a community where the cooperative farm was shared with other villages, that is, where several village-specific cooperative farms had been amalgamated during the 1970s to achieve more efficient economies of scale. These farms were often riven with struggles over resources and intervillage animosities. I also did not wish to study a model farm, known throughout the county and nation as exemplary; neither did I wish to select a cooperative farm famous for its disasters. My goal was to find a farm that was absolutely ordinary, pursuing the goals of socialist production as best it could within the constraints of local conditions and personalities.

When I mentioned my interest in studying manorial workers as well as landowning peasants, I was immediately advised to choose a village in Fejér County. My friend, an ethnographer, joked that it was a blank area (*fehér folt*)[21] on the ethnographic map. Since it had been a county dominated by manorial estates (prior to 1945, 61% of land was in manorial production, the highest rate in the nation), ethnographers presumed that the county was less populated by a traditional peasantry, so little village research had been conducted there. My friend's joke could also have referred to the settlement pattern of Fejér County. Villages were found much farther apart than in other areas of Western Hungary. In addition to village settlements, there were also small, outlying farmsteads built on large estates where manorial workers lived. These were referred to as *major* or *puszta*. These communities did not appear on the map, hidden in the bowels of monstrous manorial estates. The term *puszta* is more generally understood, by Hungarians and foreigners alike, to refer to the somewhat desertlike and barren region on the eastern boundaries of the Great Plain, known for cattle and horse breeding. The images of emptiness and barren landscape which *puszta* connotes are appropriate for these quite godforsaken communities, written about so endearingly by Illyés in his sociological memoir. This county would surely offer me a number of suitable villages to study.

When it came to selecting a specific village, I consulted with the ethnographer who worked at the county museum, the Szent István Museum located in Székesfehérvár. Studies of peasants were consigned to museums in the socialist period, as their lifestyle and belief system were considered a way of the past. He suggested Sárosd without hesitation. Although I did visit several other sites, I came to agree with the ethnographer in Székesfehérvár. Sárosd was suitable from all the perspectives I mentioned above; moreover, I was very warmly greeted by village and cooperative officials. While the political environment in Hungary was quite liberal by the early 1980s, I was still concerned that my study not

---

[21] The term "blank spot," literally "white spot" (*fehér folt*) on the map is a pun with the name of the county, Fejér County. Once spelled Fehér, it could jokingly be referred to as if it meant "white." However, the name actually derives from *fej*, meaning head, as in the head of state. Fejér County had been the site of the king's coronation until the Turkish invasion in the sixteenth century.

be hampered by intransigent officials. Though Sárosd did boast its own thick-headed party secretary, everyone else in Sárosd welcomed my interest and assisted my work.

Sárosd (Fejér County) lies seventy kilometers southwest of Budapest in the fertile plain of eastern Transdanubia (see map of Hungary, fig. 1.1).[22] The region itself was inhabited during the Roman period, as is attested by the frequent discovery of Roman artifacts: coins, stone implements, and burial sites. The first mention of Sárosd in the public record dates to 1342 (Károly 1904, 5: 262), though the adjacent manorial estate of Jakabszállás is mentioned among the properties endowed to the Catholic Church by King István early in the eleventh century (Kogutowicz 1930, 274). The estates of Sárosd changed hands several times over the next few centuries, finally being awarded to the Eszterházy family in 1700.

Before the war, the community was dominated by three manorial estates. Two of the estates belonged to Count Eszterházy, although he managed only one of them; the other he rented to the Bartha brothers during the 1920s, and then as of the mid-1930s to another set of siblings, the Elek brothers. The third estate was owned by the Widow Weisz; she lived in Budapest, and had the estate managed for her. The estate centers that surrounded the village included Jakabszállás, Csillag, Tükrös, Daru, and Andor (see fig. 1.2). The center of Eszterházy's holdings was located alongside the village proper, and was named the Puszta or Kispuszta. All estate centers, with the exception of Andor, were inhabited by manorial workers and their families, and were equipped with outbuildings for tools and stables for livestock. The village was inhabited by sharecroppers, day laborers, and a small group of landed peasants. In 1937 the combined population of Sárosd and the manors was 3,284: 2,704 people lived in the village and on the Puszta, and 580 people were distributed among the other estate centers (Schneider and Juhász 1937, 123–124). In 1980, the population of the village was 3,300, making it average in size among rural settlements in Hungary.

I employed a variety of methods and materials in this study. Fieldwork was the most important, significantly influencing how I conceptualized issues and pursued questions. I also consulted archival sources and other written materials, such as newspapers and secondary literature, to augment and confuse my firsthand observa-

---

[22] Transdanubia is the region in Hungary which reaches from the Austrian border to the Danube River.

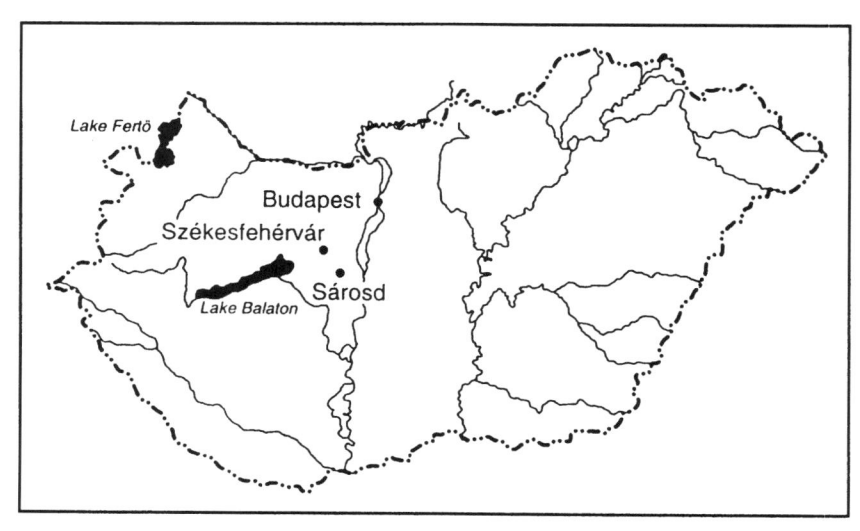

Fig. 1.1. Map of Hungary. Shown are Sárosd; the nearby county seat, Székesfehérvár; and the capital city, Budapest.

tions.[23] The fieldwork on which this analysis is based lasted twenty-eight months, from December 1981 to March 1984. The first four months in Budapest were devoted to doing preparatory archival research and interviews. I then moved to Sárosd, where I lived for the next two years. The first year of my stay was spent almost entirely in the village. In the second year, I left for shorter and longer periods, the longest being the time I was banished for two months to the city while recuperating from an appendectomy.

As my central interest was the nature of agricultural production, I took part in and examined many different kinds of agricultural tasks. I worked at the cooperative farm, primarily in the female brigade unit assigned to the garden section. Although I tried to

---

[23] The significance of literacy was brought home to me when I first started my fieldwork in Sárosd. Though I have often been praised for my ability to speak Hungarian, my ability to read and write the language played a crucial role in my initiation into village life. During my first visit to the pig farm, the brigade leader noticed the notes I had made in a small notebook. Telling his wife about our encounter that evening, he decided to ask her if he could invite me to dinner. This was my first official invitation to someone's home in the village. This relationship blossomed, and I soon regarded this man as my mentor and teacher, often consulting him on new ideas or matters that perplexed me. Book knowledge can be deceiving, however, as this villager later confided to me. While he was impressed when he first met me by how well I wrote in Hungarian, he later learned how much I did not understand. He assured me, however, that by the time I retire I might understand some of the complexities of Hungarian village life.

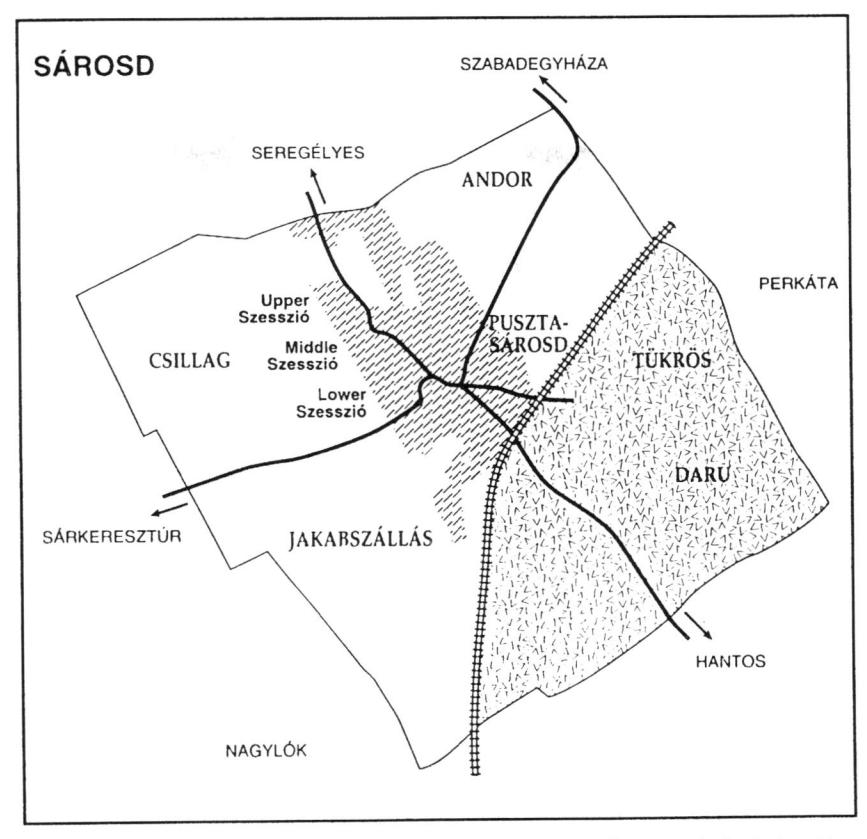

Fig. 1.2. The village of Sárosd and its surrounding manors. The central shaded section denotes the residential center of Sárosd. All lands once belonging to village residents (*szesszió*) and manorial estates became the property of the cooperative farm, except for the acreage southeast of the railroad tracks (shaded on this map). These lands were appropriated directly by the state, and donated to the neighboring state farm.

work elsewhere, I met with little success. In fact, it took me nearly a month of cautious lobbying to persuade the cooperative president that I be allowed to work at the farm in any capacity. Granted, this was an unusual request. His initial response was that I would have to take on Hungarian citizenship to be permitted to work. When I finally made it clear that I did not expect to receive a salary, the president relaxed. As I had just been bragging about my lessons in cutting hay with a scythe, he pointed to an open field before us and suggested I clear a half-acre plot immediately.

I had virtually no skills in farming, was not allowed to drive machinery, and was laughed at by machinists and other skilled craftsmen when I asked for a chance to work in their shops. My employment in the garden section made sense to everyone. Being a woman suited me to the job. Also, the skills needed for the rudimentary tasks women were given in their work at the cooperative farm made it possible for me to partake in some small measure in their activities. The garden brigade performed all the manual labor in the garden branch of production—hoeing, weeding, harvesting special crops by hand—as well as assisting other workers in particular tasks. These included a wide variety of projects, such as bagging peas, shearing sheep, picking apples, whitewashing rooms, and sweeping up after their less tidy male colleagues. I also visited all other branches of the farm, and frequently stopped off at the managerial offices to attend meetings or listen to informal discussions.

I made it a point to work with families at home, insisting that I be allowed to join in whatever task was at hand, or to be included in larger-scale projects as they arose. I had much less trouble convincing people that I was serious about participating in their work at home. The only problem came from my reluctance to take food in exchange for my time, thinking that I had been amply rewarded with good talk and informative discussion. I planted potatoes, picked beans, and cut hay. I participated in nearly all phases of tobacco cultivation, from planting and harvesting to stringing together tobacco leaves to dry before delivery to processing plants. I shucked corn, cleaned tripe, and made sausage, and spent a week cooking for a wedding. After eight months of observation and participation, I began to conduct a survey of selected residents of the village, eliciting work histories, opinions on the organization of work at the cooperative farm, information on the character of work in the home, attitudes toward work and politics, and definitions of salient terms and proverbs. Archival materials examined from the twentieth century included primary documents on Sárosd and its surrounding manors until 1947, and a sampling of county newspapers from 1948 to 1984. Primary documents included correspondence between town officials and county authorities and legal documents concerning labor complaints and housing problems of manorial workers.[24] I was not allowed to consult any primary

---

[24] I was not permitted to browse the archive files to determine which materials I wished to see. Archival officials, though always gracious, determined the sequence and content of materials I did examine. Though I am confident that I was able to examine most of the materials relevant to a study of Sárosd, I was not able to devote as much attention to those concerning labor disputes and legal battles as I would have liked, since these were presented to me only late in my stay.

materials concerning political administration or economic affairs after 1947.[25]

Although I did not know it at the time, I was particularly fortunate to have selected Sárosd when it came to the availability of archival materials. Though near the county seat, Sárosd was located in a district that was administratively centered in Sárbogárd, not in the Székesfehérvár district. During the war, the archives in that district were destroyed in the extensive bombing of Székesfehérvár, while the archives in Sárbogárd escaped unscathed. I also found out several months after I arrived in Sárosd that the doctor who had practiced for nearly twenty-five years in the community in the first half of the century had written a book advocating social welfare reforms (Kerbolt 1934). In the course of his analysis, he often mentioned anecdotes from Sárosd to illustrate his points. These materials have been extremely valuable to my recovery of earlier historical experiences. Difficulties did arise, however, in capturing the experiences of villagers in Sárosd when archival documents or memories were lacking. Other sources of information which were used to augment primary documents and secondary analyses include novels, folk songs, and statistical compilations. Building on a variety of sources allowed me to offer a more elaborate tapestry, a fuller picture of the complex relations I wish to chronicle in the following account.

This study focuses on the history of one village, Sárosd. This particular history is not shared by all communities across the nation. I wish to emphasize that the generality of my argument applies to the process of the commodification of labor in agrarian communities, not to the timing or character of the specific structure of markets in labor and goods in different regions of the nation. The culture of commodification varies in subtle ways in communities across the nation, owing to their different histories of labor, of markets, and of politics. The diversity of historical experience prompts me to be cautious in attributing the particular story of the village under study to the entire nation. However, I do consider the broad outlines of the argument to be applicable to rural com-

---

[25] Since 1989, many materials that were formerly inaccessible, particularly party documents, have been made available to the public. I have not availed myself of the opportunity to review these resources, in large part because the story told in the following pages concerns villagers' reception and reaction to party policies. Hence the use of newspapers and other materials that were published during the socialist period constitute far more valuable resources for my analysis than the insider's view offered by minutes of party meetings or debates among elite bureaucrats and leading party theorists.

munities across Hungary during the socialist period. And although the story told in the following pages is confined to the early 1980s, I do believe that particular, significant, aspects of the collapse of Communist party rule in Hungary are presaged in the analysis I offer here.

TWO

Struggling for Possession

"Rozika," he said, "listen. Do you have a soul?"

The girl smiled charmingly. She opened her mouth slightly, and said, "No, I don't."

The man was flabbergasted. "But everyone has one."

The girl was quiet, then she said, laughing a bit, "Not everyone has land." The man said nothing. . . .

"Look, the land belongs to everyone, right? Because we live from it, it produces bread. Everyone eats bread or other things the land gives us, yet one has a small garden, the other a small farm; another has property, and some have huge, huge properties, so large, that they could never wander over it all as long as they live, but still most people have no land at all."

Móricz Zsigmond, *A Lordly Spree*

Villagers were damned by landlessness. Possessions bestowed humanity; poverty stole one's soul. Few escaped the purgatory of endless movement in search of sustenance and survival: from home to manor, village to village, manor to manor, county to county. The landed, masters of their fate, chastised their poorer brethren for forsaking pride of place. Condemned to seek servitude, the landless and nearly so dreamt of acquiring land and staying home.

In this chapter, I trace the complex relations among groups within the village and at the surrounding manors of Sárosd in the unbearable calm between the first and second world wars. I argue that a description of landownership, though crucial, provides an insufficient means of delineating patterns of social hierarchy and inequality among members of the agrarian community. More important to distinguishing communities within the village were the various relations generated in work, relations that were understood and explained locally as the crucial ability to control one's labor. How one worked and with whom was of central concern; one's relative independence in work, particularly for men, was important for one's sense of social respectability and honor. Indeed, it was assumed that the ability to control one's own labor, and that of others (usually women and children), ensured one's diligence, whereas being forced to work for others was demeaning and deprived one of initiative and integrity. In other words, control of self was manifested in diligence; subservience to others resulted in irresponsibility and sloth, and so explained poverty.

The rural world of agrarian workers was elaborately fashioned according to the disposition and disposal of labor. Ownership of sufficient lands to acquire a reasonable subsistence shielded some from having to seek employment outside the home, but others, the less fortunate, frequently sold their labor, or exchanged it for goods or services. Despite the regularity with which many villagers sold their produce, or sold themselves, there was a strong moral discomfort with markets in general, and labor markets in particular, during this period. Having to seek employment outside the home forced workers to submit to the will and control of others, to surrender their most precious possession—labor—to another. Markets also represented alien forms of creating value. In the village, value was enhanced by possession, be it of one's own labor or of one's land, which ideally stayed in the family's possession for generations on end. Markets rewarded the constant movement and circulation of value, breaking ties with place and personality so central to the world of value preferred by the agrarian community, but achieved only by the privileged and wealthy.

I claim that the agrarian community of Sárosd was excruciatingly capitalist in character by the interwar period. This is demonstrated in two important characteristics. The first is the centrality of labor as the distinguishing feature of social identity and economic practice. This is paired with the active role of markets in defining the social position of villagers, differentiating and dispersing them along a steep hierarchy of respectability and disaffection. This depiction stands in contrast to the usual representation of Hungarian rural communities in the ethnographic literature, which suggests that payment in kind and nonmonetary exchange of labor constitute evidence for protofeudal or traditional relations of production and property (see Petánovics 1987; Szabó 1968). I wish to underscore the degree to which Hungarian agricultural production had been transformed in the short interval between 1848 and the end of World War I, as a direct result of quite grand changes in economy and politics which are usually described as the transition to capitalism. That these communities were not fully commoditized, had not fully embraced the fetishes of markets, of labor, and of commodities we associate with capitalism is clear in the moral proscription among villagers against reckless participation in markets and the promotion of work for its own sake.

The Village at Work

Sárosd[1] was a community dominated in the interwar period by the local manorial estates of Pusztasárosd, Jakabszállás, and Csillag. Though the soil was rich, most of its inhabitants were poor. The numbers of landowning peasants living in the village were few and their properties small. All its other inhabitants, and those of the *pusztas* that dotted its horizon, were landless or nearly so. In this respect, Sárosd was not an unusual community in the region; Fejér County was frequently referred to as "the classic homeland of large estates" (Heller 1937, 10). Prior to 1945, 61% of its lands were still held in manorial properties (Farkas 1980, 25). However, two of the villages flanking the fields of Sárosd and its manors—Seregélyes and Aba—were quite prosperous, dominated by peasants of decent or even comfortable means. The social vistas visible from Sárosd were quite diverse, however limited may have been their own horizons.

The residents of Sárosd and its surrounding manors were poor. Nearly 66% of the community (392 families) were categorized as "having nothings" (*nincstelen*), meaning that their property holdings did not reach 2 kh (1.2 ha).[2] The local manors offered possibilities for summer harvest jobs or day labor to sustain them throughout the year. Twenty-five percent of the community (151 families) had holdings of 2–10 kh, which would mark them as poor peasants, subject to the vicissitudes of day labor, harvest contracts, and other supplementary jobs to get by. Only about 8% of

---

[1] Two very important sources provided written documentation of life in Sárosd and at nearby manors prior to World War II. The first, entitled *Monograph of Sárosd* (*Sárosd monográfia*) was a general survey of village life mandated by national authorities in 1927–1930. It contains detailed discussions of agricultural production among landowners and at the manors, as well as descriptions of "folk ways" and the Roma (Gypsy) population. The second and invaluable resource is a book entitled *The Sick Village* (*A Beteg Falu*; 1934), written by Dr. Kerbolt László, who for twenty-five years had been the local doctor in Sárosd. In arguing for improved national health policies, he provided graphic descriptions of work, poverty, and disease in the village and at the estates. These sources offered valuable complements to memories I gathered during fieldwork.

[2] I use the Hungarian measurement *katasztrális hold* or "cadastral yoke" (1.42 English acres, or .57 ha; abbreviated kh) throughout, since it is still the most commonly used term in the village, and in the ethnographic and historical literature on Hungary. When possible, I also calculate land measurement in hectares, to facilitate comparison with other systems of measurement.

TABLE 2.1. Social distribution and religious affiliation of families, in Sárosd

	No. of families	% of total
Landholdings:		
0–2 kh (0–1.2 ha)	392	66.0
2–10 kh (1.2–5.8 ha)	151	25.5
10–20 kh (5.8–11.5 ha)	46	7.8
20–50 kh (11.5–28.8 ha)	3	0.5
50 kh (28.8 ha)	1	0.2
Religious affiliation:		
Roman Catholic	556	93.8
Calvinist	26	4.4
Lutheran	4	0.7
Jewish	7	1.2

Source: *Sárosd monográfia*

the community (49 families) could have been considered peasants of substance, owning 10–50 kh, truly independent and self-sustaining (see table 2.1; table 2.2 provides a comparison with the national distribution of lands in 1935). The community was strongly Roman Catholic (94%), with a few Calvinists and Lutherans and a small number of Jews also living in the village.

Manorial properties dominated arable lands. (Tables 2.3 and 2.4 show the overall distribution of properties, classified according to different branches of cultivation.) Manorial holdings equaled 78% of the total lands under cultivation; peasant holdings made up the other 22%. As one might expect, the estates owned a larger percentage of meadow and pasture lands (85.3% and 92.2%), while peasants owned close to half the lands devoted to gardening (41.4%). Manorial lands listed as garden properties (58.6%) were divided into plots and allotted to manorial workers and managerial personnel as part of their yearly contract *(kommenció;*[3] see table 2.12); vegetable gardening was not pursued at the manors as a branch of cultivation. Cultivation patterns are demonstrated in table 2.5, which provides a sketch of crops grown at the three manors and by peasant landowners. There were also differences in the character of animal husbandry between the smallholdings and the estates (table 2.6), most notably in the kinds of animals kept for draught power. In addition to the pasturage listed for the peasantry in table 2.4, the village also owned two pastures com-

[3] *Kommenció* or *konvenció* was the term used for the yearly salary, primarily in kind, for which manorial workers contracted with the estate.

TABLE 2.2. Distribution of landed properties in 1935, nationwide

Size of individual holdings (kh)	Number of landholders	Percentage of landholders	Total acreage (kh)	Percentage of all land held	Average Acreage (kh)
0–5	1,184,783	75.2	1,631,246	10.1	1.38
5–10	204,471	12.5	1,477,376	9.2	7.23
10–20	144,186	8.8	2,025,946	12.6	14.05
20–50	73,663	4.5	2,172,300	13.5	29.49
50–100	15,240	0.9	1,036,162	6.5	67.99
100–200	5,792	0.3	805,164	5.0	139.01
200–1,000	5,202	0.3	2,124,801	13.2	408.46
above 1,000	1,070	0.2	4,808,849	29.9	4,494.25
Totals	1,634,407	100.0	16,081,844	100.0	9.84

Sources: Calculated from Kerék 1939b: 298; and Orbán 172: 11

munally, one at the peasant end of the village, and another alongside the canal beyond the mill, at the "lower end" of the village. The peasant pasture was 32.2 ha (56 kh). The other was smaller—13.8 ha (24 kh)—used by the so-called new owners or land reform landowners, those who had been awarded land or house plots in the measly Nagyatádi Land Reform of 1924. By the mid-1930s the peasants and the "new owners" had each established separate pasturage associations, complete with their own stud stock.[4]

A stark distinction was made between those who lived in the village proper and those who lived on the manor, or lived there temporarily while employed during the summer months. The village proper included the areas where private homes, stores, and public offices were located; it did not extend to Eszterházy's Puszta, which though contiguous with the village, was considered outside its borders. Classic distinctions were drawn between villager (*falusi*) and outsider (*kinti*), a spatial distinction that also conveyed succinctly social differences between those who owned property and those who served others. Within the village, three categories of persons were distinguished: peasant or landowner, harvester, and day laborer. At manorial estates one found manorial workers, migrant workers, perhaps the steward and bailiff, and a few skilled craftsmen. Merchants and craftsmen also lived in the village, and although they may have been in regular contact with those working directly in agriculture, they were seen to inhabit a separate world.

Yet the greatest gulf existed between the inhabitants of the castle and all those beyond its walls. Though the Eszterházy family lived in the village, they belonged to a separate world. They maintained a separate chapel for private religious services, and quickly punished village children found swimming in the lake surrounding the young countesses' summer gazebo adjacent to the castle. People had great affection for Count Eszterházy, fondly remembering his largesse when he invited everyone in the community to his wedding feast. But these forays into the hallowed halls of the rich were quite rare.

Though the simplest division of social groups would be between those who owned property and those who did not, the more salient and subtle distinction made was whether one worked for oneself or for others. During the interwar period, the most significant social

---

[4] Previously the village had been supplied with studs by the manors. Eszterházy was particularly well known for his breeding stock, having won countless awards, including national prizes.

Struggling for Possession 41

TABLE 2.3. Manorial and peasant holdings in 1930, in Sárosd

	Peasant holdings			Manorial holdings		
	kh	ha	%	kh	ha	%
Plowland	1,523	868.1	22.0	5,420	3,089.4	78.0
Gardens	12	6.8	41.4	17	9.7	58.6
Meadow	38	21.6	14.7	220	125.4	85.3
Vineyard	25	14.25	67.6	12	6.8	32.4
Pasture	74	42.2	12.5	520	296.4	92.2
Forests	54	30.8	41.9	75	42.8	58.1

TABLE 2.4. Distribution of lands in cultivation in 1930, in Sárosd

	Peasants			Puszta		
	kh	ha	%	kh	ha	%
Plowland	1,523	868.0	88.2	2,830	1613.0	79.1
Gardens	12	6.8	0.7	12	6.8	0.3
Meadow	38	21.7	2.2	196	111.7	5.5
Vineyard	25	14.2	1.4	9	5.1	0.3
Pasture	74	42.3	4.3	466	265.6	13.0
Forests	54	30.8	3.1	64	36.5	1.8
Totals	1,726	983.8	99.9	3,577	2,038.7	100.0
	Jakabszállás			Csillag		
	kh	ha	%	kh	ha	%
Plowland	1,912	1,090.0	93.0	678	386.5	99.3
Gardens	5	2.9	0.2	—	—	—
Meadow	24	13.7	1.2	—	—	—
Vineyard	52	29.6	2.5	3	1.7	0.4
Pasture	51	29.1	2.5	2	1.1	0.3
Forests	12	6.8	0.6	—	—	—
Totals	2,056	1,172.0	100.0	683	389.3	100.0

aspect of one's work was to what degree one controlled one's own labor. The hierarchy of social identity in the village was scaled according to relative independence in work. The more one controlled one's labor, the more prestige one enjoyed. Conversely, the less one was master of one's own labor, the less respect one was accorded. Possession constituted the central distinguishing characteristic of one's identity as an independent agricultural producer. The central concern of villagers was to be one's own master, as expressed in the phrase "peasants who are masters of themselves" (*maga-ura parasztok*) (Mátyus and Tausz 1984). It has been a com-

TABLE 2.5. Distribution of crops in acreage, 1930

	Peasants			Puszta		
	kh	ha	%	kh	ha	%
Wheat	400	228.0	26.7	—	—	—
Rye	150	85.5	10.0	—	—	—
Oats	40	22.8	2.7	—	—	—
Barley	200	114.0	13.3	—	—	—
Totals, cereals	790	450.3	52.7	1,415	806.0	50.0
Corn (maize)	450	256.5	30.0	—	—	—
Potato	60	34.2	4.0	—	—	—
Sugarbeet	—	—	—	—	—	—
Poppyseeds	—	—	—	—	—	—
Lentils	—	—	—	—	—	—
Totals, row crops	510	290.7	37.0	849	484.0	30.0
Fodder	200	114.0	13.3	—	—	—
Colza (rape)	—	—	—	—	—	—
Totals, feed crops	200	114.0	13.3	566	323.0	20.0
Totals, all crops	1,500	855.0	100.0	2,829	1,613.0	100.0

	Jakabszállás			Csillag		
	kh	ha	%	kh	ha	%
Wheat	500	285.0	30.1	230	131.1	33.9
Rye	—	—	—	—	—	—
Oats	—	—	—	68	38.8	10.0
Barley	100	57.0	6.0	150	85.5	22.1
Totals, cereals	600	342.0	36.1	448	255.4	66.0
Corn (maize)	400	228.0	24.0	200	114.0	30.0
Potato	—	—	—	30	17.1	4.4
Sugar beet	100	57.0	6.0	—	—	—
Poppy seeds	200	114.0	12.0	—	—	—
Lentils	60	34.2	3.6	—	—	—
Totals, row crops	760	433.2	45.6	230	131.1	34.4
Fodder	200	114.0	12.0	—	—	—
Colza (rape)	80	45.6	4.8	—	—	—
Totals, feed crops	280	159.6	16.8	—	—	—
Totals	1,660	934.8	98.5	678	386.5	100.4

monplace in the ethnographic literature on Hungary to emphasize that during the interwar period, one's social identity was founded on the possession of land (Bell 1984, 28; Fél and Hofer 1969, 279; Sárkány 1978, 66; Vasary 1987, 35). Ethnographers have asserted that peasants valued land above all else, denying themselves

TABLE 2.6. Animals in Sárosd, 1930

	Peasants	Manorial estates Puszta	Jakab	Csillag
Horses	298	70	60	10
Foals	—	30	—	—
Oxen	—	175	20	50
Cows	—	200	140	—
Cattle	191	—	55	125
Calves	37	250	—	—
Bulls	—	60	—	—
Breeder bulls	6	—	—	—
Hogs	—	800	850	680
Sows	1,042	—	—	—
Young hogs	642	—	—	—
Boars	11	—	—	—
English Yorkshire pigs	131	—	—	—
Donkeys	—	10	—	—
Ewes	—	—	200	330

Source: Sárosd monográfia

Note: The figures are clearly disparate and not commensurate. I list them as they were presented in the *Sárosd monográfia*, to offer a general, if not consistent picture.

and their families everything to maintain and increase their holdings. Unfortunately, the almost exclusive emphasis in the literature on possessing land as the foundation of a villager's social identity obscures important aspects of identity construction during this period. Ownership of land was clearly the surest means to control one's own labor. But the criterion of possession was applied even to those who owned no land whatsoever, as in the clear demarcation between day laborers and manorial servants. Though both groups relinquished their labor to others, and so were disadvantaged, day laborers were accorded more respect, since they sold their labor only day by day, not for the entire year, as did manorial workers. Interestingly, the term generally used to refer to the landowning peasant and to the male head of household (*gazda*) meant "owner." Referring to the male head of household as owner signified that control of labor was in the final analysis a male prerogative, which extended not only to his own labor, but to that of his wife and children. Women could never fully achieve mastery of self, no matter how hard they worked as managers of the family household.

This notion of mastering self was succinctly expressed in the general term used to discuss working, *dolgom van*, which derives from the word *dolog*. When referring to work, *dolog* meant both "thing" and "activity";[5] to have things, to have activities (*dolgom van*) was to labor continuously, to be engaged in everyday affairs. The confluence of meanings—the central bond between human action and physical object—is striking. Activity presumed a focus of human action on physical objects, or, alternatively, activity was an object realized through possession.[6] The central normative quality of mastering self, possessing activity, is also found in various terms that derive from *dolog*, for example, *dolgos, dologtalan*. *Dolgos* is an adjective meaning "diligent, industrious, hard-working." The term *dologtalan*—the suffix *-talan* denoting absence or negation of the noun—may be translated as "idle, indolent, unwilling to work." It also carries the connotation of thievery.[7] Thus, this concept (*dolgom van*), which I gloss as "possessing activity," collapsed action and object into one complex form, paralleling the process of work itself. For, as one villager explained, in working, one produced one's material world and established one's honor within the community. In other words, the process of producing one's material existence was the process whereby one created social community. This was a process without end, a ceaseless project of material creation and social realization.

Ideally, then, the possession of activity constituted control over the entire work process, extending beyond production to control over the terms of exchange. In Erdei's classic treatise on the Hungarian peasantry published in 1941, he puzzled over the strength of pos-

---

[5] *Dolog* has a variety of meanings, including "task," "obligation," and "affair." (See n. 5 in the introduction.)

[6] For a discussion of similar issues with regard to the Tswana, cf. Alverson 1978; and Comaroff and Comaroff 1987.

[7] Another general term for work is *munka*. Etymologically, *munka* is defined as pain, suffering, as conscious goal-oriented human activity, as working, oeuvre, and wage (Benkö 1967, 2: 976–977). The appearance of "wage" in the list of meanings and the specific association of the word *munkás* (worker) with industrial employment suggest that the term may have generally been used in connection with nonagricultural, industrial, or craft work. The word *munka* was cited in twelfth-century Hungarian documents, so the current meaning may have been a recent innovation, which arose in the course of the nineteenth century. In the village, wage labor per se, however, has long been referred to as *napszám*, literally translated as "day-number." *Munka*, and *dolgozni* to a lesser degree, are associated with kinds of work which require regular movement outside or beyond the domestic or local domain. The inflection of *munka* akin to *dologtalan* (*munkátlan*) may be translated as "unoccupied, out of work, unemployed." However, this word carries very little of the normative censure that is so characteristic of *dologtalan*.

session as a structuring principle of action among the peasants of this region, characterized as descendants of serfs (*jobbágyparasztok*). He bemoaned the absence among these peasants of bourgeois assumptions about manipulating the market to enhance the value of one's land and labor.

> [T]his property consciousness is not the same as the ownership consciousness of the gentry or the capitalist bourgeoisie. It is manifested in completely irrational peasant behavior. . . . Property brings him absolutely no positive advantage. He lives no better and he works no less than if he had less land or he were just renting. He does not undertake creative jobs on his land or attempt individual achievements that could satisfy his ambition. He doesn't take advantage of the bourgeois benefit of property by living off the annuity. He simply enjoys owning it. It is possible that [an industrial] worker lives better; it is possible that someone who rents land produces much better. Nonetheless the landed owner considers himself far above those people because he has land. It were as if the only thing absent from a full peasant lifestyle was that he possess his land without interference, disturbance, and supervision, and that he control his produce without fixed obligations. (Erdei 1941, 93–94)

The final comment is telling: as master of one's own affairs, one wished to have complete control over cultivating and dispersing the fruits of one's labor. Thus, possession was an enclosed process, not admitting intervention by any outside authorities, be they tax collectors, banks, or local merchants. The enclosed circularity of possessing self envisioned by villagers—a bold and defiant self-sufficiency—was rarely achieved.

Dramatis Personae

Who were these people, having so much in common and sharing so little? On a balmy spring day, who would we meet as we sauntered through the village? All self-respecting peasant men would have left early for fields and pastures, not to be seen on village streets during the day. A large cart may ramble by, driven by an old hand from the manor, full of pigs on the way to market in Székesfehérvár. He would wave to the pub owner, who stands proudly in front of his establishment, in heated conversation with the grain merchant from up the street. We may see a poor peasant woman bringing eggs to the store to sell for cash. Another poor woman may be requesting, in lowered tones, that the store owner extend her credit one more time until her summer earnings come in. A Roma (Gypsy) horse vendor would be approaching the blacksmith's shop with his most recent beauty, hoping for a sale. Eszterházy's fine carriage would jog by, bringing his wife home from a visit to her in-laws in Székesfehérvár. In the evening, peasant men would walk slowly back from their fields alone, hoes propped on their shoulders, lost in quiet contemplation. Large bands of women, having spent their day weeding the fields of the manor, would march back into the village, singing all the way. A few tradesmen from the village and nearby manor would be preparing to meet their fellows at the pub, ready for a lively card game and a good night's drinking. Several young men from the Roman Catholic Youth League (Római Katolikus Ifjúsági Egyesület) would be gathering at another pub across the way, planning their roles in the upcoming program staged by the glee club. Young girls would appear at the house of one of their friends, to spend an evening spinning and gossiping, and waiting for the arrival of their young male compatriots to sing and tease them. Out at the manor, migrant workers, tired from a day's work, would fall asleep to the stories told by a visiting hobo, who would spin tales of his once glorious life as a mayor somewhere on the Great Plain, a life cut short when his underhanded dealings were discovered and he was run out of town. Manorial workers would be struggling to wrest a night's sleep in the cramped quarters of manorial housing. Living in close proximity, yet living such separate lives. This is the fate of a community divided by property and pride, love and land, work and worry.

Landowner (paraszt)

The first and preeminent category of agrarian social groups within the village hierarchy was the landowner.[8] The size of holdings for these villagers averaged 5.8–28.8 ha (10–50 kh); they comprised 8.3% of the population. The size of peasant holdings in Sárosd were small in comparison to those owned by peasants in nearby villages, and in comparison with the middle peasantry nationwide. Besides being small, holdings in Sárosd were widely dispersed in numerous locations. As in the rest of the country, a

---

[8] In Sárosd "peasant" (*paraszt*) generally means "landowner." In the literature, the term "peasant" is often used more loosely to refer to everyone living in the village. This is confusing, since it puts emphasis on land possession as a criterion of membership in an agrarian community, and hence ignores a number of groups within the agrarian community who are poor but who make their living from agriculture.

family's land was not consolidated, but scattered in five, six, or even more plots in the fields surrounding the village. Even though landowners spent nearly a third of their time traveling between plots, they were reluctant to consolidate properties, having a strong sentimental attachment to the tokens of their patrimony. The clear demarcation of a family's plots even extended to lands once owned with one's siblings. Natal families worked their lands separately, even if they were contiguous with lands owned by members of their family of origin.[9]

Plots were planted in a three-crop rotation pattern, leaving no fallow and using both chemical fertilizers and solid animal manure. However, in the three *szesszió* fields (the upper, middle, and lower *szesszió*)—the properties allotted to former serfs in 1848—the older system of crop rotation apparently had been maintained. This meant not only that families continued to leave a portion of their lands in fallow, but also that the sequence of crops and fallow land was collectively determined and maintained by the *szesszió* landowners. Acreage devoted to grape-growing was predominantly in peasant hands (67.6%). The village clerk stated that wine grapes were grown exclusively for home consumption, while Kerbolt maintained that they were sold for money or traded for corn, beans, or wheat (1934, 60).

All family members in a peasant household were involved in agricultural and domestic activities, with the possible exception of young children. Husbands enjoyed nominal status as heads of household, but the management of labor, the coordination of fieldwork, and domestic activities were primarily the responsibility of wives. Heavy fieldwork and care of larger animals fell to the men; women took care of all else. Adult men were rarely seen in the village by daylight.[10] Elderly members of the household, no

---

[9] When I asked why lands were not cultivated jointly, assuming there may have been advantages in doing so, I was told that "the women couldn't get along." As Bette Denich has pointed out (1974), the difficulty of reconciling women's diverse interests is often used as an explanation in patriarchal households for the need to keep production and consumption separate among adult siblings. Since wives are outsiders to the family, they can easily function as scapegoats. Thus potential tensions between brothers can be masked under an ideology of warring women.

[10] When asked to respond to a request for a permit to sell newspapers on the street, the mayor of Sárosd wrote: "In villages it is unproductive to sell newspapers on the street, because barely anyone stays at home, especially during work time, not like in cities, where there are constantly pedestrians on the street. If someone wishes to buy a newspaper he orders it ahead of time from the post office or the news vendor. Accordingly, we do not consider granting a permit for street vending expedient" (SJ f.i. 1734/1937).

longer able to carry the full burden of managing the farm, took on a supplementary role, often tending children and helping with domestic tasks. The task of marketing produce, of generating cash from all available sources—of ensuring the most judicious use of the family's material resources—was a woman's most crucial function within the household. She and her daughters were also responsible for doing all garden work, feeding small barnyard animals (chickens, pigs, geese, ducks), cooking, cleaning, sewing, and performing all other domestic jobs.

It was said that mothers hoped to spare their young daughters at home, in anticipation of their life of drudgery both as wives and daughters-in-law, but this seems to have been more wish than reality. The workload in even the wealthiest peasant households in Sárosd demanded a daughter's participation, if only in the lighter tasks of feeding barnyard animals and garden work. Moreover, even though young women would learn the finer points of cooking and sewing from their mothers-in-law, their hopes of a good marriage hinged on their domestic skills (in addition, of course, to the ubiquitous issue of property in land). Young men were told, "Look at the mother, then marry the daughter." A clever wife was a coveted prize in any household, and among landowning families, where the intensity and diversity of agricultural tasks demanded special managerial skills, these abilities were just as actively sought in a mate as the capacity to work hard.

Sons joined their fathers in work early, in summers during their school years, and after finishing the compulsory four years in elementary school, year-round. They learned to handle livestock, to care for the soil, to coax good yields from the earth, and to minimize the damages of natural disaster. Like all boys between the ages of twelve and eighteen, the sons of peasants were required to attend compulsory military training (*levente*). Peasant boys, however, also participated actively in the Roman Catholic Youth League, which served the joint function of instilling strong Christian values and encouraging innovative agricultural practices, a project that won Count Eszterházy's strong support.

The village monograph notes that wealthier peasants hired harvesters, and ten families engaged agricultural servants year-round. They were employed primarily to assist with stock, fieldwork, and cartage. Peasants hired girls only during the harvest.[11]

---

[11] Girls from poorer families who chose domestic service were almost always employed in the cities, and only rarely in the village, for example, by a merchant. However, in the summer months, girls would leave their urban jobs to do agricultural work, such as the wheat harvest.

As at the manor, harvesters were paid one-tenth of the portion they harvested. Though it is not stated, we may assume that harvesters working for peasant families were also fed by their employers during the six- to eight-week harvesting period, as were harvesters at the manor. But they were not necessarily fed well, or treated kindly, as the following folk song attests:

> Hey! my master, hey, my master! I leave thou here.
> From today on I don't groom either oxen or cows.
> Your courtyard should be overgrown with weeds,
> Rust should eat your pitchfork and goad!
>
> Your sour cabbage should rot in the barrel!
> Which you often made me eat in the morning and for dinner.
> Phylloxera should root out the vine-stock of your grapes.
> It'll open you up and block you up, if you drink its juice!
>
> A spider should weave a web on the door and window of your pigsty.
> Your hands should break, if you slap a servant!
> I will never set foot here again
> And never more will I eat your bitter cabbage!
>
> (Katona 1968, 33)

Agricultural servants, who were usually nineteen- to twenty-year-old bachelors from poorer families, were given full board and twenty pengö monthly. (Two pengö was the wage for day labor at the manors surrounding Sárosd. See table 2.12c.) It may have been more prestigious to serve in the village than at the manor, but that was poor consolation for servants who were abused by cruel masters.

Two crucial phases of production were mechanized by the 1920s among smallholders: sowing, and threshing wheat. (Table 2.7 lists machinery in use in Sárosd, among peasants and at the manors.) The limited number of machines in the village enabled certain members of the community to extract labor service or a percentage of the final crop in exchange for use of the machine. Labor exchange was commonly bartered for the use of equipment for sowing. Threshing, on the other hand, was undertaken for one-tenth the crop. Three threshing machine owners served the community, traveling from house to house, threshing wheat and other grains in interior courtyards.

TABLE 2.7. Machinery in use in Sárosd, 1930

	Village	Puszta	Jakab	Csillag
Threshing machine	3	—	3	21
Steam plow	—	1	1	—
Tractor	—	3	1	1
Sowing machine	10	10	—	—
Reaping machine	—	—	—	3
Harrow	—	20	—	—
Selector	—	1	—	—
Chaff-cutter	—	4	—	—
Root-cutter	—	4	—	—
Grinder	—	1	—	—
Corn-sheller	—	2	—	—
Wine press	—	4	—	—
Clover hewer	—	1	—	—
Turning lathe	—	1	—	—
Suction gas engine	—	1	—	—
Battery	—	1	—	—
Battery (accumulator) plant	—	1[a]	—	—

[a] 220 volts

According to the village monograph, fields in Sárosd were regularly plowed three times before sowing. Thus, even the poorest landowner, with scraps of a half, quarter, eighth, and sixteenth of a hectare, was compelled to procure a plow team several times in the fall and spring before planting, as only wealthier families owned a fully equipped team of horses and a plow. Regardless of one's relation to the owner of the team—close relative, neighbor, or friend—use of a plow team was to be reciprocated in kind.[12] As these complex negotiations were largely couched in the "friendly" terms of kinship reciprocity, the specific details of labor exchanges are difficult to reconstruct concretely. Yet it is patently clear that wealthier families—those outfitted with a team and plow—consistently benefited from sharing their tools and animals

[12] The text specifically reads "He who doesn't own a draught team cultivates his land with a hired team, which is reciprocated with day labor in kind." The expression "in kind," however, was generally understood as referring to the unit of time rather than to any specific task. Hence we see that concepts of task and time are only generally calculated, in line with the notion that one day's work is a generic form of activity.

with poorer villagers (Szabó 1968, 74–84). A simple calculation of families exchanging one day's labor for one day's use of a plow and team would mean that wealthier landowners could call upon nearly eight hundred days of free labor from among their poorer neighbors.[13] This is a conservative estimate. Thus patterns of labor service and sharecropping arrangements clearly demonstrated the classic process of extracting surplus through labor reciprocity, based on the unequal distribution of agricultural machinery and draught animals within the community.[14] Though studies of primitive accumulation often focus on more grand schemes of appropriating capital and property, these forms of extracting surplus must also be considered a central mechanism of developing class relations in the countryside. The old maxim that one must milk a horse by its hooves, used in modern times with reference to rental cartage, seems an apt commentary on the potential socioeconomic advantage afforded by one's possessions.

Although these families represented the ideal of peasant prosperity in the village, they did not revel in luxuries or conspicuous consumption. They were always keen to acquire more land and more stock, so nearly every penny that was earned—be it from marketing produce or other sources—was set aside for that purpose. Necessities that could not be produced at home—petroleum, salt, sugar, matches, and other staples—were obtained by exchanging grains or other produce. Their surplus stores were thus rarely consumed, but marketed to acquire either a few basic staples or cash

[13] This calculation is based on the following assumptions: fifty families (those owning 10 kh or more) would have the livestock and tools to lend to poorer villagers. Of the 543 poorer families living in the village owning less than 10 kh, it is reasonable to guess that only half would need to borrow tools from others (the poorest having no land to work whatsoever, while those with close to 10 kh having sufficient resources for their own use). If we assume that approximately 270 families borrowed a plow team three times a year, then a rough estimate of 800 days of labor exchange accumulate to be surrendered to wealthier families (working out to about fifteen days of unpaid labor for each family). This is a substantial contribution to a peasant family's labor pool. But it is also important to note other advantages of owning tools. Wealthier families are able to call up their debt when they see fit, suiting the pace and task requirements of their production schedules. Moreover, wealthier families also are able to work their own lands at optimal times. Poorer villagers, dependent upon borrowing tools, would have to work their properties when others were willing to relinquish their tools for a day—most probably at the beginning and end of the season—which could seriously erode their yields year to year. Thus the advantages of capital investments would accrue to wealthier villagers over time, while poorer villagers would always be struggling to make ends meet.

[14] Nagengast's analysis of primitive accumulation in a Polish village was an important model for my thinking here (1991).

toward the further expansion of the farm. Former landowners recalled how rarely they had ready cash to spend, in contrast to their poorer neighbors who regularly engaged in wage labor. Wage laborers may have had ready cash, but only because they were never able to accumulate enough to invest it in property or animals. Thus their habits of consumption, appearing conspicuous to wealthier families, spoke eloquently about poverty, not wealth. A woman who was a member of one of the wealthiest families in the village told me that she once witnessed a young girl buying a pair of red shoes on market day from her earnings as a day laborer. In the peasant girl's mind, that pair of shoes came to epitomize the poor laboring girl's freedom from family obligations, obligations that weighed heavily on a peasant girl's shoulders as a member of a prosperous family.

By the 1930s national programs seeking to improve the agricultural skills of the peasantry proliferated. Aimed at the sons of the middle peasantry, these programs were sponsored by groups within the Catholic Church and by independent agencies concerned with economic development. Although neither the Roman Catholic Youth League nor the Peasant Circle (Gazdakör) had membership rules excluding agricultural workers per se (they did exclude non-Catholics and craftsmen), day laborers and manorial workers were effectively discouraged from joining.[15] Thus, institutional forms disseminating knowledge about new strains of crops, new fertilizers, and innovative agricultural techniques were appropriated by the middle peasantry, widening the gap of skill and technology between the landed and the landless. When it came to marketing their produce, landowners could avail themselves of the services of the Hangya Producers, Marketers, and Consumers Cooperative. A branch was established in Sárosd to help peasants to get the best prices for their products. There was also a credit cooperative in town, associated with the National Central Credit Cooperative (Országos Központi Hitelszövetkezet). It maintained a group specializing in farm leases and rental.

Renting or sharecropping land was a means of supplementing one's holdings. Renting required cash and so was usually practiced by wealthier peasants; sharecropping was more common, but certainly not the exclusive concern of poorer peasants. Land could be rented or contracted for sharecropping from the elderly, the infirm,

---

[15] The membership of the Gazdakör in 1938 included five middle peasants (*földmíves*), twelve poorer peasants (*kisbirtokos*), one assistant clerk, one overseer, and one teacher. All the members were Catholics.

and from those families who owned land in the village but lived elsewhere. Periodically, lands owned by the village were also leased out. The miller and several merchants in town owned land that they themselves did not cultivate, primarily because their commercial activities were too demanding on their time. No ablebodied peasant would have willingly disposed of land, either to lease or in return for a portion of the produce. The need to rent out one's land or to make a sharecropping contract marked one's demise as a productive member of society, a blow to personal integrity that was feared more than anything else by aging villagers. It also symbolized social isolation, since contracting with others to work one's land meant the absence of heirs or nearby kin. By sharecropping, middle and wealthier peasants sought to expand production to the absolute limits of available labor power, in accord with an ideology built on the unquestioned value of hard and constant work. Poorer peasants, on the other hand, needed to augment their paltry holdings to be able to feed their families. Although sharecropping for half the produce was not considered as honorable as owning the land one cultivated, it nevertheless meant one performed all phases of production from start to finish. The owner paid the tax on the land and received half the yield. The sharecropper bought the seed and did all the work for the rest of the produce. The advantage for the sharecropper was that he was in complete control of the process, making all decisions about when and how to work the land. It thus most clearly approximated the ideal of agricultural work: possessing oneself and controlling the work of others. But villagers knew full well it was not an ideal situation. As one villager explained to me, the landowner clearly had the better end of the deal. "Well, in the current terminology, it would be called exploitation. In Hungarian we call it extortion."

Landowners constituted the local political and economic elite.[16] Within their ranks were the sowing machine owners, perhaps one or two machine threshers, the mayor (bíró), the market inspector (vásárbíró), the elders at church, and the leaders of the local so-

---

[16] In strongly peasant villages, the list of leading taxpayers (virilisták) was a succinct inventory of political actors. In Sárosd, there was a sharp distinction between wealthy landowners, doctors, and shopkeepers, who peopled the taxpayers' list, and the local political actors who dominated the town hall. While the wealthy virilisták may not have been viewed as integral members of the local community, they nonetheless had influence. Count Eszterházy effectively controlled national and county party politics in the village. Not so incidentally, his father-in-law was lord lieutenant of the county. Apparently, Eszterházy rarely interfered with the everyday politics or legal squabbles handled by the peasant elite.

cial and economic associations.[17] Their children were inevitably chosen as king and queen of the wine festival, and by the turn of the century, they intermarried, despite ties as close as that of first cousins. In addition to the Roman Catholic Youth League and Peasant Circle, peasant landowners in Sárosd boasted a glee club and a theater company, which performed operettas or musical pieces around the holidays.

In Sárosd, the general term for landowner was *paraszt* (peasant), or less frequently, *gazda* (male head of household, owner, peasant). In some areas of Hungary, *gazda* was the preferred term for a landowner, most notably on the Great Plain, where *paraszt* was only a term of derision. In Sárosd, the term *gazda* was more apt to refer to the position of overseer or work leader at the manor, underscoring the predominance of manorial social relations of production in the village. The verb describing a landowner's activity is *gazdálkodik*.[18] While a strict definition would read "to cultivate the land" (Benkö 1967, 1: 1037–1038), the meaning of the term *gazdálkodik* is best captured in the word "managing"; it evokes images of someone controlling the entire process of his work, and perhaps that of others.

Tradesmen, Merchants, Intelligentsia, Priests, Roma

Among the more privileged groups of the village one finds merchants, craftsmen, and the local intelligentsia (see table 2.8). Though diverse in background, skill, and social status, they were often grouped together by virtue of not being primarily identified as agrarian workers. Those considered to represent the local intellectual elite included Count Eszterházy, the pharmacist, doctor, priest, and seven school teachers.

The priest was considered a member of the wealthy elite, in the same category as the town clerk or the doctor. His comfortable ex-

---

[17] For example, when the rental contract on lands owned by the town council came up for renewal in 1938, the three bids were tendered by council members or close relatives or friends. Interestingly, the final contract was awarded to the group of three peasants who offered the least amount of grain per hectare (190 kg/kh), based on the reasoning that they chose to rent the entire 15 kh rather than just 5 kh as had the other bidders. Of the three successful bidders, one was a member of the council, one was married to a council member, and the third was father to a council member. Of the unsuccessful bidders, one was a council member and the other a wealthy peasant not a member of the council.

[18] The verb is derived from the noun *gazda*, and is not etymologically related to the word *gazdag* (wealthy).

TABLE 2.8. Craftsmen, merchants, officials, and professionals in 1927, in Sárosd

	Village	Manors
Tailor	8	—
Blacksmith	6	1
Shoemaker	6	—
Cartwright	5	2
Cabinetmaker	4	—
Machinist	4	5
Stonemason	4	—
Stable boy	3	—
Hairdresser	2	—
Baker	1	—
Butcher	1	—
Carpenter	1	—
Engine fitter	1	—
Gardener	1	—
Harness-maker	1	1
Miller	1	—
Merchant	6	—
Innkeeper	1	—
Soda-maker	1	—
Tobacconist	1	—
Manorial steward (specific manors not listed)	6	—
Railroad employee	6	—
Teacher	6	—
Clerk	2	—
Post office employee	2	—
Medical doctor	1	—
Other town employee	1	—
Pharmacist	1	—
Priest	1	—
Retiree	1	—
Other	1	—

Sources: Sárosd monográfia; and the membership list of the Sárosd Circle of Craftsmen and Merchants.

istence was sustained by a generous allotment of lands worked by his servants. Memories of priests who served in Sárosd are bitter ones.[19] One priest was known to like the bottle, and when tipsy,

[19] The current priest is a fortunate exception.

crave the favors of his female parishioners. Many women in the village complied with his demands, thinking it their religious duty to do so.

To fear the priest was common. It was not so common to challenge his authority, nor was it wise. An elderly villager recalled that when he was still a young boy during the First World War, he had been beaten quite severely by the priest, who had been told (wrongly) that the child had dropped a few church newspapers in the mud. When the boy's older brother saw that the child could not sit down for the pain, he marched over to church to confront the priest. He grabbed him by the collar, and punched him repeatedly as the priest whined and moaned. "How dare you beat my brother, when his father is off on the front fighting for you!" When the young man finally released him, the priest threatened his life; two gendarmes appeared at his doorstep the very next day. By a stroke of luck, however, the young man's draft notice was waiting for him when he returned from the church steps, a fact that his family credits with saving his life. While priests were expected to side with the powerful and have very little sympathy for their poor parishioners, one priest who served Sárosd may actually have had populist leanings. He was charged in 1904 with fraternizing with radical Christian elements. He denied the charge, and claimed he was avoiding any contact with priests who were members of the People's Party (*Néppárt*). He was eventually relieved of his post.[20]

Six merchants and an innkeeper numbered among those in commerce. A number of highly skilled machinists and craftsmen lived in Sárosd or at its surrounding manors. Though some of them pursued rather traditional trades as blacksmiths, cartwrights, or harness-makers, others had brought new skills to Sárosd, associated with mechanization and the modernization of agricultural production: engine fitters and machinists, for example. Merchants and craftsmen were often intimately tied into the agrarian community, but they were nonetheless considered peripheral to the village community. This cleavage was motivated not only by differences in occupation. Many of the craftsmen would have been recent immi-

---

[20] A priest's fortune could be enhanced by his posting, and the efforts he expended in enhancing his agricultural properties. The priest who was dismissed from Sárosd regretted having to leave, not least because his pay in Zámoly, where he was sent, was less than he received in Sárosd. More to the point, during his fifteen-year tenure in Sárosd, he had planted 3 kh of grapevines, and had built a press and a corncrib with money from his own pocket, all of which he regretted having to leave behind (FM f.b.i. 1919–43 res. sz.).

grants to the community, marking them as outsiders. (Anyone who had lived in the village less than forty years was considered an outsider.) Some of the craftsmen and several of the merchants and tradesmen were Jews, which distinguished them from the mostly Roman Catholic community, even though they had been born and raised in Sárosd.

Particular trades were more directly involved in agricultural production than others. Several of those in commerce were grain dealers, buying up wheat from local producers and selling it in larger cities. Though primarily occupied in commerce, several of these villagers also owned land, which was worked by others. They often kept animals as well, using grains from their sales to keep a healthy herd. For example, the tradesman Deutsch kept a good number of pigs in his yard, and because he was a Jew, his taste for pork sausage was often remarked upon. People joked that his only concession to religious food proscriptions was to hide out in the barn while eating sausage on high holidays.[21] The miller, who processed wheat for the local peasantry and manorial community, also had some land worked for him. It is highly probable that the wheat dealers and miller were owners of the local threshing machines. Their access to capital, and interest in acquiring surpluses of grain, would make them reasonable investors in such expensive machinery.

Other shopkeepers and tradesmen were less actively involved in agricultural production or commerce. The local shopkeeper kept a dry goods store, offering credit to villagers and *puszta* residents when their money supplies ran thin. Craftsmen such as tailors, shoemakers, and bakers served the community as well, and may have participated in the exchange of agricultural goods only when local families bartered for goods rather than pay cash for them directly.

The miscellaneous lot of intellectuals, tradesmen, craftsmen, and merchants were bound to one another in necessity as outsiders and outcasts of the community. Though many of them were far wealthier than their peasant compatriots, their skills, knowledge, or religion excluded them from many of the activities of the community. Not members of the Roman Catholic Youth League or the Peasant Circle, they founded their own social club. The Social Circle of Tradesmen and Merchants of Sárosd (Sárosdi Iparosok és

---

[21] The Jews of Sárosd do not seem to have been very orthodox in their religious practices, as were Jews in the neighboring village of Aba (Sozan 1986). This history, as so much else, perished in the Holocaust.

Kereskedök Társas Köre) met at a pub several blocks away from the one frequented by the Catholic peasantry. Here they played cards, joked, and drank. The moving picture show was also held here.

Another community even more radically excluded from village affairs was that of the Roma population. The village monograph listed seven Roma families (with fifty-one family members), who had settled in Sárosd in 1850, moving from the neighboring village of Szolgaegyháza. The town clerk claimed Roma men made bricks, while their wives panhandled. Memory records that the members of at least one family were musicians.

It is difficult to reconstruct the character of work and consumption among tradesmen, merchants, and Roma in Sárosd. Merchants and tradesmen probably lived much as did their village compatriots, while the Roma chose a life-style quite at odds with that of their agrarian neighbors. Roma crafted their lives in near stark opposition to the productivist work ethic of agrarian communities, preferring the world of trade, commerce, or very specialized occupations, such as brick-making, knife-sharpening, or musical entertainment (Stewart 1987, 1990). Virtually all the Jews from Sárosd were killed in the war; the fate of the Roma has not been documented, but I have little doubt that they also perished in large numbers in German death camps.

Harvesters (arató)

The term harvester (*arató*) was used in Sárosd mainly to refer to the poorest strata of landholders, those owning between 1.2 and 2.8 ha (2–5 kh).[22] Their meager holdings could barely sustain a family. So to augment their holdings, they sought additional income, forming the bulk of the labor force employed at the surrounding manors to reap the wheat every year.[23] Generally harvesters

---

[22] It is difficult to estimate exactly how many families fell within this category. In table 2.1, 151 families (or 25.5% of the population) are listed as having 2–10 kh, but I cannot determine with any accuracy how many of these families would have been found in the upper range of this grouping. Since the land in Sárosd was of the highest quality, families owning 7–10 kh may have been able to get by without resorting to outside income. For our purposes, I will assume that at least 12% and perhaps as many as 20% of village families could have been categorized as harvesters.

[23] Sons and daughters of wealthier landholders could also be employed at the manor for the harvest. Their participation was not compelled by poverty, however, but by the imperative to make full use of all family members during the busy summer months.

worked for one-tenth or one-eleventh of the wheat harvested, but were often expected to do additional tasks like hoeing and weeding. Male harvesters were each individually responsible for hiring and paying two attendants, usually young girls, to help gather and bind the cut wheat, but the contract with the manor represented all the harvesters of the village as a group.[24] Actual negotiations of the contract and supervision of the harvest in process were the duties of the "wheat harvest overseer" (*aratógazda*), who was selected by the harvesters to represent and supervise them.

The investment in draught animals and tools within this stratum was limited: some owned horses; none owned cows or agricultural machinery, except the basic implements of a poor peasant—hoes, a scythe, a sickle. Pigs and chickens were their primary livestock, supplies of which were scanty. Even if they owned land of good quality, it was never enough to provide them with a full year's grain supply. Accordingly, they had no surplus to exchange (except maybe eggs and odds and ends from their garden), which severely limited their commercial activities. Their contract with the manor enabled them to supplement their stores of grain; and the extra duties in tending and harvesting corn and potatoes (see table 2.12) provided additional food for the family and feed for their measly collection of barnyard animals.

Unlike among wealthier peasant families, where the character of production drew families together, among harvesters the family was just as likely to be drawn apart in work. Women often went to the fields to work alongside their husbands, but also, just as frequently, to work in place of them. Their inability to hire additional hands to assist in the fields, and the demands of reciprocal labor service and manorial sharecropping arrangements, shifted a greater proportion of fieldwork onto the wife's shoulders. Children were employed alongside their father in contractual labor projects, or they bore the additional burdens of his absence in tilling and cultivating the family's properties. The socioeconomic position of harvesters was extremely precarious; supplementing their summer gains with day labor in the winter was the only way of forestalling their eventual transformation into day laborers, the next stratum down on the village hierarchy.

Although harvesters owned land, and so were propertied members of the community, their social identity was stigmatized by

---

[24] Harvesters could save money by hiring their own children as swath-layers (*marokszedö*) and sheaf-binders (*kötözö*), rather than hiring a couple of local girls to do the job. One villager recalled that his brother never finished school, because his father insisted he help with the harvest instead.

their dependence upon others for work. To be subordinate to others in work—not to have absolute control over one's self—was normatively censured, and led the peasantry to characterize these villagers as passive and less diligent. Interestingly, movement became a defining characteristic of the work process for these villagers. The verbal descriptions of harvesting, *eljárt aratni* (he went to harvest wheat) or *eljárt az uradalomba* (he went to the manor, possibly including work done outside the harvest season), emphasized movement away from home which was habitual or frequent. In contrast, the expression describing a landowner's work in the fields, *kijárni a határba* (to go out frequently to the lands being cultivated outside the village), placed emphasis on moving from a stable center to one's fields, scattered about the periphery of the residential section of the village. One expression connoted peasant production, where the fields external to the village were bound intimately to the domestic realm; the other implied movement away from the regular productive structure of the household and domestic unit. Moreover, the use of the term *aratógazda* (overseer or supervisor) for the leader of the harvesters implied a proprietary relation. To translate the meaning of *aratógazda* as "owning the workers" conveys the sense that to submit to the control or will of others represents a loss of social identity. This is most explicitly articulated in discussions of manorial workers as animals, but the ascription of "reduced" humanity by landowners begins in relation to this stratum in the social hierarchy. People felt the pain of their poverty, and voiced it in song:

> I harvested a lot this summer,
> I didn't sleep much in bed:
> Sometimes in the forest, sometimes in the meadow,
> Sometimes in the middle of the stubble-field.
>
> Oh! my God, why do I live?
> Why was I born poor?
> If I hadn't been poor,
> I wouldn't be harvesting on someone else's bread.
>
> (Katona 1968, 28)

Wealthier villagers cast aspersions against harvesters, who were seen to have abandoned control over their own lives by leaving home to find work. Movement away from home was normatively sanctioned, understood as a renunciation of personal control over one's affairs. Clearly, specific notions of family cohesion were also at issue, better-endowed families having the luxury of working together while poorer families were forced to scatter across the local landscape in search of bread. I would point out that in combination, these attitudes suggest a strong disapproval on the part of wealthier villagers of segregating the home economy from other, more public activities within the community. As such, a private/public division—that is, an opposition between particularistic family matters and general village affairs—was alien to the hegemonic notion of a complete peasant household, a community of kin whose farming ties bound them as closely to their neighbors as to each other. The whole, undifferentiated world of peasant families—working the land, running the town council and marrying each other's children—was self-consciously contrasted with the world of poorer villagers, whose lives were ruptured by poverty.

While wealthier villagers disapproved of this fragmented existence, harvesters abhorred it. They despised those who lived off their labor, whose fields were fed by their sweat and blood.[25] Indeed, wealthy landowners were seen to gorge on the flesh of the poor.

> I harvested last summer,
> And I am still waiting for my earnings!
> Wheat is growing, barley has grown,
> The lord's granary is sated.
>
> So much of my sweat
> Streamed off me out there in the fields!
> Though what came of it?
> Someone got fattened on it!
>
> The poor man does not have any sense
> He could really live, but he still can't,
> Because he gives his blood and liver to others.
> They just squander it.

---

[25] Landowning peasants often talked of nurturing the earth with their sweat and blood, and they distinguished themselves from the landless, who, without property, were not able to invest themselves in the land. Indeed, some argued during the 1945 land reform that the landless should not be awarded property, because they had not fed the earth as had the peasantry. The poor did not share the peasants' conception that productive labor came only with landownership. As the folk song illustrates, the poor also believed that they fed the land, but they knew that their labor fed the rich as well.

> But I would like to shout,
> So that anyone could hear:
> Whoever doesn't hoe and whoever doesn't harvest,
> Should eat under the table!
>
> (Balassa 1985, 251)

Harvesters were consumed by their poverty, surrendering their most precious possession to people who squandered it away. Their humanity, their reason, slowly abandoned them, pouring off their bodies in the heat of the summer harvest.

Day Laborers (napszámos)

> The village poor are struggling
> Their whole life long.
> They have no peace on earth,
> Only out in the cemetery.
>
> His clothes wear to threads on him.
> He doesn't even have any honor.
> If he demands his rights
> Lords brand him as a tramp.
>
> Our final escape:
> Let us set out on a journey, mate!
> Perhaps foreign tyrants
> Have lighter shackles.
>
> (Katona, Simon, and Varga 1955, 159)

Day laborers constituted 66% of the village populace. Theirs was a cruel lot, living in perpetual existential uncertainty. They owned virtually no land, at most 1.2 ha (2 kh). Such holdings, divided into minuscule plots over several generations of parcellization and scattered sales, offered little more than room for a kitchen garden. They owned no draught animals and probably owned the barest minimum of tools—a hoe, maybe a sickle or scythe—necessary to be emplòyed by peasants or at the manor. No barnyard animals cluttered their courtyards or run-down stables, except perhaps for a few chickens, as all resources had to be devoted to feeding the family. The category of day laborer (*napszámos*) carries even greater connotations of subservience and existential uncertainty than that of harvester. While it was presumed that a harvester owned some land, a *napszámos* could just as well have been landless. The term used to describe those without land or nearly so was the graphic expression "have nothings" (*nincstelen*). The generic term for wage labor was *napszám* ("day number"), describing a work project that was ideally delimited by the length of one day. *Napszám* as a kind of work could be undertaken, and frequently was, by a harvester, but *napszámos* as a social group defined those with very limited sources of income or property. The expression used to describe the work of a day laborer was *napszámba jár*, "to go frequently to day labor." Similar in form to the expressions used for harvesters, *napszámba jár* reinforced the sense that one regularly sought work outside one's own domestic unit.

The day laborers of Sárosd worked almost exclusively in agriculture; rarely did they join road crews or state-funded construction projects.[26] In the depressed labor market of the 1930s, the manors surrounding the village offered relatively steady employment, discouraging fathers from leaving their families for longer periods of time to follow construction crews around the country. All members of the household sought work: children often as young as eight or nine years of age were compelled to find jobs, working long hours alongside adults in physically demanding work, such as weeding and thinning crops at the manor.

In contrast to other, more well-to-do strata, day laborers and their wives and children paid cash for goods in the village. Though they were sometimes paid in kind, they were often paid with money. Their paltry holdings did not even produce subsistence stuffs, much less a surplus, so they were constrained to pay cash rather than exchange grains or other foods for any necessities that they purchased from local merchants, such as petroleum or matches. Thus the poorer one was, the more one was forced to participate in the world of money and markets.

The constraints under which these families lived extended beyond immediate, apparently material concerns: they were even deprived of the right to choose their own political party. Only in the cities was there a secret ballot; in the countryside one's opportunities for employment at the manor were dependent upon voting for the party that was endorsed by local elites, such as Eszterházy

[26] In larger towns and cities where opportunities for day labor were not so plentiful as at the local manors in Sárosd, day laborers would gather in the town square to look for work. Overseers and others employed to recruit workers would select their crews from among the crowd. Appropriately enough, these gatherings were called "people fairs" or "people markets" (*embervásár* or *emberpiac*) (Katona 1958, 33).

and his manorial staff. Thus, the political cleavage in the village was defined not solely by the privileged access enjoyed by wealthier peasants to council offices and administrative positions in local organizations denied to the poor, but to the very choice of political affiliation. The poor had neither material privilege nor the right to determine the character of their political representation.[27]

None of the estates owned combines for harvesting wheat, even though they were widely available at the time. This was the direct result of national social policy promoted by the Agricultural Ministry. Although not illegal, the use of agricultural machinery that would deny work, at however paltry a wage, was actively discouraged (Szabó 1968, 21). Memories of agrarian strife in the 1890s, and widespread discontent in the 1920s, led interwar regimes to promote the use of labor-intensive methods at manorial estates and by wealthy peasants, especially in the case of the wheat harvest—the symbolic high point of the Hungarian agricultural calendar. For the villagers of Sárosd and its manors, the problem of unemployment was absent from otherwise keen memories of hardship in the interwar period; I was even told by one elderly informant that unemployment had been a problem only in industry. This memory of secure local employment was not shared by the villagers from the northern counties, where people were nearly always forced to migrate to find work. But as Dr. Kerbolt made clear (see table 2.9), relative unemployment was a serious problem in the winter months in Sárosd, despite its favorable labor market during the summer months. His figures apparently refer to harvesters and day laborers (Kerbolt used the term *munkás*), as manorial workers and even most peasants had little respite from their labors in what were considered by many to be the slow winter months. Dr. Kerbolt made a plea for greater distribution of industrial and crafts work in the region, as well as the rest of the country, to alleviate problems with winter employment. Local industries, especially food processing, would greatly improve social conditions.

---

[27] Szabó described the same disparity in voting rights in Tard. Peasants and others not dependent upon wage labor consistently voted for the Smallholders' Party (Kisgazdapárt), whereas anyone wishing to secure employment at the Coburg Manor was forced to vote for the Government Party (Kormánypárt) (1937, 193). Some have suggested that the political culture in formerly manorial villages has been strongly influenced by these patterns of de facto disenfranchisement in the interwar years, resulting in a general lack of knowledge about political action or interest in political affairs (Juhász, personal communication; Vági 1981).

TABLE 2.9. Employment in three villages, 1927

	Mezőkövesd	Mezőnyarad	Sárosd
January	1%	3%	5%
February	5	9	10
March	20	20	25
April	40	35	35
May	98	72	90
June	92	76	99
July	95	80	99
August	99	75	95
September	95	48	92
October	89	40	90
November	15	15	25
December	0	0	5

Source: Kerbolt 1934: 15

Note: Mezőkövesd and Mezőnyarad are communities in the north populated primarily by migrant laborers.

Not only would it provide our people with income in the winter too, but it would set our agriculture, which is struggling for its existence, back on its feet. However, since neither cottage industry nor manufacturing exists, agricultural workers spend the winter in almost complete idleness and without an income. They consume their wheat acquired through day labor or at harvest time, running up debts in advance against their income from next year at the store, butcher shop, mill, pub, if they have credit. Both material and moral damage derives from winter idleness, because the idle man, dissatisfied with his and his family's fate, easily listens to the words of those who promise much, endangering the future of the country and society, and in the final analysis, his own future and that of his family. (Kerbolt 1934, 14)

Fears among the wealthy of Communist agitators and other embittered ne'er-do-wells were not unfounded, as Dr. Kerbolt knew too well, probably having met suspicious vagrants at the manor on more than one occasion. By all accounts, poor villagers in Sárosd were tempted only by the prospect of employment, which they readily found at the manor in the peak summer season. The surrounding manors absorbed 90–100% of all agrarian workers for a good six months of the year, even in the depths of the depression, and offered odd jobs for some in quieter seasons as well.

Migrant Laborers (summás, hónapos)

Migrant laborers were considered a separate group from day laborers, although they also sought work away from their homes. The distinguishing feature here was the length of time that characterized their absence. A further critical feature distinguished migrant laborers from day laborers: while the *napszámos* was a local resident, migrant workers were outsiders. For the villagers, especially landowners, migrant laborers were complete strangers; they never met. Migrant laborers were essentially restricted to the estate, their work schedules too rigorous to permit excursions off the manor. Among manorial residents, migrant workers were treated as a separate group. And even though they may have owned as much or more land than their fellow day laborers or harvesters, the fact that they had to travel so far away from home to earn a year's keep stigmatized them.

> Mother, why did you bear me,
> If you raised me to be a migrant girl?
> You should have borne a marble stone,
> Mother, rather than me!
>
> (Borsai 1968, 134)

Food and housing provided to migrant workers were scanty and often deficient. Migrant workers frequently pooled their resources from their wages in produce, hiring a cook to feed them while they lived at the estate. If they were unlucky, the cook, or sometimes the steward, would short them on rations, leaving them poorly fed and with little recourse for compensation.

> The bailiff's assistant pocketed the price of the lard,
> And used it to buy a pocket watch.
> The pocket watch tick tocks in his pocket,
> That's why the top of the noodles is dry.
>
> (Ibid., 118)

Housing was also rudimentary, if not inadequate. At Jakabszállás migrant workers were housed in the sheep barns, emptied for the summer; manorial wives would whitewash the barns shortly before their arrival. No accommodations were made for privacy or comfort. Heller, an avid proponent of manorial production, reluctantly cited violations of the Geneva accords of 1921 which stipulated that certain basic provisions must be given to a resident workforce, none of which were fulfilled as a rule for migrant laborers in Fejér County (1937, 71).

Two terms were used to describe migrant labor: *summás* and *hónapos*. Both terms emphasized a temporal quality. The term *summás* was derived from the Latin word *summa* and referred to the fact that workers were recompensed at intervals longer than a day, usually on a monthly basis. *Hónapos* was an adjectival form of the word for month (*hónap*).[28] *Summás*es were migrant workers who were employed at the manor, generally on a six-month contract. The *hónapos*, also a migrant laborer and almost always female, was usually employed for two months, sometimes for two months in the spring and two months in the fall.

> I must go far away as a *summás* girl,
> I must leave my sweetheart here, and there's no one
> to leave him to.
> I leave him to you, my dearest companion,
> Enjoy your life with him, I don't care.
>
> I've come home, my dearest companion,
> Give me back my lover, I don't care.
> You enjoyed your life with him until now.
> I will live with him until they close my coffin.
>
> (Borsai 1968, 107)

Both groups, the *summás*es and *hónapos*es, performed a major share of the work in grain cultivation, with the general exception of seeding and reaping wheat. The *hónapos* often specialized in one crop, such as sugar beets, but the work itself was similar to that of the *summás*: hoeing, thinning, weeding crops. The work of *summás*es and *hónapos*es increased dramatically with the intensification of agricultural production toward the end of the nineteenth century, as they were the labor force most directly responsible for cultivating feed and cash crops introduced in this period (Lencsés 1982, 14). As table 2.10 also shows, Fejér County consistently employed the highest number of migrant workers during the 1930s.

In 1937 Szabó published his excellent sociographic account of a northern village of poor landholders, *The Situation in Tard* (*A Tardi Helyzet*). Two-thirds of the workforce were forced to seek day or migrant labor to feed their families (1937, 44), some of them traveling as far as Jakabszállás (1937, 39). Szabó depicted the

---

[28] In some areas *summás* and *hónapos* were interchangeable categories; not so in Sárosd.

TABLE 2.10. Trend in the number of migrant workers, by county

County	1926	1932	1933	1934	1935	1936	1937	1938	1940	1941
Baranya	—	1,814	2,145	2,521	2,551	2,717	2,946	3,392	2,560	4,523
Bars-Hont	—	—	—	—	—	—	—	—	—	2,111
Békés	—	—	—	2,051	1,394	—	—	1,040	1,239	1,834
Bihar	1,592	1,269	—	—	1,140	1,130	1,792	1,654	1,096	1,295
Borsod	—	—	—	—	—	—	—	—	—	1,097
Csanád-Arad-Torontál	3,218	—	—	2,187	2,123	1,683	2,646	2,775	1,901	3,225
Csongrád	—	—	—	—	—	—	—	—	1,714	2,194
Fejér	—	—	1,878	5,266	5,343	5,595	6,476	6,671	6,245	10,329
Győr-Moson-Pozsony	1,802	1,802	1,308	1,053	1,169	1,123	1,765	1,509	1,624	2,771
Heves	—	—	—	—	—	—	—	—	1,058	—
Jász-Nagykun-Szolnok	2,782	—	—	1,618	1,420	1,323	1,274	1,709	2,575	2,376
Komárom-Esztergom	—	—	—	—	—	1,076	1,705	—	3,010[a]	3,575[a]

County	1926	1932	1933	1934	1935	1936	1937	1938	1940	1941
Nógrád	—	—	—	—	—	—	—	1,116[b]	1,027	2,972
Pest-Pilis-Solt-Kiskun	2,258	—	—	1,268	2,213	1,536	1,384	—	3,483	3,701
Somogy	6,523	1,276	1,289	2,856	4,127	4,220	5,942	5,323	4,401	7,068
Sopron	1,606	—	—	—	—	—	1,079	1,107	1,024	1,371
Szabolcs-Ung	1,969	—	—	1,370	—	1,327	1,582	—	1,932[c]	1,832[c]
Tolna	2,805	—	—	2,358	1,996	2,714	2,942	3,660	2,696	3,903
Veszprém	3,333	1,814	—	3,692	2,922	3,577	3,761	—	3,815	5,210
Zala	—	—	1,320	—	—	—	—	—	—	—
Zemplén	1,191	—	1,685	1,421	1,615	1,348	1,500	1,125	1,110	1,760

Source: Lencsés 1982: 176
[a]Komárom
[b]Nógrád-Hont
[c]Szabolcs

helplessness and humiliation felt by the young men doomed to migrant labor, forced to supplement the fruits of their parents' meager holdings. Compelled to sell his labor—a young man's most precious possession—increasingly alienated him from the diligence of peasant life, as Erdei describes.

> In every one of them the conviction is ripening that they would no longer go to work at the manor because that servile life is unbearable, so they seek some kind of freer means of livelihood. However, since nothing like that happens, in the spring he is forced anew into *summás* work and he continues where he left off the year before. Therefore, every spring the *summás* turns into a proletarian, who doesn't want to be a manorial servant any longer, but in spring he is broken again and he will be a manorial worker till the fall after all. Yet all the rebellion and conviction transform him nonetheless from year to year. Although he returns every spring to the manor, with every spring he is less a manorial worker and more and more he endures life as a proletarian, who is freed from, even protesting against, peasant forms which are so strongly tied to this role. To a degree they become worse migrant workers and at the manors everyone complains about them more and more. At the manor, in contrast to the manorial workers, they are ever more pretentious, disrespectful, and much worse workers. (Erdei 1941, 146–147)

The demeaning experience of migrant labor forced young men and women, slowly but surely, into new worlds of work and of resistance.

> In front of the bailiff assistant's house there is a tree,
> The bailiff's assistant should be hung on it!
> I'll go to his hanging
> I'll pull the rope around his neck.
>
> (Borsai 1968, 138)

Harvesters, day laborers, and migrant workers were in a particularly problematic, interstitial position: their experience as erstwhile landowners and as wage laborers incorporated both being owners and being "owned." The pain and anger of this double life seems to have manifested itself in forms of rebellion very different from those issuing from the pains of humiliation felt by manorial workers. Unlike the abiding recalcitrance of manorial workers, harvesters and migrant workers were more abrupt. They waged strikes. In June 1929, migrant laborers mounted a strike at Jakabszállás. "Migrant workers receive their pay in wheat. The price of wheat dropped, so that they don't work as much as the overseer demanded" (FM f.b.i. 1929–33 res. sz.).[29]

Migrant workers also embraced Pentecostal or millenarian religions, depicted so well in Kovács's monograph *Silent Revolution* (*Néma Forradalom* 1937). The authorities found both forms of resistance dangerous, prosecuting self-appointed ministers of the new sects as political agitators, and treating youths missing from work as outlaws.

> These [young lads] will contract this spring and if they were not to enter service, a gendarme would come for them, so that the tranquillity of production not suffer harm even from one or two workers being truant. This is how they substitute for the kind of security which in the old days they could count on just as surely with slaves, as with animals and plows. (Szabó 1937, 43)

The treatment of agricultural workers as less than human, as animals or tools, was common.

> When he goes into the barn,
> Nyári Ferenc wakes the girls up in turn:
> "Wake up, working girls, by fifteen to six;
> Because the boss, the farm boss,
> Is shouting, 'Dogs to the fields!'"
>
> (Borsai 1968, 111)

Migrant workers, like harvesters, contracted with the manor through an appointed agent. Having come from so far away, their activity could not be defined as "frequenting the manor," as in the case of harvesters or even day laborers. The verb forms used to define their relation or activity were "to contract" (*elszegődik*) or "to serve" (*szolgál*), the same terms used for manorial workers. The verb "to contract" is richly suggestive in its connotations for concepts of power relations in work. It is derived from the verb *szeg*. The etymological dictionary lists the following meanings: "to

---

[29] In much of the literature on interwar village economies, the predominant use of payment in kind rather than monetary wages is understood to reinforce the view that feudalism was entrenched in rural communities (and to undermine the argument that an incipient capitalism was taking hold). We see in this case, however, that although migrant workers were paid in kind, they nonetheless calculated the value of their wheat in terms of its market price. Clearly, one cannot assume that workers being paid in kind are entirely unaware of market prices, or did not consider these factors in calculating contracts or strikes.

break, smash, crush; cut; ruin; to break a law or an oath; to bend, stoop, submit to someone's will; to turn in some direction (thereby deviating from one's own direction); to contract to someone" (Benkö 1967, 3: 693–694). The meanings listed for the verb *szegődik* are: "to make or enter an agreement; to take up work contracted for, to serve; to attach oneself to (ally oneself with); to tend toward, be aimed at something; to bend, be supple, incline to, yield; to offer excuses, protest, remonstrate, show reluctance" (ibid., 696). The verb to serve (*szolgálni*), which has no comparable complex of meanings, is interesting with regard to the categories of subjects it allows: soldiers, manorial workers, and tools.[30]

The cluster of meanings mentioned above suggests that a contractual relation at the manor was predicated on subservience, that is, on the renunciation of control over one's own affairs. Clearly this is consistent with the hierarchical scale that categorized the lower social strata as lacking control, which when rendered in the normative terms that were used by landowning villagers became slovenliness, stupidity, untrustworthiness, and a rejection of responsibility.

> His Excellency stood out on the balcony,
> From there he watches what migrant workers are doing:
> Do they work, or are they just loitering?
> Have they earned their sour beans?
>
> (Borsai 1968, 122–123)

The final series of meanings under *szegődik*—"to offer excuses, protest, remonstrate, show reluctance"—points up the tensions or anger that pervaded a relation based on duress or coercion, albeit the coercion of poverty. The sayings "A manorial worker/servant is a paid enemy" (*A cseléd fizetett ellenség*) or "You have as many enemies as you have servants" (*Mennyi szolgád, annyi ellenség*) convey similar attitudes (O. Nagy 1966, 122, 648). The problem of discipline and the broader issue of power struggles at the manor will be discussed below; here I wish only to note that the forces of subjugation and rebellion were mutually implicated in the very concept of contracting at the manor.

---

[30] I asked a friend to confirm my list of possible "subjects who serve," wondering why animals were not included. "I guess it's because they have a will of their own, unlike manorial workers."

Manorial Workers (cseléd)

The life of the day laborer and migrant worker was miserable, fraught with uncertainty, and characterized by powerlessness. Yet despite all the pain and humiliation suffered by these families, their lot was considered superior to that of the manorial servant or worker (cseléd) and his family, who lived beyond the confines of the village. The manorial community was a world apart: highly differentiated internally, the society of manor employees was perceived by the villagers as a self-sufficient body alien in all respects to their community. Table 2.11 lists personnel employed at Pusztasárosd, Jakabszállás, and Csillag. They include overseers, granary supervisors, tradesmen, gamekeepers, and common workers. Also listed are occasional workers, such as harvesters, migrant workers, and day laborers. Table 2.12 describes the kommenció or yearly contract provided by each of the manors surrounding Sárosd.

Earlier I suggested that centeredness and movement were an aspect of the ideology of control or possessing activity. In local moral categories, centeredness was correlated with wealth and material security, whereas movement was associated with poverty. Accordingly, as one moved downward on the social scale, the productive relation defining specific social groups was characterized by ever more movement and greater and greater distance from the domestic unit.[31] Harvesters and day laborers moved back and forth daily between the manor (or peasant households) and their own productive units. Migrant workers abandoned their villages for six months and crossed counties to earn a winter's keep. Manorial workers, who were bound to the isolation of manorial estates, nonetheless epitomized "uncenteredness": not only did they not possess their own labor (activity), they did not even have a hearth or home to call their own. In fact, the phrase used to convey the act of contracting at the manor as a worker (cselédnek ment or elment cselédnek), an act undertaken most often in desperation, was analogous to the phrase used to describe a man's moving to his wife's house at marriage. Permanence or irrevocability is implied in this phrasing, a meaning clearly absent from the expressions of movement for harvesters and day laborers which focused on frequency (jár, eljár). Central to this image was the notion of permanent subordination and humiliation: servitude at the manor was as miserable as uxorilocal residence in this strongly patriarchal soci-

---

[31] The term for vagabond, vagrant, or shady character, jöttment ("came and went"), succinctly expresses the common disdain for the transient.

TABLE 2.11. Personnel at manors surrounding Sárosd, 1930

	Pusztasárosd	Jakabszállás	Csillag
Overseer	8	3	1
Granary supervisor	2	1	1
Tradesman	16	7	3
Caretaker/ gamekeeper	4	4	3
Viticulturer	1	n.a.	n.a.
Distillery worker	n.a.	6	n.a.
Herdsman	22	n.a.	n.a.
Common worker	97	82	22
Harvester (pair of workers)			
from Sárosd	60		
from elsewhere	20[a]	60[b]	20[c]
Migrant worker (*summás*)			
from Sárosd	40	n.a.	n.a.
from elsewhere			
in spring	110–130		
in summer	30–40		
in fall	100		
Day laborer (yearly average)			
adults	3,625[d]	12,000	1,000
boys and girls	4,771		
children	1,454[e]		

Source: Sárosd monográfia

[a]These harvesters were recruited from Heves and Békés Counties.

[b]Recruited from Sárkeresztúr, a neighboring village.

[c]From Sárkeresztúr.

[d]Specifies male workers.

[e]"Boys and girls" presumably refer to those above the age of twelve, while children fall under that age.

ety where virilocal residence was the norm. The permanence, however, refers to the status of servitude rather than to a permanent association with one specific estate. Unlike the days of serfdom, when movement was restricted, here certainty of place is not the assumption; rather, it is the certainty of homelessness. The bondage of homelessness was perhaps most poignantly conveyed in images of families roaming the frozen roads of January each year when manorial contracts were renegotiated, and some families would be sent away to find employment elsewhere.

The term that I have translated throughout as manorial worker, *cseléd*, literally means "servant." Domestic servants at the

TABLE 2.12a. Salary contracts at three manors in 1930, Sárosd

	Wheat kg	Rye kg	Barley kg	Feed land mh[a]	–öl[b]
Overseer					
Puszta	800	800	600	2.0	—
Jakabszállás	1,067	1,067	333	—	—
Csillag	(Total = 1,800)			—	1,400
Tradesman					
Puszta	1,000	1,000	1,000	3.5	—
Jakabszállás	929	929	314	—	—
Csillag	(Total = 1,800)			—	1,400
Caretaker					
Puszta	800	800	600	2.5	—
Jakabszállás	950	950	300	—	—
Csillag	(Total = 1,800)			—	1,400
Common worker					
Puszta	700	800	500	1.5	—
Jakabszállás	732	829	220	—	—
Csillag	(Total = 1,800)			—	1,400
Half-wage worker[c]					
Puszta	400	400	400	0.75	—
Jakabszállás	—	—	—	—	—
Csillag	—	—	—	—	—
Granary supervisor					
Puszta	—	—	—	—	—
Jakabszállás	1,200	1,000	400	—	—
Csillag	(Total = 1,800)			—	—

Source: Sárosd monográfia

[a] mh (*magyar hold*): three-fourths the size a standard *hold*; a standard *hold* is 1,600 square fathoms, the Hungarian *hold* is 1,200 square fathoms.

[b] –öl: abbreviation for square fathom, equaling 3.57 m^2.

[c] Half-wage worker (*kisberes*): position held by adolescent boy, may also be translated as apprentice ox-driver.

Notes: Manorial workers at the Puszta were also each issued 24 kg of salt yearly, firewood as needed, and free medical care.

manor were distinguished from manorial workers by the designation *házi* (of the house), but someone working as a farmhand for a peasant family, despite his frequently intimate association with the family, was simply called a *cseléd*. The etymological dictionary explains that until the sixteenth century, *cseléd* was synonymous with the term for "family" (*család*), but by the nineteenth

TABLE 2.12b. Animals allowed at the Puszta, 1930

	No. of cows	No. of pigs
Overseer	2	2
Tradesman	3	3
Caretaker	2	2
Common worker	1	2

Source: Sárosd monográfia

Note: The *Sárosd monográfia* lists specific figures for the numbers of animals allowed each household at the Puszta, although I have no information that would indicate that the other manors in Sárosd did not allow manorial families to keep animals.

TABLE 2.12c. Wages for harvesters, migrant workers, and day laborers at the manors

	Pusztasárosd	Jakabszállás	Csillag
Harvester			
Grains per pair	1,084 kg	20 q[1]	1,500 kg
Migrant Worker			
6-month	13–15 q grain/pair	—	—
2-month	30–40 P[2] per capita	—	—
Day laborer		2 P (average)	2 P
Category I	2.00 P		
Category II	1.50 P		
Category III	1.00 P		

Source: Sárosd monográfia
[1] q = quintal
[2] P = pengö

century the meaning of the two words had diverged (Benkö 1967, 1: 493).

Isolated socially and spatially from nearby villages, manorial communities had their own, highly stratified hierarchies of social position and prestige, based upon job categories. Descriptions of manorial ranking attest to the rigidity of these boundaries.

> The hierarchy was articulated according to the character of work completed and the rank of service, and so too did the ranking among manorial families develop. . . . The *kommenció* was handed out according to rank: first to the overseer, after that to the tradesmen, then to the forester, gardener, to the lord's liveried coachman and finally to the common workers, among them first to the leading farmhands and only then to the rest of them. The value

assigned to different kinds of work was also reflected in the differentiation of incomes. The salary of managers was one and a half to two times that of others; the overseer, gardener, and the herders were given a portion of the produce or some animals as incentive. The pig herder could keep every tenth piglet from among the weaned piglets and he could feed them together with the manor's animals. Shepherds got 20% of the offspring and also received a share of the curd. Wagoners and fieldhands were not only at the bottom of the hierarchy in power and prestige, but also in income. (Mátyus and Tausz 1984, 133–135)

The social hierarchy based on the division of labor also determined the future careers of children on the manor.

> Their job was handed down from father to son; the most inflexible caste-system prevails on the puszta. The steward's son would sooner or later become steward or someone of equal rank; the puszta foreman's son, even though he went through the stages of boy labourer, ordinary labourer and carter, would certainly become a foreman by the time he was an old man. . . . It would be impossible to imagine a cattle-foreman being down-graded; he would sooner leave. Nor could a carter be demoted to ordinary labourer without causing a tragedy. And as for putting a horse-driver among the ox-drivers, it would be like trying to make a Negro out of a white Yankee. Even the very rare marriages between their families are regarded as a mild form of race-degradation. (Illyés 1967, 94)[32]

The organizational segregation of different work groups also isolated people working on the manor for shorter periods of time, such as harvesters and migrant workers.

Choice of a marriage partner among manorial workers may have been structured by one's relation to the hierarchy of labor organization, as Illyés suggested. In this sense, marriage patterns would have paralleled alliances among social groups of equal standing and property ownership in the village: landowners with landowners, harvesters with harvesters, and so forth. Memory in Sárosd records that these subtle distinctions of rank at the manor were not important. Freed of the burden of property (not needing to

---

[32] Illyés Gyula, author of *People of the Puszta* (*Puszták népe*), grew up before World War I on a manor a couple hundred miles south of Sárosd. His monograph was one of several published in the mid-1930s to counter the conservative regime's romantic depiction of the peasantry. Son of a manorial shepherd, he and his family fought their way off the manor, and he eventually became one of Hungary's leading poets.

consolidate holdings among families) and less concerned with social hierarchies, manorial workers could follow their hearts in selecting mates. Nor were they confined to the spatial boundary of their *puszta*. Moving about year to year, manorial families had social ties that spread across the entire county. Marriages were struck on one *puszta*, but as the family moved from one estate to another, children were born in parts far and wide. Birth records of manorial families thus contrast sharply with those of villagers, whose successive generations were born in one community. The pride of place villagers espoused, not least because it connoted to them an honorable work ethic, was denied manorial servants. Manorial workers were forced to roam from *puszta* to *puszta*, sparse colonies hidden behind proud villages and tucked underneath grand castles.

Manorial communities were isolated: physically isolated from villages and towns, socially isolated from the village hierarchy of property and pride. Nonetheless, manorial estates were often quite busy places: new families moved onto the *puszta*s every year and migrant workers invaded in the spring. In addition to these regular guests, occasional visitors would also frequent the *puszta*s, bringing news and notions foreign to the manorial community. Wandering figures, vagrants, and itinerant workers (known as *vendel*s in this region), would appear at the manor, performing small jobs for manorial families in exchange for money or food. In the evening, many would gather around the vagrants to hear news of other *puszta*s or listen to colorful stories. The *vendel*s were suspicious characters, often déclassé nobles and disgraced public officials, who because of scandal or love of the drink, had lost the right to remain among their peers. Still others were political refugees, who roamed from manor to manor hoping to engage the disenfranchised in a struggle for social justice (Lukács 1983).[33]

Manorial workers and their families were forced to live in wretched hovels. Four families were housed in a room, each family living in one of the four corners. A chest with four drawers, one

---

[33] When I asked in Sárosd whether there had been vagrants at the *puszta*s, I was informed that one old man had met József Attila in this way. József Attila, one of Hungary's greatest poets, came from a working-class background, which was highly unusual for a literary figure in Hungary, then or now. He was also very actively involved in left-wing politics in the 1930s, before his suicide in 1937. The story about the greatest poet of his generation loitering around Jakabszállás seems extremely improbable, but what is more important, it conveys much about the image of *vendel*s as complex figures—potential artists, probable politicians—whose lives could be seen as both noble and tragic.

per family, sat in the middle of the room, the only piece of furniture besides the bed each family had tucked in the corner. Another four families were housed in the room opposite, separated by a middle room, which served as the kitchen. It contained a large stove with eight separate burners. The buildings were of poor construction, and were rarely insulated. People said that death lived in the walls, believing that the dampness of the walls caused serious illness.[34] As one villager recalled, manorial workers were assigned places by the *gazda*, with no say whatsoever about their living quarters. "Only if there were a problem would the *gazda* intervene. It was a wretched life. People would attack each other with scythes." Heller offers us a grisly picture of the absolute squalor of these "homes" when he argues against the concern voiced by some observers that there was not enough oxygen to breathe in these crowded flats. The true problem, he explains, was the dampness in the air.

> Naturally the unhealthy character of the living quarters depends strongly on how many people live in its rooms. But contrary to general belief, the reduction of the oxygen content of the room (21%) and the increase in carbon dioxide (0.04%) through expiration is not the most important question in this respect because a decrease of oxygen to 10% is bearable. One person consumes approximately 20 liters of oxygen per hour and the natural ventilation of walls, window, and door openings is also to be taken into consideration. However, the water content and the higher temperature of the air have a much more injurious effect on health, the primary cause of which, in addition to cooking in the room, is exhalation. Beyond the possibility of infection, the curse of so many people living together is the obscuring of all possible hygienic consequences. (Heller 1937, 60)[35]

The description from Dr. Kerbolt's pen was quite different.

---

[34] When the family discovered that a young child who was sleeping next to the wall had fallen ill, the mother would then sleep in that same place, succumbing eventually to the same fate. Once a child reached the age of ten, the family could expect him or her to survive. In Europe, tuberculosis used to be called the "Hungarian disease."

[35] Heller's scientific treatment of the question of sufficient oxygen was characteristic of a whole school of Tayloristic number-crunchers bent on calculating the biological constraints on labor, in order to set the boundaries within which the most effort could be extracted from agricultural workers. By the 1940s an institute had been established, the Agricultural Work Scientific Institute (Mezőgazdasági Munkatudományi Intézet), to study questions such as caloric intake and work performance; Heller was a frequent contributor to their journal.

> [T]he home of agricultural workers, with a few exceptions, could be called everything but a home or healthy. I remember well houses dug into the ground, hovels, which had only one advantage: they were warm in the winter and cold in the summer, but they were damp, with a dirt floor, and not even by accident did the sunlight ever shine through the hand-sized window. These houses were not eliminated as much by the law as by growing social and hygienic concerns. However, common apartments where four, five, even six families lived in a large, shedlike room—at least in our region—were abolished completely only in the recent past. In the old days each corner of the room was a "suite" for a family. The stench and dirt in such rooms, where 25–30 people lived together, was inconceivable. I recall once visiting a sick patient in this kind of four-family room: the patient was a woman confined in childbirth, the second family had a dead child, the third family was slaughtering a pig. Luckily nothing out of the ordinary was happening with the fourth family. (Kerbolt 1934, 33–34)

Laws were brought in 1907 and 1913 to remedy the situation of poor manorial housing, but their implementation was hampered by the First World War. Heller cites the depression of the 1930s as an additional impediment to serious capital investment in housing at manors in Fejér County (1937, 67–68). Eventually some manors built new houses, or overhauled former barracks. The new style consisted of a kitchen and adjoining room, and only one family was assigned to each unit.

Production at the manor was divided between discrete branches, and as the description above attests, there was no horizontal movement among personnel between these units.[36] Yet horizontal rigidity was not matched by vertical complexity. The overseer and branch foremen or farmhands comprised the full complement of managerial personnel. Although some manors also employed bookkeeping personnel in addition to the overseer (as did Eszterházy), the following description is probably characteristic of manors generally.

> [T]he overseer (*gazda*) . . . directed the work. He received instructions at first hand. Every evening he had to go up to the castle, at which time they briefed him on the coming

---

[36] This paucity of experience in different kinds of agricultural work clearly contributed (but did not wholly determine) the difficulties experienced by former manorial workers in establishing viable family farms after the land reform in 1945.

day's tasks: what must be completed the next day, how many day laborers he must employ. . . . The overseer gave instructions to the farmhands and the forester, who then informed the horse-herders, cow-herders, shepherds, swine herds, wagoners, and fieldhands. They then directed the shepherd boys and farm boys. The masters directly commanded the gardener; the overseer was in charge of the craftsmen, machinists, and cartwrights. (Mátyus and Tausz 1984, 134)[37]

Understandably, the practice of running a manor differed, owners and renters varying considerably in the degree of their participation. Some owners, like Eszterházy, took an active role in supervising production; several renters were also immediately involved in production, like Bartha at Jakabszállás. Other owners and renters had absolutely no interest or inclination to become involved, and hired a steward to run the enterprise, as was the case with the Widow Weisz.

The pace and character of work for fieldhands varied from season to season. They plowed, fertilized, and conditioned the soil in slow months. In peak seasons they prepared the soil for planting, assisted the migrant labor force and harvesters in tilling the land and reaping the harvest, and carted crops to the estate center and produce to market. The work of manorial workers assigned to the stables was less variable, as the needs of the animals were fairly constant the year round. The only major difference in their regimen was occasioned by the stabling of herds in the winter and the pasturing of herds in the summer. Dairy workers (cowherds) would be at the stables as early as 3:00 A.M. to begin cleaning and milking the herd. Workers in other branches of animal husbandry would be at the stables by 4:00 A.M. to begin feeding and cleaning. The hours of a herder were long. One man who had been a cowherd told me he rarely saw his children awake, as he left before they arose and arrived home after they had gone to bed. On the rare occasions when he was home, he was so happy to spend time with his children, but they pulled away from him, crying. "It hurt me more than them. They didn't even know who I was."

The workers in the animal husbandry sections were further restricted by the constant care needed by the animals: they were rarely given holidays off. Other members of the manorial community were allowed to take Sundays off, but not those who worked

---

[37] This description bears strong similarities to descriptions of the delegation of tasks by brigade leaders (*brigádvezető*) at the cooperative farm in the early years of its existence in the 1960s.

with animals. All manorial contracts stipulated that the management would specify what constituted a day off, and that manorial responsibilities superseded all else in importance. For example, manorial workers were allowed to attend the village market fair only twice a year (the first fair in May and the last one in November), though the market was held four times a year. It is reasonable to assume that the majority of common workers at the estates were employed in the animal husbandry branches, as most of the work involved in grain production—hoeing, thinning, weeding, and harvesting—was carried out by the day laborers and harvesters.[38]

Women's primary responsibilities on the manor were caring for the children, tending the private stock of animals owned by the family, and working the kitchen garden, potato fields, and cornfields allotted to them as part of their contract with the manor. They essentially ran a minifarm, keeping a few chickens, pigs, and a calf, if allowed in their contract. Manorial workers bought fruit from a traveling peddler—a local Roma woman—exchanging their grains for fruit, weight for weight. Heller's description of women's work on the manor suggests that on occasion manorial women may have entered into sharecropping contracts with the manor.

> The fate of the family is put in the wife's hands. Her husband is overwhelmed by manorial work; the entire burden of their petty commodity household falls on her shoulders: the worry of keeping animals, the "farmhand's land" and its cultivation, work on fodder crops undertaken for a share, and exploitation of available possibilities for day labor. In other words, life is increasingly eclipsing the type of manorial wife "who sits in the doorway." (Heller 1937, 130)

From his account it is not possible to determine whether the sharecropping work mentioned was that stipulated in the husband's contract (i.e., potato and corn plots), or whether additional sharecropping was undertaken by the wife on separate contract. It does appear, however, that women engaged in day labor if the opportunity presented itself. One informant spoke of working in the

---

[38] The pronounced disinterest among current residents of the village toward regular church attendance is generally explained by the overwhelming amount of work that needs to be done around the house on the weekends, especially garden and cash crop work. Despite the memory of frequent church attendance in the interwar years, I strongly suspect that patterns of poor church attendance had already been influenced by manorial restrictions on Sunday activities. Illyés corroborates this observation (1967, 47–67, 112–117).

dairy stables in the wintertime when she was free from garden and field work. Heller's phrasing, "life . . . eclipsing . . . in the doorway," suggests that women engaging in wage labor were a more recent phenomenon. Heller clearly presented their foray into wage labor in a positive light, since as an advocate of manorial production, his overriding concern was the full use of residential labor power. But it seems clear that women's resort to day labor at the manor was related to the serious devaluation of manorial salaries between 1925 and 1934 due to the depression. Between 1928 and 1933, the monetary value of the *kommenció* suffered a loss of 55–60% (Kovács 1937, 138). Additional labor duties, then, were not the sign of new-found diligence; rather, they represented the crucial need women felt by the 1930s to augment their families' plummeting income.

Manorial workers were not immediately influenced by daily and monthly fluctuations in the labor and commodities market, in contrast to both poor and wealthy villagers. Their salaries in grains, the ability to keep animals, and usufruct rights to garden and fodder plots provided them a minimal, but fairly dependable living (barring a major epidemic in animal barns shared communally by manorial workers).[39] Nonetheless, the worsening situation in the early 1930s also touched manorial families. The monetary value of their salaries dropped precipitously, as did the monetary value of the items they brought to market. This had serious consequences for manorial workers, since their sole means of acquiring cash was to sell livestock and any surplus produce at the village fair. As Kovács describes, the category of surplus had to undergo revision as the value of marketed agricultural goods deteriorated.

> Until 1929 there was no problem because industrial and agricultural prices developed proportionately, but in 1929 the price of agricultural products fell away from the price of industrial goods and descended increasingly out of all proportion. The "agrarian scissor" [parity] opened up. Until then the manorial worker could do well taking his *konvenció* goods to market. With the opening of the "agrarian scissor," however, he also had to take proportionately more produce to the market, that is, he had to deny his own mouth to satisfy other types of needs (clothes, shoes,

---

[39] Juhász has suggested that the isolation of manorial workers from the labor market through, in part, the provision of basic subsistence needs, was repeated in the practices of state and cooperative farms, which guaranteed employees and members usufruct rights to small plots from the 1950s on, thereby assuring a stable, but less well paid workforce for agriculture (Juhász 1983, 118).

etc.). He had to choose between having his full, and having decent clothes and shoes; there wasn't enough room for both. Either he ate enough and then he had to go around in rags, or he dressed decently and then stayed hungry. (Kovács 1937, 199–200)

The market bottomed out in 1933, but good incomes were not assured anyone doing wage labor until the war boom in the early 1940s.

The separation of spouses in work characteristic of harvester families, day laborers, and migrant workers was also evident among manorial families. Wives and husbands labored independently of one another. Children also worked within the same sexual division of labor: sons alongside fathers in manorial jobs, daughters alongside mothers in pigsties, barns, gardens, and fields. The hours were long and the work debilitating. Manorial families had little recourse but to endure their lot. Without enough capital, they could not buy a house in the village to escape into the uncertainties of day-to-day labor contracts. The depression had seriously reduced the numbers of industrial jobs in the cities, forcing many back to villages to find work.

The issue of power relations in the productive process has been implicit throughout this discussion of work. The degree to which one controlled one's own activity and possessed one's time and one's produce has been shown to be a crucial element of self-esteem and to be central to constructing social value. Manorial workers represented in all respects the antithesis of managing and control to the villagers. The surrender of one's labor power to the manor—the full alienation of self—necessarily entailed or, more accurately, presupposed a repudiation of the human qualities of initiative, independence, knowledge, and reasoning. Villagers considered manorial workers to be lazy, stubborn, and stupid. It should not be surprising, then, that Illyés's discussion of work at the manor was prefaced with a detailed description of how to discipline the workforce.

Up to the age of thirty to thirty-five, the people of the pusztas are generally struck on the face. After this they usually receive blows on the back of the head or the neck and then just one blow as a rule. . . . The use of canes, sticks, and riding-whips on those over twenty is to be avoided as far as possible. For while they usually take direct punishment administered by one living body to another with resignation or even with a smile, seeing in it some kind of human contact, punishment inflicted with the aid of some external appliance arouses in them incalculable reactions. Similar observations have been noted by researchers into animal psychology. (Illyés 1967, 122)

Illyés continued, commenting on the thickness of air on the manor, which forced the stewards to shout their orders repeatedly (ibid., 125). He described the tortoiselike slowness of all actions taken by manorial workers, from plowing to blowing their noses (126) and even discussed regional differences in the efficacy of punishment (123–124). In short, he offered us an excellent portrayal of the patterns of dominance and resistance. It seems clear from numerous accounts that manorial workers were actively engaged in pacing production at the estate. What to the peasant mind was a rejection of an agrarian work ethic, or to others the survival of feudal work habits, may be called a permanent work slowdown. The most frequent charge in legal suits over labor contracts was that the worker avoided his duties, and was a bad influence over other workers. Thus, the battle waged by workers at the manor had to take the form of resistance within compliance, to maintain the appearance of servility while effectively reducing the inhuman demands of long and difficult work.[40]

Erdei (1941, 136–137), Tausz (Mátyus and Tausz 1984, 126) and Illyés (1967, 126–128) all describe work patterns that appear to be work slowdowns, but their explanation is that manorial workers wished to spare themselves the crippling drudgery of the workload. Essentially, they defined these actions as a means of coping with the difficult circumstances at the manor. This perspective underestimates the manorial workers' very active role in attempting to control the productive process. Although the manorial worker had sold his labor to the estate, he continued to put constraints on its use. The manor literally had to provoke work at every turn.

> Behind every little band of three or four folk hoeing there stands an overseer with his stick; his only job is to spur them on. His work is not easy, nor is it very successful. I

---

[40] It was said by some that as long as there were 100 manors in the country, they were undisturbed by the threat of termination, since they were sure not to live to the age of 100. They could always move on to find work elsewhere. (This attitude suggests an attenuated sense of place among manorial workers, so different from villagers, who clung to the earth.) The fact that someone had been dismissed from another estate apparently did not influence his chances of being hired again, except in the most extreme cases of insubordination. The quip about there being more manors than the number of years in one's life was also frequently used to express indifference toward complaints against a worker by management at the cooperative farm: as long as there are 100 cooperative farms . . .

would not go so far as to say that he would sometimes get better results by himself if he were to turn into direct labour all the strength he squanders on perpetual encouragement, bickering and quarrelling. This excessive supervision has at least one result—the moment they realise that they are out of range of an overseer, the folk immediately stop working. The officials on the puszta allege that they have developed a sixth sense for this. (Illyés 1967: 125)

It was as if the overseer (*gazda*), as "owner" of the servants' labor, and the servant, as the object of his possession, were required to interact continuously to complete any specific productive act. Because the labor of manorial workers was truly alienated, it was the manor's job, not theirs, to put it to use. This was another manifestation of the ideology of "possessing activity" in the agrarian community, the onus of sustaining activity falling on the shoulders of supervisors, not on workers.

The problem of theft at the manor, described by all as endemic, also sheds light on manorial workers' alternative views of their relation to the manor. Although more generous accounts considered theft by manorial residents an expression of atavistic communalism (Illyés 1967, 36), estate owners and peasants regarded it as a further indication that manorial workers had no respect for private property, clearly a direct result of their repudiation of being owners themselves. Manorial residents did, however, distinguish between different kinds of property, for anything owned by the manor was fair game, while the possessions of other residents, such as chickens or garden vegetables, were strictly off bounds. The primary resource coveted by manorial workers was feed and fodder used to supplement their own stores of feed for their private stock. Endemic theft would be more accurately represented as the manorial residents' stubborn refusal to identify their interests with that of the manor. For them the alienation of their labor did not require an alienation of family interests or needs. Actually, the needs of manorial workers and their families had to be asserted against those of the manor, as attested to in the maxim "If you don't steal from the manor, then you're stealing from your family." Thus, what appeared to peasants and stewards as the absence of a "property ethic" was an affirmation of the social identity of manorial workers, defined first and foremost by their allegiance to family over and above submission to the manor.[41]

---

[41] Manorial personnel often colluded in theft, seeing it as the healthy expression of personal initiative and creativity. One of the Jews who rented a manor in Sárosd was known to chastise his workers if they did not avail themselves of his grains. If he saw

Officials and county authorities engaged in a battle for the hearts and minds of manorial workers, but with little success. To counter the poor climate among workers at manorial estates, the lord lieutenant of the county established an award in 1938 to be given to manorial workers or construction workers (*kubikus*) who had served for a long time in one place and had thus demonstrated loyal service to their masters (SJ f.i. 711/1938). It is interesting to note that loyalty and the quality of service were demonstrated to the lord lieutenant by a worker's staying put, demonstrating that a concept of morality built on immobility was held by wealthy county officials, not only by landowning peasants. Seeking a better contract or more humane conditions by traveling from manor to manor—that is, taking advantage of a possible labor market—was seen by the wealthy to be a sign of disloyalty and unreliability.

Many town clerks responded to the lord lieutenant's call for nominees, saying that there were no such men in their community. In Sárosd, they succeeded in finding a loyal servant, Auer János, who was given eighty pengő and a certificate of merit. The political significance of this event was not to be lost, the lord lieutenant giving instructions on how to orchestrate the ceremony.

> In accord with customs hitherto, the presentation should occur in such a way that the moral value of the award be expressed before the prize winners. This should be done with the collaboration of the farming community and in consideration of local conditions. Particular stress should be placed on the prize winners as examples for the manorial servants and working people to follow.
>
> Be so kind, furthermore, as to exert your influence on the local press to mention the celebration and to send those decorated a separate copy of the paper.
>
> Székesfehérvár, October 10, 1938 (ibid.)

The lord lieutenant's campaign to instill greater industriousness among manorial workers by distributing certificates of merit seems a shallow response to the depression and despair of the late 1930s. Kerbolt's diagnosis of the sick village identified the culprit as an untreated economic malaise.

> If we recognize that the large majority of agricultural workers, dwarf landholders, even small farmers cannot

---

that someone's livestock looked thin and poorly fed, he would ask "Sleeping on your ear again?" implying that his employee should be up stealing grains, instead of sleeping through the night.

even earn the bare necessities, then we understand too why they eat poorly, why they are in rags, why they live in damp rooms with a dirt floor, why tuberculosis ravages them, why they do not turn to a doctor when sick, and why in this country so many infants die before their first year. (Kerbolt 1934, 21)

The Politics of Poverty

In the preceding discussion, I have described the various social groups living in Sárosd and its *pusztas*. The bonds between these groups varied, some quite fragile and ephemeral, others strong and long-lasting. Deep divisions existed between those with and without property, their differences expressed vividly in the work they performed day in and day out. But there were bonds built of shared assumptions, of common hopes and aspirations. Every man wanted to become independent, to be his own master and master of his family. Every woman aspired to provide a decent living for her family, sharing the joys and sorrows of building a household with her husband. If denied the chance to own property or even their own labor, people would still fight for their dignity, as the history of strikes and disturbances in Sárosd during the interwar period attests. Indeed, struggles over property and production were common following the end of the First World War. The ensuing decade was a restless time: first came the tumultuous, revolutionary flip-flops of 1919, then the skirmishes over the land reform, a series of strikes, and numerous disturbances at the local manors. By the 1930s, however, resistance waned, crushed by the heavy pall spread over the community by the depression.

The Soviet Republic of 1919

Memory records, and local documents in Sárosd support, a view of 1919 in which very little of significance occurred.[42] In the words of the village clerk, writing in 1930, "The revolution and commune did not cause a shake-up in the community to any extent. Generally, with a few exceptions, the inhabitants gave evidence of their patriotic feelings" (*Sárosd monográfia*). In early March 1919,

---

[42] On a national scale 1918–1919 was a momentous transition. In October 1918 the first Popular Democratic Republic in Hungary was declared, growing out of the ashes of the defeated empire. Its fate was short-lived, as was that of the Soviet Republic which followed it in March 1919.

during the last days of the Popular Democratic Republic, a limited land reform was initiated. Local committees were established to determine those who wished to receive lands, and those lands that could be distributed. Despite great interest among the poor for land all across the country, the committees designated to oversee the reform postponed its execution, not wishing to interfere with production already underway. So too in Sárosd.

> The president of the committee for redistributing land in Sárosd, at the urging of the steering committee of the land redistribution committee, protested most emphatically against the immediate distribution of land. He justified his position by stating that "the practical execution of the land reform—the most significant achievement of the revolution—is conceivable only if property be distributed in autumn, so as not to impede the continuity of production." (Jenei 1969, 91)

Different sentiments were expressed among the poor living on manors around Sárosd. Indeed, they demonstrated much greater interest in revolutionary activity than the village clerk reported. "The 480 claimants for land in Sárosd and farm servants of Csillag *major* attached to the village insisted on individual land grants and the immediate distribution of property" (Jenei 1969, 98). The workers at Jakabszállás sent a delegation on March 10 to the Agricultural Ministry to request that they be allowed to take possession of the *puszta* and organize it as a cooperative (ibid., 97). According to county documents, workers proceeded to do so on March 16.[43] Manorial workers clearly believed that they had legitimate rights to the land they had been working. This notion of possessing land through working it differed from views of peasant landowners who believed that possession had to be inherited.[44]

The historiography of the 1919 Soviet Republic in Fejér County, developed most thoroughly during the socialist period, made much of documents that testified to revolutionary ambitions.[45] Why

---

[43] The government commissioner's report to the the lord lieutenant of Fejér County read: "The manorial estate of Jakabszállás which forms the property of the Widow Mrs. Weisz Berthold and is attached to the community of Sárosd has been seized as a cooperative by the servants living there" (quoted in Buzás 1969, 262).

[44] This difference of opinion would come to a head during the land reform after World War II, when landless agricultural workers called for land for those who worked it, while peasants believed that only those should receive land who had already inherited it.

[45] I suspect in some cases that the significance of revolutionary sentiments may have been exaggerated in postwar socialist-period historiography. I would expect this particularly to have been the case in documents of peasant support of the Soviet

then does little memory remain of the first cooperative farm of the village? Recalling the events of that heady spring, one elderly woman denied to me that a cooperative farm ever existed. "There was no cooperative farm in 1919. It couldn't be set up. There was no group work." Work at Jakabszállás continued on as before. At this estate, as at so many others across the country, no noticeable changes occurred in the character of production or in the structure of responsibility. "[T]he Republic did not follow the Leninist strategy of distributing land among the agrarian proletariat, but socialized all estates over 100 acres, often leaving the previous owners on the premises as commissars responsible for production" (Janos 1982, 199). For example, Dr. István Elek was charged as commissarial deputy and production commissar to initiate a cooperative farm at Csillag *major*. Elek, a wealthy landowner, had been selected along with other estate owners for positions of authority, after having passed a rigorous scrutiny of his past activities as a landowner in the county. Elek and his fellow commissars were authorized to use armed force if necessary to achieve their goals (Jenei 1969, 99–100).

Neither the Soviet Republic nor the Popular Democratic Republic that preceded it saw the agricultural proletariat as able to guide their own affairs. The need for informed managerial personnel to guide the hapless agricultural proletariat was also made clear in instructions given to the man charged with investigating the request of manorial workers at Jakabszállás to form a cooperative, a request made prior to the establishment of the Soviet Republic.

> Rácz Gyula, the undersecretary of the Agricultural Ministry, immediately charged Kutas János, the ministerial envoy, to grant the request of those at Jakabszállás urgently. He should organize their cooperative farm on 1500 *hold*. He should have them select a steward, and then have them begin productive work. At the same time he directed Kutas János to stay at Jakabszállás until the former agricultural workers and farm servants master cooperative farming. (Ibid., 97)

Under such conditions, it is not surprising to learn that those living at the estates did not experience a radical change in the character of manorial farming. The owners of Jakabszállás protested the appropriation of their property to the local authorities (ibid.). By

---

Republic, since agrarian issues were so poorly addressed by this short-lived regime. These inadequacies would have been concealed by the post-1945 socialist regime.

fall, it would return to their hands, virtually untouched by revolutionary practices.

Nagyatádi Land Reform

During the brief tenure of the Soviet regime, some villages were visited with red terror squads, punished for refusing to surrender their grains (Janos 1982, 197). Once the Soviet Republic had been defeated by invading armies from the east and west, the victorious sought to castigate erstwhile reds. The White Terror in 1920 brutally punished any suspected collaborators of the 1919 Soviet Republic and incited widespread pogroms (Pamlényi 1973, 456). Sárosd escaped both these scourges. But the glaring problem of land reform could not be solved simply by brutalizing the peasantry, especially as the regime of the Regent Admiral Horthy Miklós was far from consolidated politically. More than half of the population, that is, approximately 3.8 million people, earned their livelihood from farming, of whom 3 million people were completely landless or owned dwarf holdings of .5 to 2.8 ha (1–5 kh) (ibid., 460). In other words, nearly two-thirds of those engaged in agriculture were landless or owned only one to two acres, not enough to provide a family with a year's subsistence. In a phrase attributed to his 1920s writings, right-wing populist Oláh György called Hungary "the land of three million beggars" (*három millió koldus országa*), a phrase frequently quoted by sociologists and writers across the entire political spectrum to describe the dismal life of rural communities. The meager effort at a land reform did not solve this dilemma.

In his new role as minister of agriculture Nagyatádi Szabó István, himself a peasant and former representative to Parliament, enacted the land reform in 1920 (Act XXXVI), the substance of which was severely compromised in ensuing negotiations among the competing interests in Parliament (Pamlényi 1973: 461; Macartney 1956, 40–41). Large landowners prevailed in limiting the transfer of properties to only 546,441 ha (948,682 kh) of the total national lands of 9,308,160 ha (16,160,000 kh), arguing that a larger transfer would endanger the credit structure of the country and cripple agricultural production (Macartney 1956, 41). Szabó had hoped to continue transferring properties in later stages, but his death in 1924, and the intransigence of wealthy landowners, left Hungary with one of the most pitiful postwar land reform programs in all of Central Europe. A comparison of the proportion of small and medium properties to large holdings between 1921 and

1935 illustrates the nearly inconsequential impact of the Nagyatádi Land Reform (Kerék 1939b, 201):

	Properties less than 58 ha (100 kh)	Properties more than 58 ha (100 kh)
1921	46.5%	53.5%
1935	52.0%	48.0%

Pruned to the barest minimum, the land reform was also long in coming. In its final form, it was eventually passed into law in 1924 (Act VII), but it took several more years to be implemented nationwide (ibid., 195).

Eszterházy, Weisz, and the other renters of manors surrounding Sárosd eventually distributed lands for cultivation (559 ha; 980 kh) and 219 house plots; the size of plots issued for cultivation was approximately .5 ha (1 kh). (Table 2.13 lists the average sizes of plots, explains who was awarded land, and compares the social distribution in Sárosd with national figures.) According to the village monograph of 1930, all house plots were redeemed for 10,000 korona in 1925. Remuneration for lands under cultivation was collected more slowly, as indicated by the impatient request for full payment sent to the village council in March 1927 by the Hungarian Land Credit Institute (Magyar Földhitelintézet) empowered to collect payments for transfer. As can be seen from table 2.13, lands were not distributed solely to the landless, but also to poorer landholders, public servants, craftsmen, and World War I heroes. A plot of 8.6 ha (15 kh) was even parceled out to the village council, the land to be rented regularly on a five-year competitive contract by one of the local peasants (alluded to earlier). In addition to the lands in the Andor properties owned by Eszterházy and properties at the far end of Weisz's estates, portions of the "redeemed lands" (*vagyonváltsági földek*) were also designated communal pasture.

Yet the most significant social change wrought by the distribution of lands in the village was represented by the collection of house plots apportioned beyond the mill on the road to Jakabszállás. Based on a survey of housing conditions conducted during the Soviet Republic, large numbers of villagers in Sárosd were without adequate housing. "The community of Sárosd reports that its population of 2300 persons 'are in large part proletarians without land or house property, and who are not occupied in any sort of cultivation on their own behalf. Their abodes are overcrowded, they do not receive yards or gardens'; . . . they have applied for 200 fam-

TABLE 2.13. Land reform in 1924

Recipients of land	National percentages	Sárosd percentages
War hero	0.65 %	1.3 %
War cripple	7.30	4.0
War widow	6.20	—
War orphan	0.25	6.0
Agricultural worker with no land	45.30	53.6
Dwarf- and smallholder	27.80	16.9
Public employee	0.90	—
Craftsmen and industrial workers	9.90	17.9
Agricultural manager	0.01	—
Soldier continuing to serve	0.50	—
Other	1.00	0.3

Notes: Under the land reform 228 ha (400 kh) were awarded to "This Side of the Danube Lutheran Church." This institution, which later crops up in the list of highest taxpayers of Sárosd, is a mystery. I was unable to determine who they were and why they became so prominent an institution in a village that has an inconsequential Lutheran population.

The average size of plots from the different manors is as follows:

Puszta Jakabszállás	1.0 ha	(1.8 kh)
Redeemed lands	2.5 ha	(4.3 kh)
Purchased lands	0.5 ha	(1.0 kh)
Csillag	0.9 ha	(1.5 kh)

ily houses" (Móra 1969, 82). The availability of affordable house plots opened the way for manorial workers to escape the oppression of the manor, to seek better working and living conditions. Even though little material security was gained, private dwellings symbolized the workers' new social independence. Peasants no doubt looked askance at the growing neighborhood of poor day laborers in their community, but they were unable to prevent its development. In the final analysis the land reform did little to alter the social distribution of land and wealth in Sárosd or the nation. Nonetheless, the village clerk wrote confidently in 1930, "After dividing the land, moods calmed down."

Social Strife in Sárosd

Difficulties surrounding the land reform galvanized discontent throughout the 1920s. Several confidential reports from the Sárosd village council to the lord lieutenant of the county, dated in February 1925, complained of constant incitement to divide up the land. Local agitators were resented for their interference in village affairs. "Vörös József, Hekkel István, and Andrási Gyula have hampered the affairs of the land reform. Thus the matter has unhappily dragged on now for three years. The forenamed travel constantly to Budapest, although they have no business of any kind there" (February 17, 1925; FM f.b.i. 1925–62. res. sz.). Glimmers of more organized political activity, such as these trips to the city, were apparent during this period. Vigilant civil defense authorities "confidentially requested the person selling a subversive periodical in Sárosd that he should stop disseminating the paper" (HLHM k.p.o. 1923, 500.002/2). The left wing of the Social Democratic Party, known as the Vági branch of the Socialist Worker's Party of Hungary (Magyarországi Szocialista Munkáspárt), initiated a membership campaign in Fejér County in the summer of 1925 (Farkas 1980, 81). By 1926, they had established an organization in Sárosd. Since the Vági branch was also considered to be the legal branch of the Communist party, the chief constable immediately requested a membership list (FM a.b.i. 1926–13). Officials were anxious to keep an eye on political agitators, and any other figures who would be able to wield undue influence over the working public. Labor recruiters, who traveled so broadly, were regularly required to submit a character reference, informing authorities in other communities of their personal histories.

Village certificate—from the prefecture of Mezőkövesd

Csirmaz József: agricultural day laborer, 33 years old, religion—Roman Catholic, widower, Hungarian citizen.

The forenamed is not under paternal authority, guardianship, trusteeship, or bankruptcy. He has not been punished for a crime, either for the sake of gain or a misdemeanor offending the 1921 law, paragraph III, concerning the more effective protection of state and social order.

No complaints have been brought against him with respect to morals or politics. He did not participate in the revolution. To our knowledge no complaint has been made against his activities as a recruitment boss.

January 28, 1927 (FM m.t. 559/1927)

Fear of Communist activity was rampant, leading to apprehension and mistrust.

In Sárosd there is a group of 60–70 construction laborers, who are not willing to sign onto work under the most favorable of conditions. They do not even do day labor. The chief constable assumed that they formed part of an agrarian Communist organization. He asked for a detective to find this out. However, it turned out that some of the workers are employed in railroad repairs. Thus this could only have been included in the report because of misrepresentation and naturally a fear of Communists.

May 20, 1927. (FM f.b.i. 1927–17)

The Communist menace was clearly exaggerated.

Never did the Communist party, or its various surrogates, ever find much sympathy among villagers. Families were preoccupied with their own difficulties, and were isolated from wider political movements. The desperation villagers felt, and sympathy they may have had for an alternative world of social and economic justice, were illustrated to me by the following story, told to me while I was watching the evening news one night in Sárosd. The anchorman had just finished commemorating the hundredth anniversary of the birth of the famous literary critic Bölöni, who, among other things, was being remembered for his early interest in socialism. The man I was watching television with found this claim preposterous.

> They didn't even know what socialism was then. They are always pushing it back, saying this one and that one was a socialist. Even Petőfi [the Romantic poet]. Before this they barely talked about Petőfi, but now he's a socialist thinker, right? Perhaps if they had heard one or two comments my father made, they would say he was a socialist. When my brother Laci was born in 1928, my father had to go to the steward to report that a child had been born. The steward needed to know how many children each manorial worker had. Labor power, you see. My father went to see Girgli [who received the announcement with scorn]. "So, the heir to the crown has been born." My father came up with a quick response: "If I knew that my son would become a manorial worker, I would break his neck right now." This outburst infuriated Girgli, but he could do nothing because Eszterházy really liked my father. He worked hard [and] . . . wasn't a drunkard. Also Eszterházy was a hussar in the war, a captain, and my father was a hussar too, although they hadn't served together. So Girgli couldn't pick on my dad. Now that was a socialist speech!

The following year this man succeeded in leaving the manor for good, buying a small house on the edge of the village. But his strong convictions never translated into political organizing. He taught his children early on that they should not get involved in politics, since they had to work for a living. In the final analysis, the clerk's evaluation of villagers seems credible. "They don't care much for public affairs; they do not like to engage in politics. They discuss the truncated character of our nation, but more in economic perspective" (*Sárosd monográfia*).[46]

The economic worries of villagers in Sárosd were pressing, as were the fortunes of agrarian workers across the country. "At the beginning of the counterrevolution [after the Soviet Republic was defeated] and in the years of the drift toward fascism, property relations essentially did not change. The earlier laws remained in force, so that the position of harvest workers did not improve, but in fact worsened" (Balassa 1985, 257). Many found themselves without work, despite proximity to large manors that regularly employed villagers as day laborers or harvesters. In Sárosd, 230 workers were without employment at the start of the 1920 harvest season (Farkas 1980, 75). A series of strikes and other forms of labor unrest in the 1920s expressed the anger and dissatisfaction of workers with contracts, wages, and living conditions. In August 1923, thirty pairs of harvesters were unable to find work in Sárosd as a consequence of the land reform. The manors surrounding Sárosd agreed to distribute these workers at their respective estates, but when the workers who had already been hired learned of the agreement among the manors, they refused to let the new workers participate in the harvest (HLHM k.p.o. 1923, 50.002/9).[47] On the first of July 1924, thirty-seven harvesters contracted to work at Szilfamajor refused to work because of the poor meals the manor provided. On the twenty-second of the month thirty harvesters held a one-day strike at Világosmajor, finding their wages insufficient (HLHM k.p.o. 1924, 50.126/51.209). While conditions at manorial farms were rarely pleasant, there were extreme cases of abuse and neglect. Bartha, the infamous renter of Jakabszállás, consistently appeared in legal documents during the 1920s and early 1930s, cited for his cruel treatment of workers and arbitrary methods. In one complaint lodged against him for unlawful dis-

---

[46] After the war, Hungary lost nearly two-thirds of its territory to surrounding nations. Conservatives and nationalists of all stripes complained constantly about the "truncated" nation.

[47] As table 2.11 indicates, both Jakabszállás and Csillag regularly employed harvesters from the neighboring village of Sárkeresztúr. Presumably the harvesters from Sárkeresztúr considered the incursion of harvesters from Sárosd to have threatened their privileged position at these manors, and so to have warranted extreme action.

missal, Bartha was quoted as saying, "If you can't work while you're hungry, then you can leave" (SJ f.i. 6374/1931). Bartha bred anger and dissatisfaction among his manorial workers, which came to the attention of the local military authorities.

> [A]n agitated mood predominates among the servants of Bartha Pál's rented estate of Jakabszállás, because the landowner has docked two quintals from the yearly *kommenció*. Moreover, he redeemed the servants' corn lands with five quintals of corn. August 1923 (HLHM k.p.o. 1923, 50.002/9)

Bartha was known to abuse his workers to such extremes that the chief constable saw fit to report his activities to the deputy lieutenant of the county (FM a.b.i. 1929–44). In one complaint filed with the lord lieutenant of the county, a series of accusations was made. Some workers demanded that they be paid the wages they had been promised, while others simply requested that they be released from their contractual obligations. In support of their complaint, they explained that they had been treated inhumanely. In particular, one worker requested that he be allowed to break his contract, as specified in the second paragraph of the 1898 law on manorial workers: "Szász Jenö, likewise a seasonal worker at the manorial estate of Jakabszállás, complained that during work the steward named Takács slapped his face for a harmless remark" (SJ f.i.cs.b.sz. 4891/1928). The lord lieutenant responded by demanding that the recurring squabbles and discord sown at Bartha's estate be stopped. In 1929 the lord lieutenant had to intervene to force Bartha to pay his workers their wages (ibid., 8051/1929).

Conditions at Jakabszállás had to have been quite extreme, for the consequences of bringing legal action against one's employer could be quite grave. This was voiced by a minister in the neighboring village of Aba, who had been requested by members of his church to bring a complaint against Csillag *major*. A rash of stomach problems among the residents had bred worries about the quality of water and the possibility that wells had been contaminated. I quote from his letter, dated June 18, 1930: "I would note that the poor man fears along with his fellows—most of whom are also from Aba—that if Elek learns that they lodged a protest, they would lose their livelihood. This worry is justified in today's world of unemployment and inadequate income" (SJ f.i. 3904). In another case, workers raised a complaint against Bartha concerning their two-month contract to harvest beets. "When work proceeds slowly owing to rainy days, we earn very little. Our employer wants to dock our meals from our wage on rainy days when we cannot work. Moreover our lodgings are not suitable: the door is bad and wind blows in. We request that you order cancelation of the contract; we want to go home" (SJ f.i.cs.b.sz. 6486/1928). Even though the official charged with overseeing these complaints demanded that Bartha increase the wages offered to the migrant workers, a week later the workers rescinded their complaint, explaining that they had been incited by one of their fellow workers to demand a higher wage (ibid.). They hoped to continue working, despite the poor treatment they had received.

In some instances, workers simply left for home. This could have serious consequences, since estate owners could enlist the support of the gendarmerie to return workers unjustifiably absent from their jobs. Legal documents requesting the return of workers considered delinquent were frequently filed with county authorities, as were countersuits explaining justifications for leaving, such as illness, nonpayment of wages, or the need to attend to a sick relative. So too, complaints were filed for unjustified cessation of manorial contracts or insufficient pay. Manorial officials countered with charges about poor work habits, at times asking Kerbolt to judge the severity of illnesses claimed by manorial employees. Manorial workers who had been terminated were often quite desperate. Having lost his contract (*kommenció*) with the manor at the end of the summer, one former manorial servant was unable to find another position so late in the season.

> Honorable Chief Justice!
> I respectfully request your kind order for me to receive my servant *kommenció* from Bartha András's great estate, which the chief justice of Székesfehérvár awarded to me. I am here with seven children completely without food. I cannot send my children to school, since they have neither bread nor clothing. We have been going hungry for weeks and have been completely driven to despair. I would gladly work if only I could get day labor, but I no longer have any strength left. Please be good enough to arrange that I receive it as early as possible, so I may save my family.
>
> Your humble servant,
> Fülöp János
> Seregélyes, November 9, 1931

Charges of stealing were used to dismiss employees. In their suit, manorial workers countered that they had been denied firewood, as specified in their contract. "And anyway, the other servants did the same thing" (SJ f.i.cs.b.sz. 3113/1935). If workers refused to follow orders, estate owners could solicit the help of local gendarmes to throw them off the premises. These documents paint a grim picture. One exceptional example of generosity was the town council's decision in 1924 to grant a yearly sum to two blind villagers to ease their poverty (SJ f.i. 2532). These men, however, were members of the village community, not the aliens who lived at the manor, excluded from the bonds of a village social community.

Although strikes had been waged throughout the 1920s, their numbers decreased in the 1930s (Balassa 1985, 258). The plight of the agrarian poor clearly worsened in the late 1920s and 1930s, as the burden of world depression sat heavily on everyone's shoulders.

> Decelerating industrialization, the growing gap between prices of agricultural and industrial products, increasingly difficult agricultural export—the result of all these could only mean one thing in Hungary between the two world wars: the stagnation of agricultural production, the renewal of tendencies toward extensive production, the slowing down of the technical development of agriculture. And actually this is what happened, not only in the period of the world economic depression or following it, but to a certain degree between 1924 and 1928 also, at the time of the grain boom which meant a relatively favorable situation. (Gunst 1975, 51)

The difficulties experienced were spread across the entire community, though some were more able to bear these difficulties than others. As the village clerk makes clear, the boom of the war years was no more.

> Before the war approximately 40% of the population was in debt. At this time they withdrew money for 6% from the savings bank of Fejér County and for 4% from the orphans' court. They did not make use of sinking loans. During the war and the boom they paid off their debts, but once again they have run into debt to a great degree. Now 80% of the population is indebted. The reason for this is the heavy rates and taxes, and the deviation in prices between produce and industrial goods. (*Sárosd monográfia*)

In a few short decades, twice as many villagers were in debt, victims of the local economy and the world depression.

Possessing Activity

The social relations of production in Sárosd and at the surrounding manors differed markedly. Landowners worked with family and kin, while harvesters, day laborers, migrant workers, and manorial servants worked in large bands and complex, stratified communities. It should be emphasized, however, that the process of social differentiation characterizing the entire community in the interwar years was also discernible within families. The ideology of diligence and reward hardened the boundaries between success and failure, and between emerging social strata—despite strong consanguineous ties among wealthy and poor landholders in the village.

The disparity among different work patterns is often opposed as "protocapitalist" (peasant) and "semifeudal" (manorial). In this view, peasant habits of diligence presaged a modern economy of production and property, while manorial workers were mired in the dusty world of the feudal past. But peasants and manorial workers lived in the same world; they were not isolated in different historical eras.[48] The differences observed between groups within the village community and at manorial estates must be understood in the specific context of production. Common to both groups was an ideal of mastery and of control. Yet the social form of these ideals necessarily diverged as the political and economic relations of production differed. Though all sought to possess themselves through control of their labor, the manifestation of possession took very different forms. While landowners competed with one another in diligence, manorial workers slowed down the pace of their work, resisting their bosses and stealing from the manor.

Women's experience of possessing labor varied, but not in any simple or direct relation to property and wealth. At the two extremes of the social hierarchy—as wealthy peasants or as wives of manorial workers—women were relatively independent, managing the affairs of gardens and livestock. Though nominally subordinated to their husbands or fathers, peasant women exercised extensive control over their own labor, and that of their children and daughters-in-law. Manorial women were perhaps even more independent of their husbands, who left for the animal barns of the manor early in the day, only to return late at night. Any monies collected by a manorial family from the sale of livestock,

---

[48] Fabian discusses the use of historical tropes to convey relations of social and political inequality (1983).

which constituted the sole means of freeing the family from the hardships of life on the *puszta*, were garnered through a woman's success in making ends meet. The fate of a woman living at the manor may have been influenced by the price she could get for her goods at the local market fair, but neither she nor a peasant woman participated actively in labor markets. The relative independence peasant and manorial women enjoyed did differ, insofar as peasant women were actively engaged with their husbands in managing a farm, whereas manorial women were excluded from their husband's everyday activities. This relative control over decision-making and disposition of time evaporated among those women forced to seek day labor in the village, or a summer's job at the manor. The pressures to seek supplementary sources of income, combined with the difficulties of just getting by, took a heavy toll on poorer women not sheltered from the labor market.

It bears emphasis that the degree to which alienation of one's labor to others was anathema, especially for men, is illustrated in the imperative to abandon security for insecurity, to leave behind a manorial contract to seek day labor. If manorial workers could accumulate enough capital from the sale of their own livestock to buy a house in the village, they would choose to leave the manor and live by day labor, despite the fact that their contracts at the manor offered a secure and regular income throughout the year. It mattered little that one's subsistence on day labor was far more precarious and uncertain than manorial employment. Reviewing the gradations of rank and prestige, then, one discovers that the loss of control entailed the loss of social being, so much so that the working habits and mannerisms of manorial workers—who were on the bottom of the social hierarchy—were constantly being compared to those of animals. It follows that the boundaries of the social community varied, depending upon one's ability to possess or control one's labor, and one's position within the social hierarchy. In the most extreme example, manorial workers lived "outside" (*kint*), beyond the borders of village society, in accordance with their less than human identity.

Just as important to mastery of self, however, was the *realization* of that control through the process of ongoing activity. The construction of self through work manifested itself among landowners in daily toil, sunrise to sunset, and gave rise to what may be called a "rivalry in diligence." Wives and mothers urged their men to start for the fields early, fearing the shame of being the last family to leave home; it was even shameful for men to be seen in the village by daylight. In contrast to landowning families, among manorial workers mastery of self was lived by slowing down the pace of work. As described above, bosses had to prod workers to act.

Villagers had very ambivalent attitudes toward markets. The ideal situation for a villager, as Erdei noted, was to "possess his land without interference, disturbance, and supervision, and that he control his produce without fixed obligations" (1941, 93–94). Though landowning peasants regularly engaged in transactions with local merchants or traders at the fair, they considered value to be created in the act of possession, a value that could be enhanced by withholding one's possessions from the market and other impersonal forces. However, the ability to avoid markets was not shared equally by all in the community. Landowning peasants with decent holdings could easily survive on their own, only rarely exchanging surplus for products they themselves could not produce, such as petroleum or matches. Poorer villagers, on the other hand, were forced to sell their produce and their labor to meet their obligations—taxes, debts, or needs of the farm. The ideal of possessing one's activity and resisting markets was thus a privilege few could enjoy.

The staunch refusal to see value deriving from circulation also influenced attitudes toward movement and stability. Village social hierarchy, scaled according to the relative control of self, was also articulated in terms of relative stability or mobility. Those who were completely stationary, who did not move to find work, were those villagers in complete control, masters of their own affairs. In the final analysis, only landowning men could fulfill this role. Even women from landowning families were expected to leave home upon marriage, and move the site of their domestic activity to an entirely different house in the village. This fine gradation in social mobility and honor differentiating men and women among the peasantry was repeated on a far grander scale by those less fortunate. Those who sought work outside the home to supplement their meager subsistence, such as villagers who contracted with the manor to harvest wheat, or migrant laborers from other counties who worked in manor fields all summer long, were considered socially inferior. Manorial workers were considered to be completely mobile, for even though their families may have lived at the manor for generations, they had no village to call their own. This merely reinforced the view among village dwellers that manorial workers were less than human. Mobility, then, was also the expression of a loss of control or possession. The near universal disdain for Roma has surely been based on their refusal to remain in one place, to possess objects and activities so fiercely. The very image of a Roma has been the antithesis of the moral values of the village community.[49] So for villagers, to circulate one's labor on the market, to alienate control over one's labor, was an inferior means of realizing self, and so was demeaning. Humbling oneself by journeying to work nonetheless remained the fate of most of those living in Sárosd and at its surrounding manors.

The Rise of Capitalism: 1848–1918

Working in Sárosd took different forms, with social conditions dictating various degrees of alienation among the landowning and landless, among men and women, adults and children, Jews and Roman Catholics, Roma and peasants. How had the move toward this alienation begun? How had labor acquired so central a role in village life? And how had it come to acquire so material a character? These questions prompted me to consider the history of Hungarian rural life during the century that had preceded the interwar period. The transition from feudalism to capitalism in the nineteenth century is the subject of a forthcoming book, a companion volume to this study. Let me sketch the argument here briefly.

As I mentioned in the introduction to this chapter, interwar village life has long been described as traditional, as sharing characteristics with the social and economic relations of late feudalism. The dominance of large manorial estates, the supremacy of the aristocracy and gentry in politics, nonmonetary labor exchanges, and subsistence production among the landed peasantry were all cited as factors very little changed since the abolition of feudalism in 1848. And yet the pace of work in interwar agrarian communities—described as an unremitting struggle from sunrise to sunset (*látástól vakulásig*)—appeared to me to be quite different from the lazy tempo of late feudalism, when adult men worked only intermittently in their fields, women were occupied elsewhere in do-

---

[49] Michael Stewart's recent work among Roma in Hungary has shown that Roma take pride in their participation and identification with exchange, in stark and perhaps defiant contrast to the focus on production for peasants (1987, 1990). "Being landless and without other means of production, the Gypsies defined themselves as 'sons of the market' (Romany: *foroske save*), people who lived off their wits, their ability to talk and deal with different sorts of people through their trading. Attachment to land and place, accumulation of wealth for investment, and autonomy via self-sufficient production were all rejected by the Gypsies. Instead, they constructed an image of autonomy and liberty through being able to imagine 'evading the law of reciprocal exchange' on the market" (Stewart 1993, 194).

mestic chores and crafts, and young men were often idle. Indeed, the moral universe of village life between the two wars, which I have described as a rivalry in diligence, contrasts sharply with the value system embraced by rural inhabitants in the decades following the abolition of feudalism (1850–1870). Recently freed serfs happily refused their lords' pleas to perform harvests, and regularly shunned any efforts among reform advocates to have them intensify labor inputs in agriculture. Labor was not seen to be a crucial factor in agricultural production; the possession of land, and its unhurried cultivation, was sufficient for the full achievement of social honor and respectability (Vörös 1966). In light of the contrasts in village work patterns between the immediate postfeudal era and the interwar period, "possessing activity" appears not to be a timeworn traditional notion, but a very specific product of the intensification of capitalist relations of production at the turn of the century. Mastery of self so crucial to interwar village affairs was a product of the transition to early capitalist relations in this community.

Indeed, a simple review of the historical record demonstrates that rural communities in Hungary underwent quite extensive changes in the latter half of the nineteenth century. The April Laws of 1848 not only abolished feudal services, but also introduced new forms of taxation, altered laws of inheritance, and eliminated restrictions on the partition of landed properties. While the abolition of feudalism was greeted happily by former serfs and the free peasantry, it did not substantially alter the political and economic advantages of the landed aristocracy. It must be noted that a sizable number of former serfs were not given rights to the land they had worked, but were recategorized as servants of large estates. Tempers flared, as serfs who had anticipated acquiring land in the 1848 legislation found themselves relegated to the lowly status of manorial servant. Indeed, the process of disencumbering properties following upon the changes in property and labor stipulated in the April Laws paralleled in many respects the process of enclosure familiar from British history. The landed aristocracy appropriated large tracts of formerly common properties, such as pastures, forests, and waterways, and also maintained important monopolies in milling, wine-making, the selling of alcoholic beverages, and other commercial activities. Acrimonious battles raged for years over legitimate title to lands and the legal status of workers employed on large estates. The legal confusion that issued from the transition to new relations in property and production seriously crippled agricultural production for decades.

The crash of 1873 hit agricultural production hard. The agrarian community, and most every economic sector across Europe, was profoundly affected by the ensuing depression, which lasted nearly twenty-five years. Wealthy landowners in Hungary frequently complained about the unfair competition mounted by wheat producers from the United States, Argentina, Russia, and India, and regularly demanded customs duties be imposed to prevent their financial ruin. The politics of duties and custom excises were continuously at the heart of economic debates among governmental officials and economic critics in the 1880s and 1890s, confusing the vigorous policies of laissez-faire economics publicly advocated by many politicians of the period. While grain production in Hungary did transform in important ways as a consequence of new grain markets, predictions of the demise of wealthy manor owners had been premature, not least because of changes in livestock-raising, which provided both peasants and manorial estates with alternative sources of income.

A shift to more active livestock-raising, spearheaded by the peasantry, was accompanied by a range of new agrarian practices. Crop production intensified. Fallow was abandoned and pasturage reduced to expand acreage. The loss of pasture to livestock herds was supplemented by the growth of new feed crops. Cash crops, most notably sugar beets, were introduced alongside more traditional grains. Stabling herds enabled peasants to put manure to good use as fertilizer, a practice long encouraged by agrarian reformers and stubbornly resisted by landowners. New tools also appeared. The shift from wooden to steel plows occurred in this period, and probably constituted a far more significant technological innovation than the introduction of more elaborate and expensive machinery, such as steam-driven combines or tractors.

The intrusion of the state in local affairs came in the form of taxation, increasing substantially after the 1870s as the institutionalization of bureaucracies grew rapidly. But the state was also involved in local affairs as it initiated a number of important modernization projects: building railroads, rerouting rivers, and founding schools to promote new agricultural techniques. Confronting an increasingly hostile workforce, wealthy landowners and politicians worked hard to outlaw strikes and craft new legislation stipulating the contractual obligations of agrarian workers. Laws were passed proclaiming the cultivation of agricultural property to be free of outside intervention, legislation designed to eliminate communal restrictions on the individual use of private property. However, this same legislation also set down in precise detail the degree and extent of state intervention justified to ensure the improved quality of agricultural production, including measures to build up forest nurseries and vineyards, protect pasturage and animal stock, eradicate noxious pests and plants, and establish a constabulary to protect private agricultural properties.

In the midst of these momentous innovations in agricultural production, profound changes occurred in the conceptual world of wealthy landowners, the peasantry, and the rural proletariat. Money, once seen by wealthy landowners as a pernicious influence eroding the solid bonds of patriarchalism and paternalism, was begrudgingly recognized as a possible tool in motivating workers. Slowly but surely, the wealth that flowed into the aristocracy's coffers from their extensive estates—once spent on balls in Vienna or new hunting lodges in the Tyrols—could now be seen as a productive resource, to be set aside to invest in machinery, or a new stud farm, or to build a distillery. While most of the aristocracy were slow to embrace the reform visions of agrarian specialists, a few landowners with large estates did pioneer important changes in production and commerce, joining their more innovative peasant colleagues in transforming the vistas of modern agriculture.

The identity and role of workers in agricultural production underwent important shifts. The availability of "workers' hands" (*munkáskéz*), once the sole concern of wealthy landowners, was replaced with an interest in the ethnic identity of laborers. By the turn of the century, nationalist sentiments had become a significant factor in agrarian politics. Hungarian peasants were to be given special consideration in the distribution of lands in settlement projects and in employment contracts (as well as prevented from emigrating to the New World). Conservative nationalist politicians rallied against Jewish renters and landowners, who were seen to undermine the inherent tie of Hungarian blood, body, and soil by engaging in agricultural production (see Lampland 1994). Rural workers came to identify themselves in terms of the labor they performed, often at the behest of others.

Agricultural production was undergoing a significant transformation, visible throughout the agrarian community. The domestic and international environment had shifted following a number of economic, legal, and political changes. While a number of estate owners and a few landed peasants flourished in the new agrarian regime—consisting of new crops, new techniques, and greater labor inputs—most in the agrarian community fell slowly or even precipitously into debt and penury. The burdens of taxation, the consequences of partitioning properties, the loss of lands, livestock, and tools to foreclosure undermined the ability of landowners to continue farming independently, shoving them into the ranks of landless farmers and long since proletarianized agrarian laborers. As the fortunes of many in agrarian communities worsened by the 1890s, wealthier peasants and estate owners were able to extract more and more labor for less and less reward. In fact, landowners were able to require workers to perform a number of tasks for free, forms of labor extraction reminiscent of feudalism and named the new corvée (*új robot*) by belligerent workers. The mood in many agrarian communities was angry; the popularity of agrarian socialism grew rapidly, drawing many to its cause. The decade of the 1890s is well-known in Hungary for agrarian unrest, harvest strikes, and violent clashes with authorities over the rights of workers to organize. Unlike ever before, labor had become an economic and moral category. Labor was now seen to be a factor in production, prompting peasants to work harder and harder to keep their farms. It was also a social and moral category, defining one's identity within the community according to one's devotion to diligence. The rise of a movement organizing villagers as laborers—a politics of identity through work—was unprecedented, and provides further evidence that the political and economic landscape of rural communities had changed markedly.

Central to my argument about the transition from late feudalism to early capitalism is an analysis of the shift in the source of value: from service to land in the mid-nineteenth century, to labor at the end of the nineteenth century. In other words, the privileged social relation that was seen to generate value and structure social life shifted over a period of a hundred years. Feudal service and protocapitalist labor may appear to be similar. I would argue, however, that feudal service, though rendered primarily (though not exclusively) through labor contributions, had a character very different from the concept and character of labor at the end of the nineteenth century. The central component of a feudal bond was the hierarchical relation between lord and serf, between aristocrat and commoner. This relation was realized through reciprocal services rendered—the lord providing land to cultivate, the serf labor and goods in return—which reinforced a hierarchical bond of privilege and servitude. This is very different from a notion of labor which ascribes identity to social actors through the activities they pursue in work. With the shift to land as the source of value in the mid-nineteenth century, the significant relation generating value and structuring social communities revolved around the possession of land. Not everyone could make claims to land, but the predominant conception of social value was seen to derive from the intimate relation of free citizens with their property and the fruits it bore. At the end of the nineteenth century, the concept of labor came to dominate social debates and agrarian communities. Villagers were evaluated in terms of how they worked, particularly how hard and long they worked. The process, the ongoing *activity* of work, had now come to be seen as the most significant component of one's social identity, and the primary means by which one's economic fortunes were secured. Crucial to this transformation was the development of class politics: organizing communities around who worked and how they worked became possible, indeed, became necessary for many disenfranchised workers. The rise of labor as the source of value, and as an organizing principle of politics at the end of the century, was also correlated with the development of an ideology of individualism, demonstrated most succinctly in new laws defining the contractual obligations of agrarian workers. In short, ideologies of individualism and collectivism (be they the collectivist sentiments of nationalists or agrarian socialists) arose simultaneously, clearly implicated in each other's genesis.

As a result of these momentous changes, we find communities in the interwar period defined and transfixed by labor. By the 1920s, labor was the central, overarching, and determinative source of value. It was not commensurate with other sources of value, although there were practices that regularly defined labor in monetary terms, for example, day labor. The dominant view of the construction of value in agrarian communities, however, was by and through manual labor, through physical effort, through industriousness and diligence. Markets were resisted precisely because they insinuated alternative means of determining value, such as price, into a moral universe that did not equate labor with other potential economic categories, such as time or money. Thus the capitalist world of the interwar years—centered around labor as both a moral and economic category—was not characterized by a strongly utilitarian, individualist logic of rationality. The equation of time, labor, and money was foreign to this universe, even though in so many respects rural communities had already been significantly influenced by markets. The changes that follow from the ever broadening experiences of villagers in labor and commodities markets would not occur until well into the socialist period. It is to this series of changes that we now turn.

THREE

Severing Ties

Introduction

After the war new social forces began mobilizing to transform relations of production and property in Hungary. Through the various coalitions of political parties and economic interests, a land reform was accomplished favoring the poorest families in the country. The ostensible victory for the downtrodden was short-lived; the divisions between landed and landless, between peasants, day laborers, and manorial workers in Sárosd were not erased by the land reform. The process of increasing social stratification begun in the period between the wars continued apace into the postwar period.

The land reform of 1945 strengthened the property consciousness of the poorer peasantry, who had been vindicated in their belief that working the land required ownership. But by 1948, the inherent relation between property and production was to be challenged by the Communist-led regime, which now wished to break forever the bond between personal property and labor identity. Agriculture became a target of a broad industrializing project, and large numbers of agricultural workers were to be moved into industry. The state proceeded to nationalize small shops, large industries, church lands, and manorial estates, and commenced the initial phase of collectivization, in order to destroy pride in small-scale ownership and the integrity of family production.

Former landless villagers faced numerous difficulties in establishing viable farms. Former landed families, equipped with scarce tools and draught animals, forced the disadvantaged to work far longer hours in return for use of their plows and horses. New landowners found it easier to abandon private production and to join the new cooperative farm in Sárosd, reinforcing the landed peasantry's prejudices about poverty and laziness. Forced deliveries of food and produce instituted by the state, intended to eliminate the wealthy peasantry and to promote collective farms, resulted in the impoverishment of all families engaged in farming, and seriously crippled agricultural production. Nagy Imre's reforms of state policies lightened taxation only temporarily, postponing the abandonment of private production by many for only two years. Peasants fought against the appropriation of their land, while the state worked hard and fast to instill a sense of labor identity, hoping thereby to ensure its own, new form of legitimacy. The brutality of the changes wrought by party and state policies pushed the nation to revolt in 1956. Workers took over their factories, establishing workers' councils; former manorial workers stood by their cooperative farms, while former peasants repudiated state control by taking back their lands.

At the heart of the struggles around the land reform and early collectivization was the relation between property and production. While the strength of property consciousness would be increased during the land reform, it was the socialist state's firm objective to sever those ties. The party and the state wished to elevate the status of laborer above that of property owner. Policies toward wealthy peasants and property owners in general were harsh, expressing the state's disdain and distrust of those clinging to private wealth, while industrial workers and cooperative peasants were lauded in the press and on the radio. One's value as a member of society was now to be determined solely by one's labor for the community. The value of one's labor was not to be determined by the market, however. The state prohibited the development of a market in labor during this period, seeing it as anathema to the ideological and economic goals of socialist development. The value of labor was seen to be a natural quality, expressed in one's commitment to collective production and collective ownership by working in cooperative farms, industrial factories, and state-owned shops. The ideology of labor, of identity through labor, which had characterized agrarian communities before the war was thus strengthened by party policies. On the other hand, the party worked strenuously to undermine the view that property ownership constituted honorable independence so common in rural communities. Struggles over collectivization demonstrated that for the landowning peasantry, property ownership and independent farming remained the final arbiters of integrity and honor in village life.

The Damages of War

The war took a heavy toll on Sárosd. Not a house remained unscathed; not a family escaped privation, pain, injury. Fighting reached the village only in late 1944, but before then many families had sent fathers and sons into battle. Jews were under assault, initially through laws depriving them of their property and civil rights. Estates rented or owned by Jews outside the village were taken over by Gentiles. Despite great odds, the Elek brothers contested the terms of the law depriving them of their estates, but without success. However, when the new owner of the Csillag estate also tried to have Dr. Elek István removed from the list of prominent taxpayers (*virilisták*) in Sárosd, his request was rejected. The judge declared that Dr. Elek could appear on the list, despite being Jewish, since he was in possession of documents from the National Office of Heroes (Országos Vitézi Szék) exempting him from the exclusion. As measures against the Jews intensified, the Mandls, a Jewish couple who owned a store in the center of the village, were told that their children, a daughter and son, had been rounded up in Budapest. Their former servant recalled that upon learning of her children's fate, Mrs. Mandl dressed up in her best clothes, borrowed a prayer book from a neighbor and refused to leave the house, praying incessantly. Eventually, the Jews in Sárosd were taken to the neighboring town of Sárbogárd, transported on wagons across the estate of Jakabszállás. People watched them leave, but did nothing to stop the deportation, fearing for their own lives.[1] The Mandls, still racked by grief for their children, committed suicide by taking poison when they reached Sárbogárd. Both their children survived the war. Recounting this story, the Mandls' former servant recalls watching people on the street from the Mandls' window, amazed at how they could live as if life were proceeding normally. She particularly remembered being shocked at young children laughing in the street: "I couldn't imagine how anyone could laugh, what kind of person could laugh in such times."

The front settled on Sárosd for three months, as Russian and German soldiers fought bitterly over inches of soil and scores of lives. On ten separate occasions fighting became so intense that everyone had to be evacuated from the village and surrounding manors. Each time the village changed hands, families lived in

---

[1] I do not know of any Roma families who may have been deported, but since they were also targeted by the Germans for extermination, we can assume that they also were rounded up and sent to death camps.

new fear of the soldiers' appetites and demands. Valuables such as barrels of lard, meat, and clothing were buried in courtyards and hidden in attics. Women desperately attempted to hide or to mask their identity to escape rape and physical abuse, often to no avail.

The front broke on March 15, but this did not end the suffering. Young men, many not yet conscripted, were picked up by the Russians and deported to Siberia; soldiers from Sárosd were also liberally represented in the prisoner-of-war camps in Germany and the Soviet Union. Nationwide, 850,000 to 900,000 Hungarian soldiers were prisoners-of-war; 550,000 to 570,000 ended up in the Soviet Union (Pető and Szakács 1985, 18). Those in American camps returned within a year to a year and a half, fattened up and rested. The sojourns in Siberia were lengthier and more taxing, many dying from disease, hunger, and exposure. Forty thousand prisoners of war died during captivity (ibid.). Of those who returned, most came home within two to three years, but some were not released until 1956.[2]

The damage inflicted on the nation touched every community and every family. The loss of life was staggering.

> Loss of life . . . projected on the current area of Hungary . . . amounted to, at the least, 420,000–450,000 persons. Among those more than 220,000 were Jews, and approximately 150,000–180,000 died as soldiers. The extent of the casualties is indirectly indicated by the fact that by 1949 the population should have grown by natural increase to 9.65 million, whereas there were 9.2 million inhabitants in the country, 450,000 fewer. While in the decades of the First World War and revolutions . . . the average yearly population increase was 0.94 per thousand, between 1941 and 1949 the number of inhabitants decreased at a yearly average by 0.15 per thousand. (Pető and Szakács 1985, 17)

Loss of property, goods, and transportation was extensive. The proportion of war damages experienced by different sectors varied; the agricultural and transportation sectors each suffered 16.8% of the damages, manufacturing 9.3%, private housing 8.4%, and mining 0.3%, while small factories, commerce, banks, private companies, health services, education, and other public buildings comprised 48.4% (ibid., 18). "All kinds of goods were destroyed at a value of approximately 22 billion pengő (at its 1938 value). This

---

[2] The pains of imprisonment far exceeded hunger and cold. As one villager recalled his return from a camp in Siberia, he burst into tears as he crossed the Hungarian border because he heard children's voices for the first time in three years.

sum equaled somewhat more than fivefold the national income of 1938, and totaled roughly 40% . . . of the national wealth" (ibid., 18–19). War damage was inflicted primarily in the areas of goods and produce, rather than in tools and equipment: "From the 3.7 billion pengö losses in agriculture the value of the damages to produce and livestock came to 2.8 billion (75.7%)" (ibid., 19). While industry could rebuild, for agriculture the loss of crops and animals had more serious consequences. Tools such as plows, threshing machines, and hoes may have survived the conflagration, but the limited numbers of animals (which affected both draught power and provision of animal products), the lack of good seeds for quality crops, and even the shortage of manure constrained the growth of agricultural production after the war (ibid.). Finally, the costs of reparations were also quite substantial, as were debts still weighing on the country which had been assumed by the Hungarian government during the interwar period. Although approximately half of reparations were eventually canceled by the Soviet Union, these costs still constituted 30–39% of state expenditures between 1945 and 1947, dropping to 15%, 7%, 4%, and then 2% between 1949 and 1951 (ibid., 23).

The village was in a sad state in the spring of 1945. Every house was damaged from bombings, artillery fire, and the general ravages of war. Still other houses were left empty and abandoned, silent reminders of the Jewish families who had been sent to the gas chambers.[3] Fields surrounding the village were in ruins, laced with mines and cluttered with the debris of battle. Local Russian troops continued to demand food, and the few men left in the village were frequently commandeered to assist the Russians in dismantling dangerous explosives and in transporting goods and personnel. The task of clearing the fields for cultivation overwhelmed the weary villagers, and it took several years to remedy the situation. "Of 600 dwellings 35 have been totally ruined, 500 seriously damaged. From the village's 6,989 kh of arable land, only 1,380

---

[3] Nearly all of the rural Jews in Hungary perished in the Holocaust (Sozan 1986); the first were deported in the spring of 1944. The fate of Jews from Sárosd during the war is rarely mentioned in the literature from the postwar period. The following memorandum is a stark and quiet statement about their disappearance, discussing in officialese an "inventory of personal effects in abandoned dwellings": "I return two inventories and declare that the personal property to be found in Heiszler Géza's house is being held by Gál Mihály, a resident of Sárosd; among the things in Lehner Ernö's house, a table stove is being held by the Widow Mrs. Fehér Imre. The houses of Medák Nándor and Deutsch Oszkár are being rented out, though they have yet to pay rent, since they repaired the dwellings in exchange for rent" (S n.i., 1859/1946; December 28, 1946).

kh could be worked in 1945 because of fifteen kilometers of trenches and fire-bays, while in 1947, 100% could be tilled" (Virág 1947, 27). Virtually all livestock (90–95%) had been destroyed (ibid.). In 1942, there had been 401 horses, 451 cows, and 6,666 pigs in the village; after the war, 38 chickens were left, 38 horses, 34 cows, and 25 pigs. The villagers' coffers were empty and their livestock decimated.

The confusion born of devastation was further complicated by uncertainties surrounding the land reform. In regions liberated earlier than Transdanubia, such as the eastern half of the country, local committees had already been organized by late 1944 to divide up and redistribute land. The provisional government, established in December 1944 and based in Debrecen, had advocated a land reform program, but no legislation had been drafted. Local committees did not concern themselves with legislative stalemates at the national level, and formulated their own policies for what plots of land were to be appropriated and to whom it would be given. The decisions made were thus an expression, in both positive and negative respects, of local needs and problems. Despite the effort by all parties to seek representation in village-level committees, and so foster strong coalition politics, the committees were stubbornly local in their loyalties, leading to constant clashes with national administrative bodies and even their own national party representatives (Donáth 1977, 47).[4] As legislation was developed at the national level to guide the land reform, a struggle ensued. Who should determine which interests were legitimate, with regard to agricultural policy and to new forms of government?

Land Reform

Legitimate claims to land and problems with agricultural production were not new issues in the postwar climate. The land or property question had occupied writers and politicians throughout the interwar period. In 1919, the Communist regime failed dismally in its agrarian policies, despite the pressing need to address the desire for land reform within agrarian communities. In the ensuing years, especially after the equally dismal failure of the Nagyatádi Szabó initiative to distribute land among the poor, the prob-

---

[4] These committees essentially formed local governments in the political vacuum created by the Hungarians' defeat, particularly on the Great Plain, where a long time elapsed between liberation and the formal organization of national administrative bodies. Their clashes with increasingly centralized institutions covered far more ground than just land reform decisions alone (Donáth 1977, 46–50).

lem of landless peasants and the need for improved agricultural efficiency provoked continuous debates. Two questions—property rights and the problem of economic rationality in agriculture—dominated discussion. With regard to property rights, disagreements surrounded interpretations of the rights of ownership. For wealthy peasants and large landowners, property rights meant the inalienability of inherited wealth and the privileges of one's birthright. For the poor and their advocates, property rights meant the right to own the land one cultivated. Thus debates over the right to own property clashed over different principles: social justice versus the sanctity of ownership. To address the questions of property distribution, a wide range of policy options were aired. Some conservative voices argued against distribution in any form; others suggested that otherwise fallow lands, such as large tracts of entailed estates not being cultivated, should be distributed, preferably through rental contracts for smallholdings (Szekeres 1936). Arguments against distributing lands among the poor included resistance to intervening in the natural stratification of classes within society (Némethy, B. 1938). Moreover, it was argued that the breakup of large estates would be an irresponsible move; it would deprive the agricultural proletariat of employment and the opportunity to improve their lot by being elevated to the status of manorial worker, who—it was asserted—enjoyed a decent living and guaranteed income (Heller 1937). In stark opposition to this position, social democratic theorists in the 1920s argued for small farms because they would employ more people than large estates, which were primarily engaged in extensive cultivation and mechanized production. In this view, unemployment among the agrarian poor would be solved by land reform (Solt and Szigeti 1990, 92). Others argued for a program combining distribution of properties laying fallow with rental programs, consolidation of holdings, and settlement projects distributing the population more evenly throughout the country (Kerék 1934; Bajcsy-Zsilinszky 1938). Still others believed that appropriating lands owned by Jews would satisfy land hunger among the poor. Finally, there were those who argued simply for the full appropriation of large estates for distribution among the landless (Károlyi 1931; Nagy 1946).

Often, the viability of these plans was judged on the basis of economic rationality. Many argued that manorial production represented a higher form of rationality, guaranteeing productivity and ensuring the modernization of Hungarian agriculture (Kornfeld 1935). Some argued that economic rationality would be found in intensive, small-scale production, described in their slogan "Toward a Garden Hungary" (Somogyi 1942). Many feared that distributing lands among the poorest landowners would result in lost productivity and impoverishment. In plans put forward by the Social Democratic Party, these potential problems were to be remedied by state-supported initiatives to train and provision smallholders, as well as efforts to found cooperative associations (Dániel 1931). A middle-of-the-road position was advocated by Éber, who suggested that economies of scale varied with crops and animal-raising. Hence well-equipped manorial estates should be supported, and small farms—specializing in areas best suited to their scale of production—should be organized in cooperative organizations to promote their own interests within society (1932).[5]

In the debates about productivity, one finds quite odd bedfellows. Aristocrats and Communists both argue for the higher value of large-scale agriculture, though in one case on privately owned land, and in the other case on state farms or collectives (Kerék 1939a).[6] Moreover, cultural and racial politics also motivated positions within the debate. Populist groups with strong attitudes about the integrity and inherent value of peasant communities sought to realize a policy that benefited peasant families. So too, anti-Semites saw policies targeting Jewish property for appropriation by Christian peasants as just. Welcoming legislation during the war which expropriated land from Jews, one writer proclaimed: "Land should belong to those who in peace work it with their sweat, and in war protect it with their blood. Hence we salute the transfiguration of life-generating, producing, and nation-protecting sweat and blood in this extremely significant measure of the government" (quoted in Ferge 1986, 132). Those with low opinions of the working poor—who were considered backward and unproductive—fought against the measures distributing land among them, as did those who held a positive opinion of the modernizing role capitalist owners and renters played in agricultural production. All these factors—diverse attitudes toward peasants and ethnic groups, various theories of rationality, differing concepts of

---

[5] One might suggest that this policy would find its realization in the distribution of tasks between state-controlled production and second-economy activities in the late socialist period.

[6] This accounts for the fact that once collectivization was under way, many agrarian specialists—former stewards and other managerial personnel who had been employed at manorial estates prior to 1945—sided with the Communist program in the 1950s, seeing large-scale production to be superior in all respects to peasant family production.

justice—rendered the debates quite complicated and forestalled their ultimate resolution before the end of the war.

By the end of the war, the notion that settlement projects would solve the problem had been abandoned. So too had land distribution schemes that would have left manorial estates untouched. Essentially two opposing conceptions of the land reform were proposed, represented on one hand by the Smallholders' Party (Kisgazda Párt), and on the other by the Communist party (Magyar Kommunista Párt) and the National Peasant Party (Nemzeti Parasztpárt).[7] While the motivations of the Smallholders' Party were based on a specific plan for the development of capitalist agriculture, the Communist position was founded on an explicitly political concern: to win the allegiance of dwarf holders and the landless. Both plans did share one crucial point: agreement on the necessity of appropriating and redistributing manorial estates.

Smallholders, the party that represented the middle and wealthy peasants, advocated distribution of properties among peasants who were already equipped with tools, draught animals, and service buildings, and who had experience in managing a "farm." Distribution of lands to the poorest landowners and the landless, it was thought, would result in lower levels of production initially, and would burden the national economy with the costs of start-up capital, which could not be spared.[8] Finally, it was important that the pool of cheap wage-laborers be maintained (Donáth 1977, 40). This conception had strong support from the middle peasantry.

> The land reform program of the Smallholders' Party reflected above all the thinking of the upper strata of the Hungarian landed peasantry. Socially and economically they also had been oppressed by the large estates. They approved of and supported eliminating the manorial system, because it blocked their own embourgeoisement. They

---

[7] My discussion of the land reform period is based on the excellent scholarship of Donáth (1965, 1969, 1976, 1977). His critical voice has always spoken from an intimate knowledge of the philosophical and pragmatic goals of Communist party policies. He was in the Communist underground during the war, and was repeatedly jailed during the early 1950s and, after 1956, for his participation in the 1956 revolution.

[8] During the debates in 1957 and 1958, prior to the final process of collectivization, similar sentiments were voiced. A strong minority pushed for establishing cooperatives devoted solely to grain production. They argued that since every family already owned stables and barns to house livestock, creating animal husbandry units at cooperative farms would duplicate existing resources and drain coffers, thus crippling the overall development of cooperative agriculture (Donáth 1977, 154). Their predictions were to come true.

declared: the aristocratic world that looks down on the peasant should perish, the estates that grab every kind of advantage to the peasants' detriment should be abolished. But that which is good for the peasant who wishes to profit and grow richer should be maintained: respect for property, and cheap labor, which he well knows is the primary source of wealth. (Donáth 1977, 40)

The Communists and the National Peasant Party, on the other hand, argued for a land reform plan that awarded properties to the landless and poor peasantry. The National Peasant Party represented poorer peasants and the agricultural proletariat, whose position was represented by the slogan "Land to those who work it!"

Donáth characterized this plan, which was eventually implemented, as explicitly antifeudal, antifascist, and anticapitalist (1977, 41–42). Its antifeudal objective, shared by the Smallholders' Party, was the redistribution of large estates owned by aristocratic landowners. Strong antifascist principles were also incorporated into the reform. "[The legislation] decreed the confiscation of lands belonging to war criminals and enemies of the people, to representatives of Fascist parties and organizations, and the supporters of German imperialism, irrespective of the size of their holdings" (ibid., 42).[9] The anticapitalist thrust of the land reform was expressed in the appropriation of agrarian properties—land and fixed capital—owned by banks or other capitalist enterprises, and, more significantly, by choosing extreme parcellization of properties over the concentration of land among the middle peasantry. Divergent motivations prompted this aspect of the reform. The overriding concern of the landless and poor peasantry was that they receive land, despite their lack of proper tools and draught animals. In this sense, their position was also anticapitalist. "They did not consider someone a peasant if he did not participate in production with the work of his own two hands" (Donáth 1977, 42). The Communist party's interest, on the other hand, was solely in establishing a wide base of support, in anticipation of intense power struggles among postwar coalition parties.

---

[9] Perhaps the most consistently abused policies in the land reform were those relating to war criminals and German sympathizers. In the heat of the postwar land reform, the charge of war criminal or sympathizer was easily believed and not necessarily substantiated. Agelong animosities in many villages against the German ethnic minority, compounded by simple greed, influenced decisions made. Their lands expropriated, thousands of ethnic German families were eventually deported to the Federal Republic of Germany.

> The poor peasant conception of the land reform . . . was not directly linked to the Communist party's political strategy, which aimed at the abolition of capitalist relations of production, thus the abolition of the private property of the means of production. Surely it was not logical; in fact, it appeared theoretically mistaken and opportunistic (see the position of R. Luxemburg and others on the 1917 Russian Revolution) simultaneously to fight for the abolition of social classes and to increase immeasurably the number of smallholders. But then, in politics the geometric rule that a straight line is the shortest route between two points does not always prevail. (Donáth 1977, 41)

Similar strategic considerations led the Communist party to advocate a higher ceiling on landholdings than considered reasonable by the poor and landless. Unwilling at this time to alienate irretrievably a large portion of the middle peasantry, the Communists fought for a ceiling of 57.5 ha (100 kh) for manorial properties and 115 ha (200 kh) for peasant holdings (ibid., 43).

Despite the strong differences of opinion concerning the land reform, no true debate among opposing parties was ever staged. The middle peasantry had hoped to fight the battle in the calm of postwar parliamentary circles, preferably in a favorable political climate. But several factors persuaded the Smallholders to join the other parties in announcing the land reform plan advocated by the National Peasant and Communist parties in March 1945. First and foremost, it was hoped that the announcement of a land reform would encourage Hungarian soldiers still battling Russian troops in Western Hungary to abandon the Germans and return home. Secondly, the Smallholders feared a radicalization of demands among embittered agricultural workers awaiting action, possibly incited by the Communists. The radical climate among local groups on the Great Plain undoubtedly confirmed their worries. Finally, spring planting carried its own, time-bound urgency, an urgency heightened by the poverty of a nation devastated by war (Donáth 1977, 44–45).

Nearly 35% of the country's 9.3 million ha (16 million kh) was expropriated, of which 60% was distributed to individuals (equaling 1,890,068 ha or 3,258,738 kh). Another third of the properties—primarily forests and pastures—were transferred to state and cooperative ownership (Pető and Szakács 1985, 37–38). (Table 3.1 lists the numbers and acreage of properties expropriated, and the legal means by which these properties were appropriated in the land reform; table 3.2 shows the persons and agencies who

TABLE 3.1. Number and size of landed properties appropriated in the course of the land reform (1945–1947)

	Landed properties		Redeemed from properties < 100 kh		Appropriated landed properties		Total
	Confiscated	Redeemed	War acquisition	Vineyards, orchards	Forests	For house plots	From necessity	
Number	43,245	8,479	2,629	590	763	12,372	7,427	75,505
Acreage (kh)	539,171	4,835,348	89,808	8,682	18,302	19,243	89,091	5,599,645
Percentage of entire acreage	3.3	29.9	0.6	0.0	0.1	0.1	0.6	34.6

Source: Orbán 1972: 38

TABLE 3.2. Grantees in the land reform (1945–1947)

	Acreage (kh)	Percentage
Persons	3,258,738	58.2
Model farms and experimental farms	51,079	0.9
State fish hatcheries and reed farms	46,542	0.8
Communities (forest)	83,643	1.5
Joint tenants (forest)	15,932	0.3
State (forest)	1,360,592	24.3
Common pastures	422,496	7.6
Churches	19,739	0.4
For settlements	91,809	1.6
For house plots	110,387	1.9
For public purposes	26,152	0.5
Acreage in reserve	112,536	2.0
Totals	5,599,645	100.0

Source: Orbán 1972: 41

were given land during the land reform process.) In addition to acreage, land reform legislation also expropriated equipment, buildings, and produce.

> In the then 25 counties of the country more than 1,500 manor houses and mansions, along with the almost 8,000 kh which belonged to them, became state property. More than half a thousand distilleries, mills, various agricultural trade workshops, as well as 175 other small factories (cartwright and blacksmith shops, soda-water works) became partly the property of cooperatives forming then, and partly the property of other public institutions. Many thousands of large agricultural machines (tractors, threshing machines, etc.) fell into the hands of cooperatives. (Ibid., 38)

Small landowners could clearly not use the large machinery once employed at manorial farms. In addition to large machinery, however, another 160,000 tools and smaller tilling equipment were distributed among poor peasants, as were 250,000 other machines (Donáth 1977, 79).

Eventually 642,342 persons received land in the reform (Orbán 1972, 42). As planned, the agrarian poor predominated among new landholders: 93.3% of the land apportioned to individuals went to agricultural workers, manorial workers, and semiproletarians (ibid.). (See table 3.3 for exact figures on the occupational identity of those receiving land and the average size of plots portioned out

TABLE 3.3. Distribution of new landowners, nationwide (1945–1947)

Occupation	Receiving land Persons	%	Land apportioned Kh	%	Kh per capita
Manorial worker	109,875	17.1	922,255	28.3	8.4
Agrarian worker	261,088	40.6	1,288,463	39.5	4.9
Dwarf holder	213,930	33.3	829,477	25.5	3.9
Smallholder	32,865	5.1	143,131	4.4	4.4
Farmhand and craftsman	22,164	3.5	53,866	1.7	2.4
Trained agriculturalist	1,256	0.2	14,548	0.4	11.6
Forestry worker	1,164	0.2	6,998	0.2	6.0
Totals	642,342	100.0	3,258,738	100.0	5.1

Source: Calculated from Donáth 1977: 76; and Orbán 1972: 42

to those groups). Favoring the poorest families altered social differentiation within the agrarian community to a significant degree.

> Within the agricultural population the numerical proportion of landless in 1941 figures dropped from 46% to 17%, the number of small-scale producers increased from 47% to 80%, while the ratio of wealthy peasants and remaining lordly landowners (former large and middle landowners) fell from 7% to 2.8%. As a result of the land reform, the rocketing increase of the numbers of petty commodity producers as a proportion of the entire population increased to over 43%, while in 1941 they had not reached 27%. (Pető and Szakács 1985, 40)

Yet some were still left out. Orbán has calculated that despite the allocation of lands favorable to the poorest groups of peasants and agricultural workers, more than 300,000 agricultural workers did not receive a land allotment in the land reform. This was due in part to industrial workers returning to agriculture following the destruction of urban industries during the war, and in part to the arrival of 350,000 Hungarians who had been deported from neighboring countries after the war. Their numbers were only partially balanced by the 170,000 German and 70,000 Slovak deportees from Hungary (Orbán 1972, 43). Roma were also excluded entirely from the land reform process.

> Earlier, Gypsies had been disenfranchised and marginalized to such a degree . . . that—according to all indica-

> tions—they could not even realize that perhaps they could have rights now. Presumably part of the reason for this was that (with the exception of horse traders, one of the "highest level" groups in the hierarchy) earlier they had little to do with land or animals, and also that a significant proportion had been caravan dwellers. All of this however is not a true cause, only a partial explanation or excuse. *If* the conditions had been different, if—above all—there had existed sensitivity toward the Gypsy question, then perhaps a number of later, current difficulties could have been prevented. However, the country did not take advantage of this possibility. There was *no one* who encouraged the Gypsies to demand land or for whom the idea would have occurred. Thus they were *entirely* overlooked in the revolutionary transformation. (Ferge 1986, 150–151)

So although no legal provisions excluded Roma from participation, centuries of exile within village and urban communities ensured their disenfranchisement.[10]

Regional differences also characterized the distribution of properties. The historical differences in settlement patterns and the distribution of large estates varied across the country, so that both the numbers of claimants and the amount of land to distribute were often at odds. In the Great Plain and the northern counties, there were high numbers of claimants and very little land to distribute, whereas in Transdanubia the average proportion of people to land was more favorable, owing to large tracts of scarcely populated regions being dominated by manorial estates. So too, the number of middle-sized farms (e.g., above 25 kh) was proportionately greater on the Great Plain than in Transdanubia.

> Since from the three million agrarian poor, only one million (wage-earners and dependents together!) could be awarded land and those were in diverging proportions in various regions in the country, not only did the number of those awarded land and their proportion develop differentially, but also the regional distribution of the remaining two million agrarian poor also was somewhat modified. Before the land reform . . . the dispersion of the agrarian poor was greater. But now, instead of the earlier 45–46% in the eastern part of the country (with the exception of Sza-

---

[10] Stewart offers figures that contest Ferge's representation of Roma as not engaging in agricultural work. "Gypsies had made up one-quarter of all agricultural wage-workers prior to the war and would therefore have been eligible for land in any land reform. But when this came in the wake of the liberation in 1944–45 they were left out of the redistribution" (1993, 199, n. 2).

bolcs, Szatmár), 52–53% were to be found in the counties of the Great Plain, and in Borsod-Abaúj-Zemplén County, which was later merged. But their polar situation also was altered, since in the aforementioned counties of the Great Plain the farms of more than 25 *hold* developed in certain constraints in proportions exceeding that of the country generally. (Orbán 1972, 45)

Thus, the social tensions between poorer and wealthier peasants, between those who got and those who were left out, were much more pronounced in the eastern and northern sections of the country than in Transdanubia, even after the land reform (ibid., 44–45). An article entitled "There Is an Agricultural Labor Shortage in Fejér County" in the *Fehérvári Népszava* (March 19, 1946) illustrates how differently the problem of land and labor was lived in Fejér County than in other regions of the country.

For spring and summer tasks one must reckon with an absence of 6–8,000 agricultural workers in the county. Therefore worker demand is evident, but until now supply has been only from Zala County. Work morale has seriously deteriorated. . . . In connection with the land reform and the resettlement of Germans, there is an urgent need to bring settlers, because it is feared that some of the lands reserved for them will remain uncultivated as a consequence of tardy measures.

Thus different regions in the country would also have different experiences with class warfare during the early phase of collectivization. While in Sárosd the category of kulak was relatively unknown, in other regions the wealthy, exploitative peasant was very much a member of the community: easily targeted and frequently abused.

Opinions differed concerning how to administer the land reform, the two options advocated following directly from the divergent policies of the Smallholders and the National Peasant and Communist parties. The Smallholders proposed that officials from the state and wealthier peasants oversee the process; agricultural workers fought for locally elected committees, which were to be composed of those who had claims to land. Efforts were made to create national- and county-level organizational structures for the reform, but in many communities local committees effectively took matters into their own hands. On the national level, the National Council Supervising Land Reform (Országos Földbirtokrendezö Tanács) was established to clarify and adjudicate legal problems arising from the redistribution of properties. The more immediately administrative functions of implementing the land reform, that is, material and technical concerns and questions over property registration, were handled by the National Land Office (Országos Földhivatal). On the county level committees were set up to supervise directly both the expropriation and the allocation of lands (Megyei Földbirtokrendezö Tanács), but their role was severely limited by village land claims committees. Legally, the village committees were empowered to make recommendations to the county level, but they did far more.

> In practice, to wit, everywhere in the country supervising bodies from higher levels were confronted by *faits accomplis*. The Land Claims Committees not only inventoried the lands to be expropriated, and not only did they take a position on the confiscation or redemption of properties, they actually expropriated them and distributed them immediately. They did not merely judge who was entitled to land, but they also determined who would receive how much land. They did not wait for their plan on apportionment to be sanctioned. Thus they stepped far beyond the sphere of authority established in the law! (Donáth 1977, 52)

The local control of decision-making led to charges of abuse, patronage, and nepotism, which were not entirely without foundation. In many instances land claims committees judged a claimant's case not only according to the size of his family and previous holdings, but also in terms of his merit, his reputation as a peasant (ibid., 73). The inclusion of such subjective factors undoubtedly complicated the process. In regions where land was scarce, such as the Great Plain, contested claims led to bitter fights.

Two independent factors strengthened local control of the land reform. Centuries of land hunger emboldened the landless and poor peasantry to wrest land for their families, despite fears of retaliation or government interference.[11] Among the committees first established on the Great Plain, these sentiments were particularly widespread, building on the strong tradition of political organizing among the agricultural proletariat in these regions. The other major factor was the ever present concern for the state of agricultural production. The sole preoccupation of the county agricultural supervisor, and of his superiors, was the condition of local agricultural production. In areas where war destruction had not been so

---

[11] The exalted aims of peasants and workers for a new kind of society were paired with an intense hatred of the aristocracy and gentry. All across the country, wealthy houses were sacked and property was demolished, reminiscent of peasant wars in the past.

extensive, the rhythm of the seasons had been only briefly interrupted. In Fejér County, and all of the northwestern counties—site of the heaviest fighting outside Budapest—it was imperative to get people farming again, despite the obstacles of land mines, heavy metal debris, and labor shortages. On more than one occasion the county supervisor threatened imprisonment, and candidly stated that problems would be handled by the Russians if not attended to immediately. The level of desperation is clear in the following communiqué, which states in no uncertain terms that unresolved issues concerning the land reform must not represent an impediment to the wheat harvest.

> The time has arrived to complete the tasks of the wheat harvest. I direct the Community Production Committee to begin immediately with all the power at its disposal and to continue in full force without interruption. . . .
>
> I serve notice to the Committee that the above directive must be satisfied without delay and without regard to circumstances. The beginning of work cannot be prevented by the possibility that the Committee, or more properly the Land Claims Committee, is unclear about the results of implementing the land reform, e.g., who gets how much land, what the former owner is to receive, how to start up the work, and how to divide up [the work] among new landowners, and possibly among others from outside the community who are without land but are willing to participate in the harvest, all of which is the job of the Land Claims Committee.
>
> Insofar as the least negligence affects implementation of this directive, I will begin proceedings for internment against the remiss.
> Koltai János
> Székesfehérvár, June 23, 1945
> (S n.i., 20/1945)

Thus, lengthy deliberations over this and that parcel of land, over the legality and propriety of properties distributed, were actively discouraged.

Sárosd came late to the problem of land reform, but was no less fraught with confusion than other communities. Rumors reached Sárosd of land being redistributed on the Great Plain. Stories of renegade committees frightened villagers. Many remembered the promises made in 1919 to provide agricultural workers with land, and they had not forgotten the retaliations against presumed supporters of the Soviet Republic during the White Terror. When I asked the head of the land reform committee whether there was much haggling about where and how much land people were given, he explained:

> No, they were afraid that it would cost them. . . . Well, they remembered what had happened in 1919. They began, they had set about starting up [in 1919], and by then the situation had changed. They tied people up behind wagons, they beat up many people. [In 1945] people shouted at me that I would be hanged. "What will happen when the lords return?" My wife didn't speak to me for three weeks, because she was afraid, her too, she didn't like what I was doing. Well, we had all these children.

Kovács, the former owner of Jakabszállás, came back to the village and threatened the people working on his land, giving substance to their fears about reprisals. He even went so far as to force people to remove the potatoes they had planted. The head of the Land Claims Committee had to go out and intervene, explaining to Kovács that he had no more rights to the land. Some refused to be intimidated and, with encouragement from Budapest, they proceeded to arrange elections for the Land Claims Committee in mid-April. The membership was predominantly composed of former manorial workers and harvesters, and was headed by a former construction laborer who had only recently joined the Communist party. Although the mood in the village was generally skeptical, the Land Claims Committee acted quickly and Sárosd was declared the second village in the county to execute the land reform.

Land was parceled out according to the size of the family and the amount of land already owned. The initial allotment of properties satisfied every claim, but left no surplus land. Later in the year, however, modifications in land reform legislation forced the Land Claims Committee to redistribute properties. In Sárosd, families without land were given at least 2.85 ha (5 kh), the amount increasing, depending upon the size of the family, to an upper limit of 8.6 ha (15 kh). Families with limited holdings were also given small parcels of land. The village clerk counted 310 middle and wealthy peasants (2.85–57 ha; 5–100 kh) and 200 owners of smaller properties (0–2.85 ha; 0–5 kh) after the land reform was completed.

Land was apportioned under the direction of the president of the Land Claims Committee, whose experience as a roadworker had taught him to measure tracts of land skillfully. Every claimant brought a stake 60 cm high, with his name prominently displayed, to mark the allotted plot.[12] (Those fearing the return of their former lords brought short, anonymous sticks lest their complicity in the reform later be revealed.) The second allotment left 228 ha (400 kh) undistributed from the properties of the Daru estate center, which were then apportioned to several families from the Great Plain.[13] The families from the east worked their farms of 57 ha (100 kh) each until 1948, when the state appropriated their lands for the state farm.

Since there was plenty of land to distribute among villagers and residents of the former estate centers, the land reform was not marred by overt conflict as it was in villages to the east. Nonetheless, some problems did arise. Technical difficulties issued from the redistribution mandated by the state after changes were made in land reform legislation in mid-1945. Some sections of land had been left fallow; other portions had been sown with crops. If someone's cultivated land was to be reapportioned to another claimant, and the land to be received lay fallow, then a compromise was reached. The claimants would share half of each plot until the crops were harvested, and then the lands would go to their new, rightful owner. Though amicably reached in all cases, such compromises on labor invested and land readied for cultivation surely complicated recovery in the year following the land reform. A more serious problem, never resolved to some people's satisfaction, was the disagreement over how claimants were defined. Generally, a claimant was defined by the Land Claims Committee as a married couple, preferably with children. Young adults hoping to establish a farm independent of their parents were not awarded land. In larger families this doomed young men and women to continue day labor and other wage employment in agriculture, as the land could not support them, and urban industry was in a shambles.[14]

---

[12] There is no indication that attempts were made to divide up each claimant's allotted lands among several different plots, in the interest of dispersing better (and poorer) lands among many families. The only apparent advantage new landowners wielded over their peasant rivals in the postwar recovery was that their lands were consolidated, not divided up in multiple plots like those of the peasantry.

[13] The scarcity of land in many communities on the Great Plain was reduced by offering wealthier peasants lands elsewhere, in exchange for their own lands in the community. The wealthier peasant was recompensed for his expropriated lands, e.g., in Fejér County, and thus local landless and poor peasants in the east had more land to distribute among themselves.

[14] In some communities land was reserved for men detained in prisoner-of-war camps and in Siberia, but I do not know this to have been the case in Sárosd. Since

Swift recovery in agricultural production was a national priority, and a pressing concern for the new landowners of Sárosd. Obstacles were great, and far greater for new landowners than for the peasantry. Machines from the manorial estates were made available for sale as part of the land reform, but they were often designed explicitly for large-scale production. They were certainly not suited for the small plots that now dominated the landscape. The few animals that had survived the front were owned almost exclusively by the peasantry. Thus the new landowners had neither tools nor draught power, nor even the means to acquire them. The absolute dearth of basic tools in Sárosd after the war was demonstrated by the June 23, 1945, request for 200 scythes by the village council; the county made them available to needy communities at low cost. Indeed, in some cases in the county—and perhaps in Sárosd as well—local elites guaranteed that their disproportionate control of livestock and draught power would endure, as the following memorandum makes very clear.

> Subject: the delivery of cows for reparations
>
> In the course of the delivery of cows to be commandeered for reparations, complaints have arrived on my desk from many quarters that parish councils are not requisitioning cows in a just fashion. In many cases they considered taking animals from those who had only one, rather than from those who have many animals. This measure not only contradicts the decree, but it is totally unfair.
>
> Consequently I call upon the parish councils to report whether the Village Production Committee participated in the commandeering of cows in the village. I request an account of who they requisitioned the delivered cows from, how many cows those people have, and who are those who still have two or more cows.
>
> Dr. Szirbik Ferenc, Subprefect
> Székesfehérvár, June 25, 1945
> (S n.i., 694/1945)

The terms of reparations were thus quite strict; anyone with more than two cows could be targeted for expropriation.

In areas that had escaped major war damage, one could buy cows, horses, and pigs in exchange for foodstuffs and other valuables. Peasants blessed with materials salvaged from the war,

---

many of the men detained in camps east and west fell into the category of unmarried adults, I suspect they also were victims of the narrow definition of claimant employed locally and so were welcomed home to no land.

such as urns of lard or wheat, traveled to Tolna County and other regions to the south to acquire livestock and other valuable goods.[15] Outfitted with the few horses and cows in the village, peasants were in a position to extract even greater amounts of labor from poorer families and new landowners in exchange for use of a draught team than they had before the war. Labor service inflated drastically: for one day's use of a draught team, one reciprocated with five to seven days of labor. Being ill equipped, former manorial workers and day laborers were forced to exchange their labor cheaply. As one villager commented, "It was as if they had reinstated the feudal tithe!"

The landed peasantry considered new landowners to be lazy and morally weak, an opinion that was confirmed by their poor start in private farming. The charge that new landowners lacked any ability to farm—though exaggerated by years of social prejudice—nonetheless had some basis in fact. The experiences of new landowners in agricultural production had been severely limited, particularly in complex decision-making processes. Confronted with unfamiliar problems, they were forced into decidedly novel situations: to experiment, to question, and to seek the advice of their "betters." In this regard, they had little resort but to appeal to the generosity of their peasant kin and acquaintances for help and guidance. Thus the stage was set for increased socioeconomic differentiation of former and new landowners, based on differential access to the basic means of production: to tools, to animals, to sufficient labor power, and to critical knowledge. To complicate matters, two of the three crucial years that constitute the transition period between the land reform and the institution of the command economy were stricken by drought. This only exacerbated the difficulties encountered by new landowners in building and sustaining viable farms.

The ideology of diligence and possession, which before the war had explained the patterns of stratification in the village and on the manor, was further strengthened during the transition period by the growing socioeconomic disparity between peasants and new landowners. The economic advantages enjoyed by the peasantry

---

[15] Postwar inflation of the Hungarian currency reached a scale unprecedented before or since. "In the German inflation following the First World War, the exchange rate of the dollar rose to 4.2 billion (10^{12}) marks; in the Hungarian inflation following the First World War, to 100,000 crowns, and in the Hungarian inflation following the Second World War, to 5 quintillion (10^{30}) paper pengős" (Nötel 1986, 538). Until the conversion to the forint in August 1946, any exchanges were preferably transacted in tangibles: foodstuffs, clothing, precious metals, etc.

were understood to be the privileges of hard work, keen knowledge, and the ever critical status of being an owner, of controlling one's labor. It was presumed that the new landowners' complete ignorance of the experience of possessing land made them completely unsuited to working land on their own. Roma required even more direct forms of control over their labor; they were to be closely watched and schooled in new patterns of behavior. In the work plan on social issues submitted by the village clerk to his superior at the district offices, he describes local efforts to discipline the Roma population.

> We have taken care of the issue of Gypsies living in the village by having the *vajda* who represents them confirming in a daily report where the Gypsies are and what they are working on. We have ordered vagrants immediately to public works projects. This way we have forced them to work in order to insure them a living, and so not require village support.
>
> Deputy clerk
> Sárosd, September 13, 1946
> (S n.i., 1527/1946)

The difficulties new landowners experienced and the behavior of Roma workers merely reinforced the peasants' view that the fate of new landowners and Roma laborers was to be directed and controlled, to be "owned" rather than to own. This would be indisputably confirmed to the old elite when the pressures of heavy taxation and forced deliveries drove new landowners into cooperative production in the early 1950s.

The Rise of the New State

The role of the state in assisting, directing, and regulating agricultural production increased in the postwar period. However, the state's role in agricultural production and commerce had been quite substantial in the interwar period. The participation of state agencies in regulating commercial transactions, setting prices, restricting acreage, and promoting certain crops was particularly significant during the depression years of the late 1920s and early 1930s (see Szuhay 1961, 1962), not least because of the high degree of competition among countries in the agrarian market. By the mid-1930s, the production, trade, and price structure of the following crops were under different degrees of state supervision: wheat, sugar beets, tobacco, linseed and linen fiber, hemp, milk, firewood, wool, paprika, alcohol, and potatoes (Ihrig 1935, 131). In search of effective policies to deal with the pressures of a depressed market, economists and politicians compared the success and failure of the various methods employed by the states of Europe and America. They were keenly aware of Soviet attempts at regulating the economy, and often debated their merits and deficiencies. Discussing the scope and relative value of state intervention, Ihrig quoted from an editorial in the *Economist* published in September 1934 to pose the question more clearly. "The Economist, which is by no means biased, writes the following: 'today the great question of the world, for which every country searches for an answer in its own way, is not whether the state should introduce systematic plans into economic life, but rather, who should do this and how should it be done'" (ibid., 130).

In addition to the state's regulation and supervision of private production, it had long had agricultural holdings of its own. The state ran agricultural enterprises, such as experimental farms, teaching and model farms, reed farms, and fish hatcheries, as well as owning large tracts of forests. Some of these enterprises, such as horse-breeding farms, were designed to serve military needs, while others were integral to educational institutions. However, the degree to which state officials were directly involved in managing varied, depending on whether the specific institution was a joint stock company, in which the state owned the majority of shares, or an institution directly owned by the state (Róna-Tas 1990, 60). Most village communities owned pastures and forests, and also rented land to local peasants, as in Sárosd.

By the early 1940s, the war economy and demands made by the Germans on Hungarian resources increased even further the state's involvement in agricultural production. Compulsory agricultural deliveries were instituted in 1940, and during Jurcsek Béla's tenure as minister of food a requisitioning system was established to appropriate surplus grains from the peasantry (Pamlényi 1973, 520). The specific level of food rationing in the country was dictated in large part by the Germans' requisitioning of foodstuffs (Berend and Ránki 1979, 228). As the war lingered on, greater and greater restrictions were imposed on the Hungarian populace, while Germans attempted to extract more and more supplies. In order to placate the Germans, and also provide a minimum of provisions and supplies for its own population, Hungarian officials introduced a variety of different industrial crops and production restrictions.

> The area sown with the oil seeds demanded by Germany was compulsorily increased. Sunflower had to be grown on 5

> per cent of the arable area of holdings of over 50 *holds* . . . and 2 per cent more castor beans had to be grown on holdings of over 20 *holds* than before. By 1942, compulsory production was extended to holdings of over 15 *holds*, with 7 per cent of the land to be devoted to specified products. The war years thus brought considerable changes in Hungary's agricultural structure. The area devoted to industrial crops grew the most: by 1944, it was 3.2 times what it had been between 1931 and 1938. . . . Compulsory delivery orders resulted in significant growth in the production of rough fodder, too. (Ibid., 229–230)

Thus the role of the state in expropriating grains and livestock, as well as in demanding the cultivation of cash and industrial crops, was quite influential.

Since the turn of the century, the state had been actively involved in setting wages and setting the terms for contractual negotiations. Between the wars, the state continued its role in employment policies. In the midst of the economic chaos following the First World War, the income of agricultural workers had dropped precipitously, prompting the government to pass legislation in 1923 "for the prevention of the illegal exploitation of the labor power of agricultural workers" (quoted in Lencsés 1982, 109). This statute was restricted to wages for day labor, leaving other forms of agricultural wage labor unregulated (ibid., 110). By 1941, the conditions influencing wages had changed, compelling state agencies to set an upper limit for wages; the passage of the governmental regulation was explained as "wish[ing] to thwart the increase of living costs and the decline of the buying power of money" (quoted in ibid., 112).

After 1919, the state's concern with the political mood of the labor force and the agitation of undesirable political elements resulted in legislation restricting those who could participate in labor recruitment. In the 1920s, people seeking permission to act as labor recruiters, usually contracting migrant labor for manorial estates, were expected to demonstrate an untarnished political past, for instance, no association with the 1919 Soviet Republic. This concern for political radicalism faded in the early 1930s and then grew again in the late 1930s, when radical religious fundamentalism and political extremism swept large regions of the Great Plain. The Hungarian government began to be more directly involved in contracting agricultural laborers as early as 1937, specifically to work in Germany. In 1938, the National Economic Work Commission Office (Országos Gazdasági Munkaközvetítö Iroda) had arranged for 10,000 workers to travel to Germany, and by 1939 the Germans expressed an interest in 15,000 to 20,000 Hungarian workers (Lencsés 1982, 53). Nearly 23,000 agricultural and industrial workers were employed in Germany in 1942, but in following years their numbers decreased as the demands of the Hungarian military and domestic labor market overshadowed German requests for more workers (ibid., 60–61).

In the spring and summer months of 1945 the correspondence between the village council and the county agricultural officer and other government agencies grew rapidly. Despite its long history in regulating and influencing agricultural production, the state was now even more directly involved in local village affairs, in response to the urgency of postwar reconstruction. Three general problems were addressed, all assessing the current situation and the tasks that lay ahead for agricultural, and national, recovery. County and national agencies urged villages to comply with the demands of Russian troops, which included supplying provisions, housing, transportation, and labor to clear the fields of mines and debris, and to bury the dead. Queries were made about the condition of machines, tools, parts for machinery, fertilizers, seeds, and the status of labor resources and livestock reserves in the village. Finally, and perhaps most urgently, the county strove to insure that cultivation would be resumed as swiftly and extensively as possible.

The documents leave no doubt that many county and national agencies perceived the resumption of agricultural production to be absolutely necessary for economic recovery, and thus a high priority of government attention. Community production committees, established to handle problems at the village level, were coaxed, admonished, and even threatened with imprisonment if county directives were not discharged. Protection of abandoned crops, fulfillment of quotas, the prompt submission of data all came to be construed as central to the national interest. Local interests, however, took exception to these priorities, often contesting the propriety of outside agencies in determining their affairs. The battle over control of resources was raging. The following memorandum concerning an official barter campaign makes this very clear:

To the Local Subprefect:

In reports sent to me I have come to learn that several public institutions (or officials working at public institutions), public agencies, various persons, as well as parties, have been obstructing—for unknown reasons or out of ignorance—the organization of the official exchange scheme initiated by the Hungarian Minister of Public Food. In certain places public authorities are totally misinformed of the circumstances, in others they interfere in the exchange scheme. In many cases the local authorities, Production and National Committees, or political parties take the industrial goods and use the grain for seed.

I call the subprefect's attention to the fact that the bread grain collected for the purpose of the exchange scheme is an exclusively central stockpile (see law no. 104.900/1945 KM.sz.rend.7); it cannot be used or exchanged by any organ, party, production committee, national committee, etc., and cannot be issued to any of them.

Szabó István, lord lieutenant
Székesfehérvár, February 11, 1946
(S n.i., 361/1946)

In another memorandum stipulating that results be improved in the official collection of lard, county officials complained that administrators were not keeping proper records and not taking the project seriously. Accordingly, too many people were butchering pigs illegally, and those who did so with official permission were still not submitting their proper allowances of lard (S n.i., 1446/1946). Direct intervention by government agencies in production decisions, marketing, and even consumption of foodstuffs in the village thus became an integral aspect of administrative practice several years before the institutionalization of Soviet-style national planning in 1948. So too, local practices resisting state agencies in their intrusion in local affairs were also quite evident in these protracted negotiations, battles, and struggles.

The fateful Year of Transition, as it has come to be called in official forums (late 1947 to late 1948), marked the final demise of coalition politics. In 1947, the Communist party discredited, jailed, or exiled all remaining opponents in the left-wing coalition Independence Front (Függetlenségi Front), and in June 1948 it absorbed the Social Democratic Party.[16] Banks were nationalized, additional industries were appropriated by the state, and the national system of parochial schools was abolished, to be replaced with public schools. Governmental agencies were subjected to sweeping changes in their organizational structure, to institutionalize their role in planning, administering, and regulating the growing state

---

[16] Despite the virtual takeover by the Communist party in June 1948, in October 1949 three parties were still active in Sárosd: the Hungarian Workers' Party (Magyar Dolgozók Pártja), the Independent Smallholders (Független Kisgazda Párt) and the National Peasant Party (Nemzeti Parasztpárt).

sector (Pető and Szakács 1985). New policies for the development of agriculture were also implemented.

Public announcement of changes in agricultural policy was abrupt. From one day to the next newspaper editorials shifted their position drastically: support of private property in agriculture was abandoned and voluntary collectivization was declared the solution to the class struggle in the countryside. In a front-page article on July 13, 1948, the county newspaper, *Fehérvári Napló*, declared, "The government not only defends private property of the peasantry, it wants to strengthen it."

> The reactionaries have begun a new campaign of lies against the democracy . . . It is spreading the news . . . that the government wants to change its current policies toward landed property and wishes to place agriculture in kolkhozes . . . the government has no intention of altering its policies in agriculture and toward landed property.

On the following day, an editorial proclaimed voluntary collectivization to be the weapon against exploitation by kulaks and the basis of worker-peasant solidarity.

> There is no question . . . that since liberation kulaks have only gained strength in the democracy, have grown richer in greater measure than the working peasantry. . . . The instrument, the best weapon for the protection of common economic interests, is the voluntary unification of workers: the cooperative. . . . Of course, even as a joke it is boring to say the cooperative is just like a kolkhoz. To make people believe this is only in the interest of kulaks, who resist in direct proportion to their desperation. . . . The ascending, flowering peasantry assures the basis of the nation's prosperity: the worker-peasant alliance. (*Fehérvári Napló*, July 14, 1948)[17]

To paraphrase another editorial, class warfare was transferred from the city to the village. Peasant property was under siege by the state. (Table 3.4 shows the distribution of properties in Sárosd in 1949.)

The tone of articles became increasingly aggressive; vilifying synonyms for the class enemy multiplied. Greasy peasant, specula-

---

[17] The term "working peasantry" was coined in the late 1940s to refer to those members of the landowning peasantry the state deemed to be engaged in honorable labor. They were regularly contrasted with kulaks, who were purported to live off the labor of others. The opposition was meant to underscore the alliance of the state with respectable (but poor) peasants against class enemies and exploiters.

Securing Ties 137

TABLE 3.4. Size and number of properties in Sárosd, 1949

Size of properties (kh)	Number of properties
0 (Peasants without any land)	56
1–5	264
5–10	283
10–15	120
15–20	25
20–25	4
25–30	6
30–40	4
40–50	2
50–100	6
Above 100	1

Source: Chief clerk's notes on Sárosd, addressed to the Sárbogárd district October 1, 1949

tive middleman, capitalistic parasite, millers who fleece people, fatty-necked peasant: the terms reek of the awkward, forced slogans of the 1950s. Editorials and newspaper stories blamed the kulaks, and their partners in sabotage, the "black reactionaries" (the priesthood), for every possible complication in cultivating and harvesting crops in the summer and fall of 1948. The consequences of resisting state demands were severe, as the following article entitled, "They Are Proceeding against Kulaks Who Sabotage with Relentless Severity," attests:

> According to reports received *a majority of kulaks are sabotaging fall work.* In many places the kulaks *are not willing to lend their tractors to small peasants to plow their lands,* so that they are inhibiting them in their work. The government will proceed against sabotaging kulaks with the full severity of the law. They will seriously punish those who neglect fall work. And as for those who do not hand over their tractor to the small peasant, in addition to the penalty, they will have their machine confiscated. The government will prevent sabotage with every means and will insure the working people their next year's bread supply. (*Fehérvári Napló,* October 14, 1948)

Simply refusing to relinquish one's tools to cooperative production and poor peasants was construed as sabotage, as undermining the people's economy, a quite novel conception of the rights of property and control over one's personal effects. All the while the party fought to convince poorer peasants that their allies were not the

TABLE 3.5. Expansion of state property during the three-year plan

Economic Branch	July 1947	July 1948	December 1948	December 1949
Mining	91%	91%	91%	100%
Industry	38%	80%	87%	100%
Transportation	98%	98%	98%	100%
Banking	—	95%	95%	95%
Wholesale trade	—	16%	75%	100%
Retail trade	—	—	20%	30%

Source: Pető and Szakács 1985: 103

wealthy, yet familiar peasants of their village, but the urban working class.

Peasant Unity against the Peasant

It is time that the working peasantry of the village awaken to a consciousness of its class position and not permit the kulaks to take over the inheritance of the Pappenheims and the Zichys [Jews and aristocrats] in the village. They are trying to lump the peasantry together. They proclaim the unity of the peasantry and would have us believe that the interest of a 5 kh [2.8 ha] peasant is the same as the greasy peasant's, who has 50 kh [28 ha]. The kulaks' ceaseless demagoguery now incites against forced deliveries, and claims that the industrial worker takes the ration right out of the producing peasant's mouth. Of course, so many words, so many lies. (*Fehérvári Napló*, July 21, 1948)

The intensity of the rhetoric was directly related to the stubborn refusal of the agricultural population to recognize the working class as their ally, and their adamant rejection of collective production. As early as the 1945 and 1947 elections, Communist party observers noticed with dismay that even in regions known for their radicalism, new landowners voted overwhelmingly for parties proclaiming the sanctity of private property. As a communiqué issued by the Communist party explained, "their consciousness was washed away by the fact that they got land" (Orbán 1972, 63). By the end of 1949 the agricultural sector alone had escaped nationalization, whereas mining, industry, transportation, banking, large commercial enterprises, and a sizable proportion of small shops had been absorbed by the state sector (Pető and Szakács 1983, 213). (See table 3.5 for the pace and degree of nationalization.)

Age-old animosities between rural and urban communities in Hungary lay at the root of landowners' unwillingness to embrace the urban working class as comrade in the struggle. The polarization of former peasants and new landowners was certainly troublesome to poorer villagers, but this experience did not lead to a rethinking of their class allegiances. On the contrary, the offensive against kulaks, and ensuing policies of heavy taxation and agricultural quotas, only heightened the sense of community in Sárosd with fellow peasants against the city and the state (synonymous entities to many minds), and against their representatives, the party bureaucracy and the secret police. A further factor, Communist ideology itself, also reinforced sentiments estranging village and town. The worker-peasant alliance, so fondly evoked in print and oratory, was an alliance founded against the kulak, not an alliance based on social or political equality between the industrial and agricultural proletariat. Editorials hammered home the ideological supremacy of the industrial worker; agriculturalists were exhorted to follow their progressive lead. Industrial workers, on the other hand, were encouraged to sponsor cooperative farms and other rural institutions, as well as to conduct "consciousness-raising sessions" to convince peasants of the superiority of collective production. Worker agitators, called "instructors of the people" (*népnevelök*), had very little credibility. The campaigns waged in villages by urban workers struck peasants as absurd, since they were far more knowledgeable than many workers about the best way to go about farming. Furthermore, in the national division of labor, agriculture was proclaimed to be the peasants' specialty. As one villager recalled, workers were invited to join peasants in the fields, to talk while they worked. It turned out that the agitators could not even hoe, which merely reinforced the peasants' unwillingness to take them seriously. Incidentally, workers did not always participate voluntarily, being forced to campaign as part of their obligations to the factory, which probably made them far less convincing advocates for the state than was hoped. The harsh policies of rapid industrialization in the early 1950s, financed by agricultural production and designed in part to force labor out of agriculture and into industry, merely underscored the peasants' opinion that socialism guaranteed the secondary status of peasants and villagers.[18]

[18] For comparison with similar processes in Southern Africa at the turn of the century, see Palmer and Parsons 1977, Marks and Rathbone 1982, and van Onselen 1976; cf. Hart 1982 for a discussion of increased social inequalities and the character of class alliance in West Africa.

Great discrepancies existed between officials in agricultural production who were trained in the field and those who were there by virtue of being party members or industrial workers. Many former stewards and management personnel from manorial estates became actively involved in early cooperative and state farms, putting their professional knowledge to use and demonstrating their commitment to modernized, large-scale production. They easily clashed with those officials whose positions were owed to their allegiance to the party, and whose political concerns did not translate into a well-run enterprise. Yet despite the superior training and practical knowledge agrarian specialists may have enjoyed, they were still subordinated to local political officials. One inhabitant of Sárosd, who was employed as a local agrarian advisor by the county, had been the steward of a large estate owned by the Károly family in Slovakia before the war. He kept his former job a secret. Even though he had joined the party (solely for political advantage), he nonetheless was called upon at one point to prove his political reliability. Among other things, his superiors asked him what kind of political books he read. At a loss for an answer, the agrarian advisor from Sárosd half jokingly said that he had just read *Credit* by Popov. The county official leaned back in his chair, and nodded, "Yes, that is a good book; I've read it too." The agrarian advisor knew he could rely on the party official's ignorance of economic history not to recognize Széchenyi's title and the self-serving answer he gave.[19] On another occasion, having tired of the constant pestering he received from the political appointee at a neighboring cooperative farm, the agrarian expert from Sárosd decided to play a practical joke. He told the good comrade that it was time for him to plant the macaroni cuttings. The cooperative president balked, claiming he hadn't received his memo on this matter. The man ran to the granary the next day, where everyone had a good laugh at his expense. So ignorant was the political appointee about agriculture (and cooking, for that matter) that the agrarian expert finally had to explain his little joke to him, to prevent the comrade from spending any more time scouring county warehouses for macaroni plants. Though the joke was well received, conflicts between agrarian specialists

---

[19] *Credit* (*Hitel*), written by Széchenyi István in 1830, was a treatise on political economy arguing for the abolition of serfdom and articulating a plan for nationalist development. As a central icon of the reform period (1820–1850), it represents a crucial formulation of nationalist programs for modernization in Hungary, and is not quite what the party official would have had in mind, if he knew anything about political economy.

and political appointees would not always end amicably. Political reliability and commitment to party goals superseded all other considerations in the Stalinist period, often with dreadful consequences for cooperative farms and village communities. By 1957 the party would reevaluate its policy toward agrarian specialists, recognizing their significance both as economic advisors and as political leaders in collectivization.

Compulsory or forced deliveries of agricultural produce to the state were reintroduced in 1946.[20] The overall progressive rate was not great: the proportion of produce to be surrendered by the top category of landowners was a little less than twice the amount turned over by landowners with the least amount of land (those owning under .57 ha being exempt). In 1947, the proportion became 2.5, progressively increasing regularly. The category exempted from deliveries in 1947 included those with less than 4.6 ha (8 kh), but by 1948 the earlier category of .57 ha (1 kh) was reinstated. A major shift in determining delivery obligations took place in 1948, when the value of the land itself was reckoned as part of the calculation. This in itself did not result in an overall increase in quotas, since there were about as many landowners whose quotas decreased as there were owners whose deliveries increased (Zavada 1984, 141). A further innovation was the stipulation that those owning more than 14.25 ha (25 kh) or cultivating properties with a value greater than 250 gold crowns were obliged to fulfill a quota in animal fat. (Table 3.6 shows the deliveries according to size of holdings 1946–1948.)

The new 1949 regulation specifying the level of agricultural quotas represented a departure from earlier statutes. For the first time it designated an upper limit to normally progressive quotas (which had been increased drastically: 1 to 12.2 proportionally), and created a category that was subject to heavier dues. Although not officially named in the regulations until 1951, this was the category for kulaks. The legal boundary for the class enemy was 14.25 ha (25 kh) or a total land value of 350 gold crowns.[21] The new regulation was significant, furthermore, as an economic and unabashedly political attack upon the wealthy and middle peasantry, endan-

---

[20] The following discussion of agricultural quotas in the 1950s is based on the thorough and insightful article by Zavada Pál (1984), "With Full Force: Agrarian Policies 1949–1953."

[21] In Sárosd everyone remembers the gold crown value to have been 250, suggesting that the party secretary altered national specifications for Sárosd. This modification of regulations by local officials was common throughout the country (see discussion of the abuse of kulak categorizations below).

TABLE 3.6. Delivery quotas in Hungary for 1946–1948

	1946[a]		1947	1948[b]
	Wheat (kg)	Meat (kg)	Wheat (kg)	Wheat (kg)
5 kh	541	35	—	220–330
8 kh	914	62	640	448–672
10 kh	1,163	80	1,040	720–1,080
15 kh	1,788	125	1,900	1,440–2,160
20 kh	2,407	165	2,900	2,400–3,600
25 kh	4,053	285	4,000	3,200–4,800
50 kh	8,893	485	10,250	8,000–12,000

Source: Zavada 1984: 142

[a] In 1946, an additional quota was included, of kilograms of meat.

[b] In 1948, quota values varied, as shown on the table, according to the quality of the land.

gering the very viability of their families' farms. Until 1951 the quotas demanded of the kulak category grew rapidly; after 1951 owners with less land were targeted (see table 3.7).

The harshness of agrarian policy was further aggravated by local abuse in implementing regulations. Despite the simple definition of those to be targeted in early statutes, the process of determining who was to be named a kulak was universally abused. As early as December 1948 Donáth warned against the problem, in his article entitled "The Unsympathetic Attitude Being Taken Toward the Middle Peasant Is a Gross Violation of Our Party's Policies":

> We cannot set limits on the kulak . . . if we do not create an alliance of the working class and poor peasantry with the middle peasant. This compact is not a tactical question, but rather it is the indispensable link of our entire village policy. Our policy has always kept the interests of the middle peasantry in mind and this strengthened their view that the new system is not their enemy. Already during the first measures directed against kulaks we have realized that our comrades are misconstruing our policies: at every turn they are confusing the middle peasantry with kulaks. Our goal is to gather the entire working peasantry, excluding kulaks, into a new, united peasant organization against the reactionaries and exploiters and for the socialist development of the working class and the village. (*Fehérvári Napló*, December 4, 1948)

Donáth still held hopes that zealous party functionaries would not alienate an important group who should be rallied to the worker-peasant alliance, an alliance that he believed could be

TABLE 3.7. Proportion of produce and animal products per kh in relation to land tax per kh (land tax = 1), nationwide

Size of lands	1948	1949	1950	1951	1952	1953	1954–1956
2 kh (1–3 kh)	1.0	1.0	1.0	1.5–2.0	4.0–4.3	6.0–6.5	5.3–6.4
4 kh (3–5 kh)	1.0	1.0	1.0	2.5–3.3	5.2–5.3	7.4–7.7	6.5–7.5
7.5 kh (5–10 kh)	1.3	1.5	1.5	3.8–5.0	6.7–7.0	8.8–8.9	7.6–8.5
12.5 kh (10–15 kh)	2.2–2.6	2.5	2.5	6.3–8.3	9.0–9.6	10.8–11.2	9.1–9.9
17.5 kh: not kulak (15–20 kh)	2.6	3.0	3.0	7.8	10.1	11.1	9.4–10.2
17.5 kh: kulak (15–20 kh)	3	6.1	6.1	15.5	12.5	12.7	—
22.5 kh: not kulak	3.0	3.5	3.5	9.2	11.2	11.6	9.9–10.6
22.5 kh: kulak (20–25 kh)	3.3	6.1	6.1	18.5	13.9	13.2	—
above 25 kh (up to 100 kh; kulak)	3.7–4.6	6.6–12.2	6.6–12.2	12.0–16.0	12.8–13.3	11.9–12.5	10.3–11.1

Source: Zavada 1984: 144

forged among poorer members of the community against the exploiter and the reactionary forces. Unfortunately, both the policies of his party and the attitudes of the peasantry resoundingly defeated the worker-peasant alliance.

Political training during the 1980s did not clarify the process of divining and designating class enemies. I once discussed these difficult times with a young friend of mine in the village, who had been enrolled in the party's evening classes in political education for several years. He explained to me that in the 1950s "people became kulaks." His comment was very interesting, since it suggested that no local prior category equivalent to the kulak, that is, wealthy, exploiting peasant, existed prior to the war. This clearly is related to the absence of a strong middle or wealthy peasantry in Sárosd, in contrast to some areas on the Great Plain. But his comment also conveyed, inadvertently, the historical process itself. The category of kulak was very loose, and was not a given, absolute grouping, but depended upon local party decisions. Thus, one's connections with the local party secretary, and other village notables, in large part determined one's socioeconomic categorization.

There were cases across the country of wealthier peasants not being defined as a kulak, thereby surrendering proportionately much less to the state than would have been mandated. This way they also escaped ugly psychological persecution. In the most frequently cited example of such abuses in Sárosd the reverse was the case. A former manorial worker was awarded the full allotment of 8.5 ha (15 kh) in the land reform because he had a big family. But the combined gold crown value of his land exceeded 250 gold crowns, the local limit. So despite a life of toil as a true proletarian, he was branded a kulak and subjected to very harsh quotas. This corresponds to the general tendency to broaden the category, including far more families than initially had been intended. With time, such minor offenses would pale in comparison to the state's severe agrarian policies and police terror.

The First Cooperative

The first cooperative in Sárosd was established in August 1950. Based at the former estate center of Jakabszállás, it was named Progress (Haladás) and had thirty members, of whom 78% were of manorial worker background and another 19% came from the category of harvesters or day laborers. (Not until 1952 did a peasant join.) Starting with 148 ha (260 kh), the farm varied in size according to the endowments brought to or later withdrawn from collective cultivation.[22] All the tools and animals owned by the farm were accumulated from the members' private holdings, or acquired in exchange for grain supplies from neighboring communities. One informant remembered that initially the cooperative owned three horses, ten sows, and fifty piglets, but no cow. The organization of the cooperative was very simple. The members were divided into three work groups, two assigned to grain production and one to animal husbandry. Each work group had a leader to direct activities, each of whom was based at different centers of production: one at Jakabszállás, one at Tükrös, and one in the village. Members were paid in produce, their shares calculated according to the number of work units they had amassed during the year.

The primary motivation to join the cooperative was to escape the burdens of a failing farm, to avoid the growing weight of unpaid taxes and the obligations of agricultural quotas. (Coopera-

---

[22] After the state mandated consolidation of holdings, certain properties initially owned by the cooperative were turned over to the state farm, most notably the estate center of Tükrös. The cooperative was compensated with lands elsewhere within the boundaries of the cultivated lands surrounding Sárosd.

TABLE 3.8. Increase in the number of cooperative farms, their membership, and their acreage (1948–1953)

Period	Number of cooperative farms Types I, II	Type III, and indep.	Totals	Members (1,000 persons)	Total acreage (1,000 ha)	Ave. acreage (ha)
Dec. 1948	189	279	468	—	—	—
June 1949	140	444	584	13	55	94
Dec. 31, 1949	77	1,290	1,367	36	182	133
June 30, 1950	50	1,738	1,788	68	278	155
Dec. 1, 1950	36	2,149	2,185	120	444	203
June 30, 1951	1,640	2,406	4,046	260	846	209
Dec. 31, 1951	1,999	2,626	4,625	311	1,002	217
June 30, 1952	1,565	3,098	4,663	322	1,186	254
Dec. 31, 1952	1,478	3,632	5,110	369	1,501	294
June 30, 1953	1,456	3,768	5,224	376	1,620	310

Source: Pető and Szakács 1985: 181

tives were subject to taxation and agricultural deliveries, but at a far lower level than for those still working in the private sector.) Similar pressures forced some landowners to seek employment at the state farm. However, this required that they surrender their lands entirely to the state, a choice that could be avoided if one joined the cooperative. One frequent strategy employed was to enroll one family member in the cooperative, thereby holding onto one's properties, and to have other family members find work at the state farm or in industry, to ensure that there would be some cash coming into the household. Memory records that some chose to join the cooperative because of their political beliefs; several founding members were staunch Communist party members. They welcomed the opportunity to act upon their new ideals of collective production. But the large majority of members were forced to forsake private farming for collective production simply because they could not produce enough to feed their families and still fulfill their obligations to the state. (Table 3.8 shows the numbers of cooperatives, membership levels, and acreage nationwide between 1948 and 1953; table 3.9 shows the distribution of acreage among state farms, cooperatives, and the peasantry between 1949 and 1953.)

Throughout the 1950s the possibility remained open to any cooperative member that he or she could return to private farming. A parcel equivalent in size to one's initial contribution to the coop-

TABLE 3.9. Distribution of plow land, 1949–1953, in thousands kh

Year	State farms	Cooperative farms	Peasant small farms	Farms qualified as kulak
1949	113	76	7,731	1,650
1950	463	419	7,391	1,271
1951	685	1,253	6,620	942
1952	1,050	1,795	6,019	530
1953	1,300	2,478	5,067	449

Source: Petö and Szakács, 1985: 188

erative was then designated at the edge of the collective holdings of the group. However, the choice to work at the cooperative or state farm was not entirely a free one. A large segment of the formerly manorial workforce was still quartered at the old estate centers. If one was not housed at the cooperative center, Jakabszállás—but lived instead at Csillag or Tükrös—it was extremely difficult to become a member of the cooperative farm. Once the lands at Csillag and Tükrös had been absorbed into the state farm, it was virtually impossible to become a cooperative farm member, because without land or animals to contribute, the cooperative farm would not confer membership. Some became quite bitter. "They wouldn't let us join, yet they had worked at Jakab too. We grew up together!"

Movement between state farm employment and cooperative membership was often constrained by the selfish motives of cooperative members in restricting access to land. However, access to property was not the only factor restricting movement between jobs and between sectors of the economy during this period. The state also attempted to reduce movement by workers in employment, motivated by overarching ideological concerns and by immediate worries with planning and industrial discipline. The state made a concerted effort to eliminate labor markets during the 1950s as a means of defining and controlling labor.[23] The discomfort with markets clearly grew out of the Marxist analysis of the alienation of labor in capitalism; socialism was expected to rectify the indignities suffered by having one's value determined by exploitative capitalists and impersonal forces. However, attempts to eliminate markets were also intimately related to the project of state economic planning.

[23] I have benefited greatly in my discussion of markets in the socialist period by lengthy debates with Ákos Róna-Tas. For a clear exposition of his argument on proletarianization without commodification, see Róna-Tas 1990, 91–107.

> [T]he leadership felt it needed complete control over the
> labor allocation process, and preferred direct administrative methods over the indirect methods through the markets. . . . the labor market had to be abolished . . . it was incompatible with fast, centralized accumulation, the principal objective of socialist industrialization. (Róna-Tas 1990, 93)

The state initiated a variety of measures to ensure its control over workers and their actions within the workplace. Trade unions were subordinated entirely to the goals of state production, losing their role as vehicles for expressing and addressing workers' grievances. The state also passed laws that would prevent dissatisfied workers from moving to a more congenial workplace, either in industry or in agriculture.

> As of 1950 "arbitrary quitting,"—leaving a job without the consent of the employer,—again became a civil, and later, it was elevated to a criminal offense. The penalty for this criminal act usually did not exceed a fine and "corrective labor" at the work-place, but in certain cases could carry a prison sentence too. (Ibid., 85–86).

Yet these legal provisions did not succeed entirely in restricting movement, as demonstrated by the fact that in 1952 alone nearly "7000 cases of 'refusing to continue employment' were brought to trial" (ibid.). Among these cases one could find examples of workers attempting to find better jobs in a variety of different industrial workplaces, but also instances in which industrial workers would abandon their jobs during the peak of the agricultural season to assist family members in the harvest. Thus, the struggle over controlling one's property among the peasantry was paralleled by workers attempting to wrest some control over their labor. There were consequences to resisting the state's policies, but these were borne by workers who yearned after some semblance of control over their own lives.

Tables 3.10 and 3.11 describe the social background of the membership of Progress Farm and show the fluctuations in membership between 1950 and 1956. The fortunes of Progress followed national trends, both with respect to the social composition of the membership, and to the timing of membership levels (see table 3.12 for national figures). Nationwide, 70% of the founding members of cooperatives in 1949 were landless agricultural wage laborers, and an additional 29% had owned properties of less than 4 ha (Donáth 1977, 142). In Sárosd the membership was overwhelmingly from the poorest strata of the village and estates, all of whom owned

TABLE 3.10. Social background of Progress Cooperative Farm membership

	1950	1951	1952	1953	1954	1955	1956
Manorial worker	77.8%	85.0%	80.0%	53.2%	89.5%	73.7%	65.4%
Manorial worker and harvester	3.7	—	—	4.5	—	3.5	7.7
Harvester	14.8	5.0	—	29.7	5.3	14.0	26.9
Harvester and peasant	—	—	—	1.8	—	1.8	—
Peasant	—	5.0	20.0	7.2	—	—	—
Craftsman	3.7	5.0	—	0.9	—	5.3	—
Intelligentsia	—	—	—	—	—	1.8	—
Unknown	—	—	—	2.7	5.3	—	—

TABLE 3.11. New members of Progress Cooperative Farm

August 1950	29	October 1954	2
December 1950	1	November 1954	1
February 1951	10	December 1954	4
August 1951	3	February 1955	3
September 1951	4	March 1955	3
October 1951	2	April 1955	2
December 1951	1	May 1955	1
February 1952	2	June 1955	1
March 1952	1	July 1955	2
October 1952	1	August 1955	7
November 1952	1	September 1955	2
January 1953	60	October 1955	5
February 1953	42	November 1955	17
March 1953	6	December 1955	4
July 1953	3	January 1956	10
February 1954	1	February 1956	1
March 1954	1	April 1956	6
April 1954	1	May 1956	2
August 1954	6	June 1956	1
September 1954	3	July 1956	1

land by virtue of the land reform. Clearly the problems new landowners had encountered in establishing viable farms had proven too great. As the poorest equipped and least experienced landowners, they could hope to alleviate their difficulties only by banding together with others, though their problems were far from

TABLE 3.12. Number of families in Hungarian agricultural cooperatives according to size of properties ceded

	No. of landless	0–4 ha	4–8.5 ha	8.5–14.5 ha	Nonagr. jobs	Total no. of families
June 30, 1949	4,226	1,625	52	7	90	6,000
June 30, 1951	37,061	44,517	19,299	3,014	3,721	107,612
June 30, 1953	45,791	96,427	65,591	10,813	7,167	225,789
Dec. 30, 1953	40,386	57,239	37,951	6,081	4,861	146,518
Dec. 31, 1955	63,953	71,348	46,435	6,956	8,336	197,028
Dec. 31, 1956	—	—	—	—	—	76,513
Dec. 31, 1958	62,074	30,295	16,861	1,592	5,590	116,412

Source: Calculated from Donáth 1977: 142

solved. The first two years of the cooperative were very unsuccessful; later years showed only minor improvements.

The state's policies of severe taxation and agricultural quotas forced large segments of the agricultural population into collective production—state farms or cooperatives—or out of agriculture altogether and into the burgeoning industrial sector. This process of accelerated proletarianization further deepened the rift between new and old landowners. In 1945, former manorial workers and day laborers had been transformed overnight from agricultural proletarians to petty landowners. Unlike prewar landowners, they had no deep sentiments or historical attachment to their newly acquired lands. To step back into the familiar world of working in large groups—be it at the factory or the cooperative farm—was thus far less heart-wrenching for them than it would be for those whose families had owned land. Yet the apparent ease with which new landowners relinquished their properties merely confirmed the peasantry's views that the agricultural poor were lazy and stupid. The prejudices of the landowning peasantry were clearly articulated by one villager, in comments criticizing her boss at the cooperative farm during the 1980s.

They had nothing, only what they got after the war. They had one *hold* at the count's, something like the current *háztáji*. What they had belonged to the count. . . . Just ask [him], how he acquired land. He didn't acquire it, he got it. . . . Just ask the manorial worker, how he acquired the land. The ones who joined the cooperative first are those who didn't procure land, they didn't cling to it. Peasants only joined in 1959; they endured the forced deliveries, everything, just so they wouldn't have to surrender it. Mano-

TABLE 3.13. Income of cooperative members compared to that of individual farmers (1951–1955), nationwide

Year	Percentage
1951	55.1
1952	67.5
1953	79.4
1954	90.8
1955	93.3

Source: Petö and Szakács 1985: 215

rial workers didn't worry about that. What did they know how to do? They always just drove the oxen, they didn't know what to do with the land.

Not having possessed land in the past, not having planted generations of sweat and blood in the soil, it was said, rendered them incapable of being owners, of demonstrating the socially laudable quality of diligence through possession. They had not acquired their land with hard work; they received it from state agencies, only to turn it back when cooperatives were founded. Private ownership and diligence went hand in hand; sloth and collectivity were a natural pair, be it at the manor or the cooperative.

The poor performance of cooperatives and state farms across the country, despite strong ideological support and some financial assistance from the state,[24] also reinforced the prejudices of the private sector against collective production and those engaged in it. The 1950s abbreviation for cooperative farm, "t.sz.cs." (from *termelöszövetkezeti csoport*), was given new meaning to denigrate cooperative members, the letters transposed to stand for "rehired manorial worker" (*tovább szegödö cseléd*). Cooperative members earned less than peasants in the private sector, less than workers at state farms, and significantly less than workers in industry (Donáth 1977, 155) (see tables 3.13–3.15). Furthermore, unit production was higher in the private sector than in cooperative production.

In the productive structure of private farms, labor-intensive and higher valued field crops (corn, potatoes, sugar beets, etc.), vegetables, grapes, and fruit cultivation were represented in greater proportion, and livestock per unit area

---

[24] The financial support of agriculture was extremely limited in the early 1950s. Cooperatives, however, did have privileged access to tractors and other machinery through the county machine stations.

TABLE 3.14. Real income of the major population groups in Hungary (1949 = 100)

	Workers and employees[1]	Peasantry[2]
1950	102.8	112.7
1951	97.8	118.8
1952	87.5	106.6
1953	91.0	100.6
1954	115.0	111.0
1955	121.8	124.5
1956	129.3	131.2

Source: Petö and Szakács 1985: 217
[1] Per capita real income.
[2] Real value of per capita consumption, without deducting taxes and working expenses.

TABLE 3.15. Average monthly wages in Hungary, in forints, 1949 and 1955

	1949	1955	Index (1949 = 100)
Total National Economy	616	1,133	183.9
Mining, manufacturing	632	1,217	192.6
Construction	646	1,194	184.8
Agriculture	405	926	228.6
Transportation	619	1,117	180.5
Commerce	640	1,024	160.0
State Administration	606	1,104	182.2
Industry Administered by Ministries			
Worker	587	1,164	198.3
Technical	1,166	2,007	172.1
Administrative	824	1,205	146.2
Construction Industry			
Worker	657	1,081	164.5
Technical	974	1,884	193.4
Administrative	634	1,249	197.0
State Agriculture			
Worker	377	890	236.1
Technical	645	1,371	212.6
Administrative	636	1,055	165.9

Source: Petö and Szakács 1985: 222

was also higher. . . . According to calculations of the time nationwide, private farms produced somewhat more than 20% greater productive value and more commodities than collective farms. (Donáth 1977, 155)

The poor record of collective production also proved a disincentive to join. Those truly impoverished, the landless and failing new landowners, had little choice but to enter collective cultivation, but as long as they could hold out, the peasantry continued to farm privately. All these factors underscored the social prejudices of the landed against the proletarian, of the peasant against the agricultural worker.

Between 1948 and 1951 the social and economic pressures brought by the state had been focused on the category of producers designated as kulaks. As the following figures demonstrate, the antikulak policies had been effective in forcing larger landowners off the land.

> In sum total from 1949 to 1953 the number of farms of 1–5 kh decreased 17%, 5–10 kh 30%, 10–20 kh 40%, 20–25 kh 54%, and farms over 25 kh decreased 78%. Fewer than three-fourths of the 1,441,000 private farms in 1949 were in existence by 1953, albeit among these, approximately 230,000 families had taken their farms into cooperative farms. (Zavada 1984, 151)

Ironically, the success of antikulak policies significantly reduced the quantities of produce and animal products surrendered to the state. To curtail their delivery obligations, families attempted to spread their holdings among kin, or relinquished the poorer parcels of their land to the state, thereby decreasing their quota level. This enabled them to produce primarily for their own subsistence, which drastically reduced agricultural production levels nationally (Szabó and Virágh 1984, 167; Zavada 1984, 139). Only in 1955 did the level of agricultural production reach that achieved in 1938. In early 1953, the Ministry of Agricultural Deliveries calculated that, owing to the large number of abandoned farms, 399,000 ha (700,000 kh) would lie fallow, and an additional 114,000 ha (200,000) would probably not be cultivated that year (Szabó and Virágh 1984, 167).

After 1951, already severe quota obligations were widened to cover the entire peasantry. From 1951 to 1953, state delivery obligations for private farms of less than 5.7 ha (10 kh) increased to six to eight times the level they had been before 1951 (Zavada 1984, 151). By 1952 the amount of land surrendered to the state by the poor and middle peasantry was proportionately equal to that relinquished by kulaks (Szabó and Virágh 1984, 167). But as in the case of the kulaks, the increase in agricultural deliveries among poorer peasants was not solely dictated by economic imperatives. The measures were just as strongly political in nature, motivated by the overall policy of the Communist party to encourage collective production and strongly discourage private ownership. Harassment of agricultural producers peaked in 1952 with the threat of up to five years' imprisonment if they did not fulfill their quotas (Zavada 1984, 142–143). A total of 400,000 legal proceedings were brought against the peasantry, of which only 280,000 were directed against kulaks. It has been calculated that this translated into an average of four proceedings against each wealthy peasant, many of which resulted in internment (Zavada 1984, 154). The sharp rise in membership at Progress Farm in early 1953 reflects the worsening conditions of villagers, forced to submit their holdings to collective cultivation.

The state's unremitting campaign—ideological, political, and economic—waged against the peasantry was marred by serious problems.[25] All agencies followed the dictates of their planned objectives, ostensibly objectives to fulfill the national plan of economic development. Nonetheless, different agencies involved in controlling and regulating agricultural production often found implementation of their mandates in direct conflict.[26] For example, agencies expropriating produce and livestock from the peasantry came into conflict with other agencies responsible for processing, whose interests were not in bulk, but in the quality of goods available.

> In every issue of the Procurement Bulletin one can read extensive correspondence or several dozen news items telling of fines, which were imposed for not fulfilling the enterprise procurement plan or for abuses. In fact, on April 23, 1950, the workers' court condemned to death two managers of the Meat Sales Enterprise for a crime committed against public supply. They carried out the sentence that very day. (Szabó and Virágh 1984, 175)

Clashes between planned objectives at the ministerial level hampered the implementation of policies at the county and village levels, as expectations from different quarters were impossible to

---

[25] Szabó and Virágh's article (1984) is an excellent, and more extensive, treatment of the contradictions inherent in the system of forced deliveries. Zavada (1984) also deals at length with the discrepancies between the intentions motivating agricultural policies in the 1950s and their eventual consequences.

[26] Needless to say, the Soviet Union was also plagued with problems of planning. In one famous critique of planning published in 1965, Antonov "supplied dozens of instances of the irrational behavior of managers and of workers who were forced to wreck plans and destroy goods because of the contradictions in the system" (Lewin 1974, 148).

fulfill simultaneously. Contradictions in the regulations themselves which stipulated quota obligations further confused their implementation.

To complicate matters even further, the plan was frequently changed, whenever unanticipated issues surfaced or conditions changed. "In Hungary in 1952, the then current five-year plan was changed 472 times, and the yearly plan for 1952 113 times. In 1953 the Council of Ministers and the Planning Office changed the plan 225 times, and plans as passed to enterprises were changed 71 times" (Swain 1992, 71). Even if local officials had been endowed with the wisdom of Solomon, they could not have reconciled all the problems that arose from attempting to put the plan into effect.

> It is common knowledge that grave economic failures characterized the domestic period of the Stalinist exercise of power. It was an economy that was overplanned to a state of planlessness and overorganized to the point of collapse. Although natural plan indicators arranged the last producible nail and deliverable egg into its militarily rigid, plan-command system, it was, in the final analysis, disorganized, clumsily deformed, and anarchic. (Zavada 1984, 139)

The chaos clothed in planned order was all the more dreadful for the terror it fostered in every home. Local officials were just as likely to increase quotas as they were to lighten them. There were cases of village administrators trying to lighten the load of fellow villagers, sometimes out of sympathy, sometimes out of fear of the villagers' anger. But in just as many communities the overzealous party secretary or county official reigned supreme.[27] Unreasonable officials, harboring hopes to garner favor by exceeding the plan, would increase quotas, and so not be able to fulfill the plan at all, as families gave up under the pressure. Arbitrary use of police terror also characterized the methods of many party secretaries and administrators. One villager was interned for three months for singing the simple ditty (while under the influence): "I can't wait for the English/I had to, just had to join the cooperative." Yet by 1953 county officials found it necessary to explain that the use of

---

[27] The party secretary in Sárosd was even taken to task in the county newspaper for his high-handed methods (*Fehérvár Napló*, August 6, 1953). Discussing the serious problems that had arisen in party education over the previous year, the universally hated party secretary of Sárosd was singled out as having appropriated the duties of the committee in determining who would attend party classes. He also neglected to include trainees from the working peasantry.

force and internment must be directed toward stimulating agricultural production, not hampering it, as illustrated in the objection voiced by the district attorney of Hajdú-Bihar County, dated July 14, 1953:

> [C]riminal jurisdiction is not serving agricultural tasks adequately. This is demonstrated by the fact that, with regard to petty offenses, things like not sweeping the sidewalk and school problems figure primarily, instead of placing more emphasis on the neglect of crops or other agricultural jobs, and on those who do not fulfill their delivery quotas. (Szabó and Virágh 1984, 176–177)

The pervasive confusion of means and ends led to the constant abuse of local power.

Memories of the period still bring tears to the eyes of villagers. Loudspeakers were placed all over the village and droned interminably, broadcasting announcements, playing political songs, extolling First Party Secretary Rákosi Mátyás's plans for the future, and heaping vile slander upon local class enemies. One older peasant, whose hearing was so poor he bent his head to hear one speak, was described as the kulak "who doesn't dare look you in the eye." Grain bins were commonly swept clean by the state collectors; enforced impoverishment was the fate of many who chose to remain in private farming. One villager remembered cultivating 3 kh (1.7 ha) of sunflowers to pay off her taxes; as a preferred cash crop given good value in tax calculations, she had expected it to cover her full debt. Her eyes tearing, she recalled, "One came home without a penny. Everything went to taxes, not enough was left even to buy an apron." People joked about how an elderly peasant, sent to a labor camp because he was too feeble to work his land, came home fattened up. He even brought a bushel of dried bread home to feed his wife. One woman, formerly branded as a kulak, recalled how state collectors took every stick of furniture from her house when she was unable to fulfill her quota. Only a chair was left, because it had been hidden by the door. This hurt far less than when a well-intentioned neighbor suggested she not try to enroll her three-year-old daughter in the new nursery school, since she was a kulak. "A child? A poor little child stigmatized at age three?" Health problems, heart and stomach problems, and nervous illnesses proliferated.

For those willing to throw in their lot with local functionaries and national agencies, it was a time of real bounty. Collection of agricultural quotas, and occasional surprise searches, were entrusted to the leadership of a woman of ill repute in Sárosd.[28] She carried out her duties in a cleverly fashioned skirt, which could hide underneath its many folds and pleats sausages, brandy, every imaginable delicacy so easily appropriated under the state's imprimatur. Others chose to outwit collectors. Unauthorized slaughters were illegal, yet I frequently heard stories of secret, late-night pig slaughters, complicated by the pig's propensity to squeal when bludgeoned. Villagers would forewarn others of surprise searches if they could, allowing families to secure a hiding place for their small cache of food and valuables. Still others resisted with simple courage. When confronted with collectors at her door, one woman stood in their way, shouting that if they were the Lord God himself she would not allow them to cross the threshold. Faced with such boldfaced audacity, the collectors let her be. The president of the cooperative farm, who was in charge of forced collections, was threatened on more than one occasion with physical violence, in retaliation for his duties. Threats were often made after heavy drinking. In these cases, as in others when hotheaded young men wished to take revenge, villagers conspired to prevent the aggrieved from carrying out their threats, fearing that the price to pay would be too high. Cleverness and courage were in evidence day by day, but it could not unite villagers, at least until 1956. The social isolation of families was complete. As one villager recalled, hearing the secret police drive up the street sent chills of fear through the whole family, though they relaxed when they realized the police had stopped next door. "Those were times when one was happy they took somebody else."

By the spring of 1953 agricultural production was in a shambles. The quantities of produce delivered to the state had decreased as

---

[28] It is interesting to note that women and Roma are often mentioned as collectors during the Stalinist period. I would suggest that this memory says more about the changing character of social policies and the politics of the state toward women and Roma than it does about their actual role in collection procedures. In other words, once the state attempted to support women's issues and address problems with minority populations—no matter how superficially and inadequately these policies were implemented—these groups became identified with the state apparatus, especially their punitive institutions. Thus the shift away from traditional authority relations that was experienced in the early 1950s is remembered as a time when women and Roma took over official positions, even though all accounts of party membership and bureaucratic personnel would contradict these memories. (For a thorough discussion of the position of women in socialism and the problems that this raised for the redefinition of power relations, see Goven 1993. Kenedi [1986] addresses the manner in which Roma replace Jews as the target of vilification and scapegoating during the socialist period.)

a result of the state's policies of pushing kulaks and other peasants off the land, into cooperatives or state farms, or out of the agricultural sector entirely. The widespread use of police terror in the villages, and its absolutely arbitrary character, crippled its efficacy. Rákosi had brought Hungary to the brink of destruction. His policies of rapid industrialization were just as harmful for the national economy as his policies in agriculture had proven to be. The Russians were furious, and instructed Rákosi ("Stalin's best disciple") and Nagy Imre, among others, to go to Moscow for consultations in June 1953. Shortly thereafter, Nagy, one-time minister of agriculture and peasant advocate, was named prime minister. Surprisingly, Rákosi was allowed to remain in the position of first party secretary. Nagy announced his new policies in agriculture and industry on July 4, creating a stir throughout the country.[29]

The most important change in agricultural policy for the peasantry under Nagy was the newly sanctioned opportunity to abandon collective production to return to private farming. Even though the state permitted cooperative members to return to the private sector, and even for cooperative farms to disband entirely, it continued to support collective production, as the following editorial demonstrates.

Let Us Defend Our Cooperatives

Until now the enemy has tried to hurt our cooperative movement with mean slander, but after the appearance of the government program it has begun a general attack against the cooperative movement throughout the county. The government's program aims for the cooperatives to rid themselves of those who did not step on the road to large-scale production out of conviction, and moreover from those who fritter away their day instead of working, causing huge damages to cooperative members. . . . Let us make our cooperatives so strong, so secure that they become the ruin of the enemy. (*Fejér Megyei Néplap*, August 23, 1953)

As tables 3.11 and 3.12 show, thousands of families took advantage of this provision, and went back to farming their own parcels of land. Significantly, however, the overwhelming majority of those returning to private production had been smallholders; for-

---

[29] One of Nagy's most enduring political acts when he assumed the prime minister's post was his release of nearly 90,000 political prisoners. In performing the simply humane act of freeing so many wrongly accused people, Nagy also set the stage for the revolution in 1956, since it was no longer possible to believe, however naively, that imprisonment meant guilt, causing serious reevaluation of the goals of the regime by intellectuals and party members.

TABLE 3.16. Drop in membership and acreage of cooperative farms,
June 30, 1953, to December 31, 1954

	June 30, 1953	Dec. 31, 1953	Dec. 31, 1954
No. of farms	5,224	4,536	4,281
Membership	376,000	250,000	230,000
Acreage (ha)	1,620,000	1,143,000	1,082,000

Source: Calculated from Pető and Szakács 1985: 257

mer agricultural proletarians stood fast by cooperative production (Donáth 1977, 142). These sentiments were voiced by the party secretary of Progress Farm, as recorded in his comments at a countywide meeting of vanguard workers:

We Are Campaigning by Deed

Our cooperative farm was founded on the 20th of August 1950 with seventeen families. In the past we were all manorial workers. The lords got the profit, the stewards got the honor, and we got the work. . . . After the publication of the government program, several considered leaving our cooperative. We held a membership meeting on the evening of August 26, by which time there was not one member who wanted to quit; on the contrary, two working peasants asked that they be admitted. By our good work we wish to demonstrate to the individually working peasants that our cooperative farm develops and that we can show better results in collective production than as individuals. We are campaigning for cooperative production by deeds. (*Fejér Megyei Néplap*, September 39, 1953)

As the numbers show, such acts were not convincing (see table 3.16).

Nagy's program represented a respite from ever increasing agricultural quotas, but the deliveries mandated by the state were not drastically reduced.[30] In proposing the development of a three-year quota plan, Nagy clearly advocated predictability over the earlier policies of frequent and unanticipated changes in delivery quotas. Tensions were further eased by the elimination of imprisonment and fines for late deliveries. The accumulation of economic pressure continued nonetheless, and by late 1955 families were returning in large numbers to cooperative production (see tables 3.11 and 3.12). The shifting contours of the cooperative community are

---

[30] Zavada (1984, 158) calculated that the relaxation of quotas essentially meant 1,500 to 1,600 forints per family by the end of 1953, representing approximately several month's wages, which could not be construed as a significant reform of the quota system.

well demonstrated in the history of one villager's career in and out of the cooperative farm. He joined in the spring of 1953, then in the fall of 1953 he left, after Nagy's policies came into effect. Unfortunately, he was held responsible for taxes for the whole year of 1953, as if he had not been a member of the farm. This defeated his original intention of joining the cooperative, which had been to escape his tax burden. He returned to the cooperative farm in 1955 to escape his accumulating debt, and then withdrew in 1956. He finally joined for good in 1958. This individual story was repeated many times in Sárosd, and all across the country.

The state and party fought hard to transform the peasantry's commitment to private farming during the 1950s, but they were not very successful. Villagers moved in and out of cooperative farms, trying to eke out a living despite the severe measures employed to push them off the land. Still others withheld their labor and their land from cooperative production entirely, at great cost to their family's welfare and their own health.

> Thus the ever harsher collection of progressive delivery obligations, as well as taxes and other debts, had to serve not only primitive accumulation, but also political intimidation, making the existence of individual farms impossible, and forcing the peasantry into cooperative farms. (It should be noted that despite this, collectivization occurred much more slowly than would follow solely from political coercion. After the year of the greatest pressure, in 1953—the high point before 1956—the number of cooperative farm families did not reach 20% of the families living from agriculture: 230,000 versus 1.3 million small farmers.) The degree of coercion is indicated by the fact that with the lightening of the pressure by 1954, the number of cooperative families fell under 200,000; in December 1956, it had dropped to 76,000. (Ferge 1986, 48)

It behooves us to remember the degree to which the peasantry withstood the state's harsh policies, even in so dark an era in Hungarian history as the Stalinist period. Their resistance was valiant, though done quietly and deliberately. An article published in 1953, describing the new policy canceling outstanding agricultural taxes, describes the villagers of Sárosd well. "The working peasants of Sárosd know their responsibilities. Sure, they don't talk about them much, but they strive harder by their efforts for that reason" (*Fejér Megyei Néplap*, September 16, 1953). In the final analysis, the action taken by thousands was to leave agriculture for good.

A total of 360,000 left agriculture. Between 1949 and 1956 the numbers employed in industry increased by 375,000, whereas in agriculture from 1949 to 1954 they dropped by 227,000. During the years between 1954 and 1956—as a result of the change in economic policies—the numbers of agricultural employees rose again, and approximately 80,000 tried to ensure their living in this way. In total, while in 1949 54.7% of employees worked in agriculture, already by 1954 they were 43.4% and by the beginning of 1956, 44.2%. (Petö and Szakács 1985, 203)

Collectivization was after all a strategy of industrialization: reduce the peasantry and increase the size of the industrial workforce. Thousands upon thousands were alienated from agricultural production completely. In this respect, and perhaps only in this respect, did the agrarian policies of the socialist state succeed.

1956

Nagy's reign was short-lived. Georgi Malenkov, Nagy's sponsor in the Kremlin, lost control of the Soviet Central Committee by the fall of 1955. Consequently, Nagy was deposed from the premiership, and when he refused to exercise self-criticism for his policies, he was kicked out of the party. Rákosi returned to his former position, with plans for stiffened economic measures and major purges among intellectuals. The ensuing year in Hungarian history is a tumultuous one, though major events leading up to the revolution in 1956 were centered primarily in the capital and other industrial centers. But rural communities were also swept up into the storm. Gerö Ernö's radio speech on the twenty-third of October, evening of the initial demonstrations, signaled to the rural population that something was happening in Budapest. The Fejér County newspaper carried stories about demonstrations and clashes at the radio station on the following day, and continued to print news of the uprising at least until the twenty-sixth of October. Székesfehérvár saw its own demonstrations as well, serving to carry the revolutionary message even closer to Sárosd. Throughout the revolution broadcasts from workers' councils and factory committees outside Budapest were carried on national radio, further disseminating revolutionary platforms among villagers. During the fighting peasants showed their support for the uprising by bringing foods and supplies to the fighters. City folk, long convinced of the stinginess of the peasantry, were overwhelmed by their generosity and assured of their support.

When news of an uprising spread through the village, a spontaneous, heavily nationalistic demonstration was staged. One man recalled working in the fields outside the village and hearing the din of patriotic songs rising from the center of town. As these kinds of songs were not permitted to be sung, he knew something was afoot and ran into the village. Another man recounted that he was in the process of giving orders to a tractor driver on his daily duties, but was told by the driver that he would not comply, since a revolution had begun. The manager then promptly went into the village center, to find a large crowd in front of the town council building.

Unfortunately, sentiments were not restricted to benign nationalistic pride. A march had been organized in the village, which proceeded from house to house, calling out under the windows, "Come out if you are a patriot, otherwise we will beat you up!" Revenge was heavy on some people's minds. One villager, who had spent eight years in hiding (in his own cellar, unbeknownst to everyone), brandished a rifle and proclaimed to the crowd that it was time to take vengeance. Luckily, cooler heads prevailed. A man who had served as a gendarme before the war insisted on ensuring proper legal proceedings for all, once the winds of revolt quieted down. The hated party secretary had long since fled the village, to hide at a distant estate center until the revolution was over. Another well-known Communist party member and cooperative farm official was threatened with physical harm, so he escaped to another village, where several young men promptly beat him. Others were successfully protected, most notably a "true comrade," the only Communist party member in the village whose credentials actually dated back to the Soviet Republic of 1919. Several prominent citizens were entrusted with supervising village affairs from the town hall, the former gendarme among them. While some, like the gendarme, had stepped into positions of authority to prevent violence, others had been solicited to join the new ruling committee. The manager who had come into the village center when told that a revolution was brewing had called the local istrict office to inquire what should be done, and had been told to select the most popular man in the village to take over the town hall. Consensus was reached over whom to select, but the designated leader was unwilling to accept the position without the support of other village leaders. Once fortunes took a turn for the worse, the newly formed ruling council did all in their power to prevent the village from suffering any harm. Soviet tanks rolled through the center of town in the early days of November.

Several young men attempted to barricade the road, hoping to impede their movement, but the citizens' committee forced the blockade to be removed, fearing that the Russians would decimate the village in retaliation. After the revolution was crushed, several men, including the former gendarme, were tortured and imprisoned on the order of the village's despised party secretary. The one "authentic Communist" strongly protested these actions, declaring that he owed his life to the local leaders' levelheaded handling of affairs during the revolution. The men who had stood for law and order, had prevented bloodshed and possible injustice, were never to hold decent jobs again, haunted by the stigma of having been 1956 activists.

The most significant event in the village during 1956 was the complete repudiation of the cooperative. Once again villagers sought to regain control of their land and of their labor. Nationwide the agricultural workers' attitude toward collective production demonstrated the same patterns in shifting membership as had followed Nagy's ascension to power. Former landowners turned their backs on cooperatives; the core of the agricultural proletariat, the landless in 1948, stood by the cooperatives throughout the revolution. The primary motivation of those abandoning cooperative production remained the issue of control and possession, as Donáth points out. "The larger segment of cooperative peasants—the majority of former landowning peasants who had rallied by cooperatives—*strove to restore the old relationship of work and property*, with the approval of private producers" (Donáth 1977, 157; emphasis in the original). By the second half of 1956 nearly 60% of cooperative farms in Hungary were disbanded (ibid., 158). The majority of the agricultural workforce showed their unequivocal support of private production and strong disdain for collective production.

The course of events in Sárosd during 1956 follows in close outline the pattern of social upheaval in many villages in the countryside, as Magyar Bálint has described in his insightful analysis of rural communities during this period (1988).[31] He distinguishes three phases in rural revolts. The first phase occurred beyond the

---

[31] The literature on 1956 is quite substantial, but virtually all of the monographs and articles that address the period center their analysis in Budapest, and to some extent on events in several industrial centers in the countryside. While both the genesis of particular demonstrations and the eventual conflagration occur in Budapest, it is nonetheless important to investigate rural communities, precisely because their experiences differed so significantly from those of city dwellers. Magyar's analysis has done much to clarify the issues relevant to village communities.

village proper, the revolution breaking in Budapest and other cities on October 22 and 23. Villagers became aware of events from a variety of sources—including radio broadcasts and eyewitness testimony from commuting workers and young soldiers in the military—and waited to see what would transpire. In Sárosd, this included keeping a watchful eye on the events in Székesfehérvár.

> The second, perhaps shortest phase, occurs over one to two days sometime between October 25 and 27. Its import is the village's recognition that there is a power vacuum, and the actual collapse of local authority. At this point the telephone is transformed from a tool of authority once more into a mere technical toy. No longer would a cry for help [be answered] by party commands, nor would it bring the auxiliary police or AVH troops. At the same time, local symbols of power were removed, usually in the context of a mass demonstration. (Magyar 1988, 207)

The third phase occurred when local leaders were selected to form the National Committee (Nemzeti Bizottság), and a local division of the National Guard (Nemzetörség) was founded. During this final phase, new institutions of local authority and political decision-making were established (ibid., 207–208).

Two significant points distinguish the actions of villagers and townspeople during the revolution, as Magyar points out and the description of Sárosd illustrates. The first is the cautious attitude of village leaders who took over once Communist party leaders and other figures of authority had scattered from the regime in disarray. The new, often democratically elected leaders moved to consolidate power, proceeding cautiously to secure a calm environment in which to build a legal foundation for authority. Theirs was not an active revolutionary program, nor did they share the mood of revenge or anger demonstrated in other villages, which resorted to physical violence or abuse.

> Their dissatisfaction does not appear on the political scene inarticulately or as mere negation. Rather, it appears in the guise of creating order and a system. The middle peasantry and crafts strata, who grew up on traditions of associations, and not so long ago even belonged to a wide variety of reading clubs and groups, organize to establish the new institutions of the revolution—the National Committee and the National Guard—with the ease of "déjà vu." (Ibid., 208–209)

Keeping the community's broader interests in mind, these leaders prevented the wholesale distribution of food stores and other goods stockpiled in granaries and other cooperative buildings, fearing that the injustices the authorities sanctioned in expropriating goods would only be visited upon the village once again, considering the anarchic conditions and means by which redistribution had been proposed (ibid., 209).

Finally, it is important to note the different approaches toward property and production taken by various groups among industrial and agrarian workers. As is well known, industrial workers in Budapest and other cities initiated workers' councils during the revolution. Equally well known was the proclivity of former peasants to leave cooperatives and return to private farming, while poorer peasants and former manorial workers chose to keep their cooperative farms intact. Clearly, however, the motivations for these actions differed. Workers' councils wrested control over production from management and the party, asserting that as primary producers they had the right to direct the enterprise as well. In this act, they confirmed the value of collective production and management, their interpretation of socialist ideals taking a different form from that of party officials in earlier years. One could possibly see a parallel between workers' councils in factories and the continued viability of cooperative farms whose membership consisted of former manorial workers and poorer peasants. In both cases, those doing the labor insisted upon running the enterprise. However, for cooperative members the change in managerial structures and patterns of authority within the enterprise had occurred when new cooperative farms were first established in the late 1940s and early 1950s. It was during this period that the structure of large-scale production, which so many of these workers had known at the manorial estate, had broken down, replaced by the new, more democratically organized cooperative farm. It bears emphasis that many of those who chose to enter cooperative farms in these early years had been employed in quite hierarchical organizations, such as manorial farms, and had also suffered through the years following the land reform with poor tools and poor draught power. Thus, when they entered cooperative farms—which at this time were small conglomerations of ten to thirty families—they had been able to have a much greater say in decision-making and a much more immediate sense of their own empowerment as peasants collectively working the soil. Wealthier landowning peasants had not shared this experience of large hierarchical organizations. Hence, they constituted the exception to these efforts at invigorating collective production and property relations. They wished to restore earlier relations, in which private, family-based production and property were the center and core. Unfortunately, the events in the fall of 1956 would only serve to postpone for several years the state's enforced collectivization of agriculture.

Conclusion

Between 1945 and 1956 Hungary would see a major reorganization of landed properties and the institution of a Soviet-style planned economy. Yet the trends in social polarization in the village begun before the war continued apace. Former agricultural workers, manorial workers, harvesters, and day laborers were forced into collective production as their paltry holdings and lack of adequate tools and knowledge deprived them of a solid foundation for establishing a viable family farm. Wealthier peasants, targeted by the regime for extinction, held on as long as they could, often forsaking agriculture altogether as the prospect of joining a cooperative farm became a necessity. Though the level of private production varied as the fortunes of Nagy and Rákosi ebbed and waned, the overall tendency was to deprive peasants of their land and force them into wage labor, either in agriculture or in industry. This step would be finally realized with the forced collectivization of nearly all agricultural production in 1959–1960.

The socialist state had mounted an all-out campaign to alter the character of property relations, production, and social identity. One's value to society and to the community was to emanate from one's labor, not from one's property. New alliances were to be forged between the workers of the city and the countryside, against intransigent peasants and class enemies from the old regime. This assault on the split between urban and rural communities was not very successful, since the party's position on the political consciousness of the industrial working-class vanguard, always contrasted with the backwardness of the agrarian poor, reproduced the very division they were attempting to undermine. Distinctions of gender and race were to be erased, along with the pernicious influence of property ownership, in a society now defined by one's productive capacities and dedication to the planned goals of the party. Yet the result of the state's purported attempts to liberate women and to aid minority communities such as Roma was their absorption into wage labor on a scale unprecedented in previous years. Moreover, policies promoting new rights for women and minorities reinforced views among many that socialism fostered unnatural relations, endangering the family and society (Goven 1993).

The party's offensive against social relations and attitudes inimical to socialism during the 1950s targeted crucial hegemonic constructs, such as property consciousness, gender politics, and racial prejudices. These measures clashed with common sense. However, in other respects, the socialist platform did not ring false to villagers. The sanctification of labor was not a foreign concept in rural communities where the integrity of one's work had long been a significant component of one's social standing. In fact, the Marxist-Leninist theory of labor value and the philosophical pragmatism of socialist writers and policy makers made sense in village communities. Party activists and villagers thus shared a commitment to action; party activists "spoke with their actions," as the editorials proclaimed, and residents of Sárosd did the same. However, their actions differed significantly, arising from a crucial distinction in the final goals of labor integrity and diligence. For the party, labor was rewarding when it served the party's goals of a planned economy; for villagers, labor was valuable when it enriched the independence of the family from external forces, be they aristocrats, markets, or state agencies. Thus the core of resistance to the socialist state was already evident in early struggles over workplace discipline and in battles over property relations in agriculture, years before the street battles of 1956.

FOUR

Reforming Notions

Introduction

In the aftermath of the revolution, the party began to rethink its policies of economic development. Policies regarding the agricultural sector were given special scrutiny, and the suggested revisions presaged the broader economic reforms of the 1960s. In the years following collectivization, and the gradual mechanization of agricultural production, the composition of the cooperative workforce changed, including more and more young, and more and more skilled workers. The structure of the farm also altered, becoming an increasingly hierarchical and segmented institution, managed by a technocratic elite less and less accountable to the membership. Remuneration shifted from work units to money, equating time and money, as work shifts came to be represented in terms of their temporal and monetary value. The labor of villagers had become increasingly commodified, contributing to the widespread alienation of the cooperative workforce.

Agrarian Reforms

The Russian invasion crushed the revolution in a week or so of heavy street fighting; strikes at major factories lasted through early December. As one villager recalled, "when the Russians invaded on November 4th, there were so many of them that the earth nearly cracked beneath them." Nagy Imre, Lukács György, and others of the revolutionary government, who had escaped temporarily to the Yugoslav Embassy, were abducted in mid-November and taken to Romania. Two hundred thousand people fled the country, 10,000 were killed—3,000 in the fighting—and many were imprisoned and tortured. Nagy was executed in June 1958.

TABLE 4.1. Change in wages in Hungary, 1956–1957

	Average monthly income (Ft)	
	1956	1957
Total workers and employees	1,235	1,445
Industrial workers	1,234	1,486
Construction workers	1,152	1,516
Central state administration, public health	1,338	1,560
Local state administration, public health	1,136	1,238

Source: Petö and Szakács 1985: 314

The revolution was to have far-reaching consequences for domestic policies in Hungary. Of immediate consequence for villagers was the abolition of forced deliveries and the quota system. Even though the policy had been adopted during the revolution, it was not reversed by the new regime. The abandonment of cooperative production by all but its most staunch supporters demonstrated once more the peasantry's intransigence. As party theorists debated the new course for agricultural production, peasants felt free to pursue their goals of private production unhampered by the state. Meanwhile, the state took measures to improve the standard of living of industrial workers and those in the service sector and government administration. Wages were increased (table 4.1), profit-sharing was introduced in 90% of state-owned factories, and housing construction was speeded up (Petö and Szakács 1985, 313–315).[1]

In July 1957, the reorganized Communist party issued the *Agrarian Theses* (*Az MSZMP Agrárpolitikájának Ttzisei*), summarizing the party's revised policies toward the agricultural sector. In classic Communist party style the document chronicled the advances in agriculture achieved under socialism, and attributed the difficulties experienced from 1948 to 1956 to left- and right-wing aberrations and misjudgment. Simply put, "The relationship of the party, the vanguard of the working class, was not always formed appropriately with certain strata of the peasantry" (*Agrarian Theses* 1957, 56). The abuse of administrative measures was condemned. Collectivization had become an end in itself, rather than the means to increased agricultural production. Pervasive uncertainty surrounding agriculture reduced production levels and alienated the potential allies of the working class: small and middle landown-

[1] In addition to the tremendous damages on housing inflicted by the Russian troops during the invasion, Hungary had also been struck by an earthquake and flood in the spring of 1956. The rubble from all of these tragedies had to be swept away, repaired, and rebuilt (Petö and Szakács 1985, 315).

ers. Furthermore, the state had neglected to contribute investments and development costs to strengthen state farms, leading to poor performance in the socialist sector of agriculture. The goal of winning the peasantry over to cooperative production had been totally undermined.

The *Agrarian Theses* of the Communist party severely criticized the harshness and accelerated pace of agrarian policies of the Rákosi regime, but the core of the new proposals did not represent a radical shift in theory. Agriculture would eventually become fully collectivized, fulfilling the socialist goals of cooperative production and state ownership. However, the developmental period was now conceived of as much longer in duration, approximately fifteen to twenty years. In the spring of 1958, the plan submitted to the Economic Committee of the National Planning Office predicted that by 1965 the socialist sector of agriculture could reach 50% of all production (Berend 1983, 246). During the transitional phase, state and cooperative farms would represent the economic and political vanguard within the agricultural sector, proving once and for all the superiority of collective production to private peasants. The transitional character of specific forms of cooperative production, and of the cooperative sector as a whole, thus became a crucial feature of this policy.

Renouncing direct administrative force, the new policies on the development of collective agriculture placed greater emphasis upon moral and political education.

> [T]o this very day contradictions may be perceived in many cases between the socialist economic position of cooperative members and their consciousness. This manifests itself in frictions between former middle peasantry and former poor peasantry members, in the destructive survivals of the smallholders' thought patterns, etc. Therefore in the future, the party, cooperative leaders, and every class-conscious cooperative member must devote greater care to forging the consciousness and unity of the cooperative peasantry, to developing a unified, healthy cooperative spirit. (*Agrarian Theses* 1957, 58–59)

The developmental conception outlined in the document explicitly designated social strata within the agricultural populace in terms of property relations and class consciousness. The transitional position envisioned for each group was thus determined by the degree of socialist reeducation required, as well as their relation to the means of production. The agricultural proletariat, who demonstrated their progressive attitudes by supporting cooperatives in the worst of times (in 1953 and 1956), would become the wage-labor workforce of the state sector, the preeminently socialist domain. Smallholders were targeted for cooperative farms. Producing little more than subsistence, smallholders were property owners nonetheless, whose positive attitudes toward private ownership and family farms would have to be overcome by political education and through participation in successful cooperative farms.

The task of reeducating the middle peasantry was seen to be the most difficult, as their success in commodity production strongly inclined them to remain in private production. More restricted forms of cooperative production were to be encouraged, such as farms where only tools, draught animals, and machines were shared by the membership. Local officials were also advised not to oppose issuing permits for cooperative farms of which the membership was exclusively from the middle peasantry. This suggests that class prejudice among peasants against their poorer neighbors became a major obstacle to cooperative production in many areas. This was certainly the case in Sárosd. Overall, however, the thrust was to allow middle peasants to continue in private production as long as they fed the nation.

> As long as these [capitalist] tendencies serve the promotion of agricultural production, that is, they bring more food to the working people and more raw materials to industry, the state of the dictatorship of the proletariat, with expedient pricing policy and provision of industrial goods, can divert those [capitalist tendencies] into appropriate channels for itself. (Ibid., 60)

The upper limit for the size of a family farm was set at 11–14 ha (20–25 kh) to prevent the development of truly capitalistic farms, that is, those employing outside labor. The long-term perspective of the policy is most clearly evident in the treatment of the middle peasantry, as it was assumed that private farming would be abandoned only when cooperative production guaranteed higher yields, better pay, and less work. Finally, though cautioning against past abuses, the document called for the complete social and political isolation of kulaks. Although they were to be allowed to continue farming privately, and even allowed to employ laborers under provisions stipulating a heavy employee tax levy, they were to be watched carefully for any semblance of behavior hostile to the people's state, and swiftly punished if caught.[2] Ku-

---

[2] The category of kulak was carefully redefined to apply to someone within the agricultural population who exploited the labor of others formerly or at present. But the

laks and class enemies were clearly thought politically incorrigible, for no mention is made of educational rehabilitation in their case. As a concession, their children and close relatives were to be allowed to prosper if they did not display the adverse behaviors attributed to this group.

Among the specific proposals for the development of collective production which were summarized in the *Agrarian Theses*, three points deserve mention. These were (1) support for a larger role for agrarian specialists in directing production, (2) the introduction of restricted market mechanisms in regulating prices and stimulating production, and (3) a concern with local conditions and consideration of different forms of cooperative organization and auxiliary family plots. Though not necessarily highlighted at the time, all three issues became crucial components of later reform legislation, significantly altering the nature of agricultural production in Hungary.

Throughout the 1950s, agricultural specialists were employed, usually by the town council, to advise cooperatives and private producers. The lack of comprehensive farming experience among new landowners in private or cooperative production created a demand for village-level personnel able to assist the needy. Many of the specialists came from wealthier peasant families, which before the war had made it possible for them to pursue advanced training in agricultural sciences. In the 1950s, their social background was often a handicap. Yet because of their advanced training, these specialists generally advocated collective production; as proponents of "modern" agriculture, they promoted consolidated, large-scale farms or enterprises. Although they may have won favor from party officials, they were alienated from the peasantry, who wished to continue production in family farms. Furthermore, traditional antiintellectual prejudices of the agricultural proletariat, and even the peasantry at large, combined with the sanctification of political goals over economic concerns among party secretaries and local administrators, effectively undermined the experts' role in assisting or improving agricultural production. The party hoped to reverse these tendencies, since plans for highly developed collective production were seen to depend on improving the position of agricultural experts.

category of class enemy was widened to include a great variety of elements: former army officers, former bureaucrats, intellectuals, déclassé supporters of the old regime, essentially those who had actively participated in the 1956 revolution. The harsh tone taken against counterrevolutionaries in the document captures the repressive mood of the period. The crackdown did not cease until late 1959, early 1960.

> Let us give greater respect to the professionals working in the area of agriculture. We must build the professional administration of agriculture on suitable experienced professionals who are loyal to our people's democracy. The party and other bodies should hearken to the dangers of destructive, sectarian, antiprofessional behavior. At the same time, let us respect and strengthen in their positions those worker and peasant cadres without diplomas who have great practical experience, perform their work well, know the scope of their activities, and are loyal to the party and the people's democracy. (*Agrarian Theses* 1957, 68)

The position advocated here is significant for its championing of both practical knowledge and agricultural training. A crucial aspect of the ensuing collectivization drive in Hungary would be the recruitment of middle peasants to important leadership positions in cooperative farms, justified by their practical knowledge and experience. The support of a trained elite would have far broader political and economic implications, resulting in the political appropriation of cooperative farms by management as of the mid-1970s. The first steps along the road to professionalization were taken when the *Theses* suggested that advanced training in agricultural sciences be elevated to a higher educational level, jointly with a reduction in the number of students attending agrarian colleges (ibid., 68). Rudimentary course work would continue to be offered, and even required, for less skilled agricultural workers throughout the 1960s and 1970s, but the stage was set by the *Agrarian Theses* of 1957 for a clear separation between manual and professional employees even before collectivization was finally implemented.

The vision of modern agriculture set forth in the *Agrarian Theses* held the hope that agricultural exports would defray foreign debts as well as perform the perfunctory duties of feeding the nation and providing industry with raw materials. Yet with the abolition of forced deliveries and quotas, the state had to find new means of insuring an increase in the levels of production and improvement in the quality of agriculture. To meet these goals, new policies were to be devised for calculating prices of agricultural goods and coordinating the purchase of produce from collective and private producers. The policies were significant in their redefinition of the state's role: from one of directing production to one of regulating the market of agricultural produce (*az állam piacszabályozó szerepe*) (ibid., 67). This was in contrast to the state's role in industry.

> [T]he single exception to maintaining essentially the earlier economic mechanism [after 1956] . . . was agriculture, the planning-directing mechanism of which was transformed in every main feature; and the earlier direct administrative control, although it lived much longer in hidden, noninstitutional form, was replaced by indirect management built on market relations. (Berend 1983, 235)

As Berend points out, direct intervention in production decisions, primarily by local party officials and county administrators, continued much longer. At least until 1967 or 1968 the effective role of the state was as active participant in determining the kind and proportion of crops to be produced.

The point to stress here is the *concept* of a market designed to stimulate the agricultural sector. The shift from simple measures characterizing the command economy to some sort of regulated market is significant, since it signals the introduction into state policy of new attitudes toward human motivation and work. Moreover, the reason for distinguishing in policy between industry and agriculture may be that private production was primarily motivated by individual needs, whereas the state sector, by definition collective, would not require policies geared to utilitarian notions of self and action. Regulated markets are discussed specifically in the context of increasing production: agricultural production is presumed to be intimately related to the exchange of its products and so can be expanded by altering prices and easing sales. Thus, the revision of the state's mode of intervention, these inklings of the new "economic mechanism"—however theoretical at the time—may be understood as compelling moves by the state toward generalizing commodity markets and supporting assumptions that human interest is tied to exchange value; that work is determined by shifting monetary values; that labor, time, and object are comparable and causally interrelated. The conceptual shift in economic policy is important, even if labor markets would come to play a significant role in the Hungarian economy only in the 1970s and 1980s.

As a transitional phase in the development of socialist agriculture, the campaign to win the peasantry over to collective production was to include promoting simpler forms of cooperative organization, and to allow for the consideration of local conditions. "The goal of establishing simpler cooperative forms is to ease the way to large-scale socialist farms for working peasants with a strongly proprietary attitude" (*Agrarian Theses* 1957, 72). Among these different forms would be consumer cooperatives, concerned primarily with sales, marketing, and credit activities, and a variety of technical or professional associations, that is, groups that marketed produce, shared machinery, and made other group investments to aid production. Abandoning a unilateral national policy dictating cooperative organization meant that the concern with local conditions extended to classic cooperatives as well. For example, if facilities for establishing collective animal husbandry units were lacking, arrangements were to be made to house the jointly owned livestock in private barns and sheds until investments could be devoted to construction (with the additional proviso that a proportion of the manure be surrendered to the cooperative). This bow to household-based production, though a minor point in the *Theses*, presages later policies in which domestic activities would come to be seen as auxiliary to cooperative production. Household-based production, conducted with virtually no capital investment by the cooperative farm and exploiting labor not otherwise employed at the cooperative, would eventually fulfill in every aspect the paramount goal of agricultural development, as expressed in the slogan of the *Theses*: "All out for intensification!" (*Arccal a belterjesség felé*) (1957, 63).

In the year and a half following the publication of the *Agrarian Theses*, debates continued among party and ministerial officials on the appropriate form and rate of collectivization. Essentially two camps developed, representatives of which were found at all levels, from the ministries and Central Committee down to county and village officials (Donáth 1977, 166–170; Berend 1983, 246–259). One faction embraced the long-term or transitional program of collectivization elaborated in the *Theses*. This camp held that a prior condition for full collectivization should be the availability of material supports, machinery, and monetary and fixed capital investments. Accordingly, they supported the conception of collectivization which saw farms as cultivating only grains. The near universal availability of quarters for livestock in every village household prompted this camp to advocate an initial separation of production branches, that is, leaving animal husbandry to the supervision of cooperative members in their private barns, stables, sheds. This would allow investments to be focused solely on grain production, speeding up modernization in this branch. When sufficiently developed, cooperative farms could then incorporate animal husbandry units as investments permitted.

The overall conception of the transitional camp was based on a considered policy of industrial and other economic supports for agricultural production. As Donáth points out, the availability of tractors, chemical fertilizers, and building materials was limited in 1957–1958, providing no base for immediate, intensive collectivization (1977, 168). At the time the Three Year Plan was devised (1958–1960), the transitional option was still being assumed, as demonstrated by the slow rate of tractor production calculated into the plan. During the three-year period the number of tractors was to be increased from 26,200 to only 27,400; the production profile of other agricultural machines for large-scale farms was equally limited (ibid., 168). Furthermore, the transitional faction also stood firmly by the principles of voluntary membership and the legitimacy in the short-term of simpler forms of cooperative organization for the landed peasantry.

The opposing camp espoused a program of swift collectivization, which covered both grains and animal husbandry. In contrast to the economic considerations that were voiced by the transitional camp, this faction was primarily motivated by the political concerns that private production be abolished and that socialist agriculture be established as soon as possible. As presented in an editorial of the party newspaper *Népszabadság* in 1957, their position advocated direct, forceful administrative measures to reach their goal.

> [V]oluntarism is a very wise and beautiful thing, only I don't know then how there will be socialist agriculture. ... There is a need for some kind of economic coercion—precisely in the interest of the further future of the peasantry—otherwise it won't happen. (Quoted in Berend 1983, 248)

This camp clearly distrusted the landed peasantry. Their program declared that reduction in agricultural production was less worrisome than the survival of private production in the guise of simpler cooperative forms (Donáth 1977, 168–169).

After extensive debates covering a variety of proposals spanning the two extremes, the party finally brought forward their program outlining the fate of the cooperative movement in December 1958. Major principles of both camps were rejected. Forceful methods reminiscent of the early 1950s were renounced, as were any plans stipulating that the rate of collectivization be contingent upon economic factors, such as industrial supports and capital investments. Collectivization would be completed in two years, by April 1961, at which time simpler cooperative farms would be reduced to half their 1958 number. As formal large-scale enterprises, cooperative farms would include grains and animal husbandry. The collective movement would strive for two goals simultaneously: the restructuring and improvement of agricultural production.

Although aspects of the new policy were drawn from domestic debates, the final decision on the character of collectivization was made under strong Soviet pressure. Khrushchev's pronouncements on the true road to socialist agriculture in December 1958, and the intensified collectivization drives in neighboring countries, influenced the party in its final deliberations (Donáth 1977, 167–169). As Donáth notes with regard to pressure from the Soviets and neighboring countries, "These circumstances greatly influenced the peasantry, as Kádár János writes, and let us add, they naturally influenced the country's leaders as well" (168). The quickened pace of collectivization and other economic policies were modeled after the Great Leap in China, the theoretical influence of the Chinese not having been finally exorcised until 1960 (Berend 1983, 269–270).

The Drive to Collectivize

The renaissance of private farming after 1956 seriously influenced the character of cooperative production, such as it existed. With the loosening of regulations following publication of the *Agrarian Theses*, a greater variety of organizational forms sprung up among cooperatives. By the middle of 1957, two-fifths of all cooperatives were not operating according to cooperative statutes. Cultivation, and even the marketing of produce, was often pursued individually; the size of family plots within cooperative farms increased, permitting a wide-scale withdrawal from cooperative activities (Donáth 1977, 163–164). The number of cooperative farms increased within the year, but this did not contradict the trend toward an entrenched private farming community, since the majority of cooperative farm members were landless or had very little land to contribute to cooperative production (see table 4.2). At the end of 1958, 70% of cooperative farm acreage was still state-owned property (ibid., 162–163). As table 4.3 shows, neither the number of private farms nor the acreage in private cultivation changed significantly during this period.

A few economic measures were brought to pressure private producers in 1958 and early 1959. The prices of certain agricultural goods were reduced, the price of industrial goods increased, and a new tax was levied on private peasants. Inducements to join cooperatives were contained in two policies introduced nationwide: (1) the guarantee of social insurance for cooperative members, in the form of health insurance and retirement pensions, and (2) the

TABLE 4.2. Distribution of members according to the size of arable land contributed to the cooperative farm

	Dec. 31, 1958	Percentage
Landless	62,074	53.3
Less than 4 hectares	30,295	26.0
4–8.5 hectares	16,861	14.5
More than 8.6 hectares	1,592	1.4
Nonagricultural occupations	5,590	4.8
Totals	116,412	100.0

Source: Donáth 1977: 163

TABLE 4.3. Number and size of cooperative and private farms between December 1956 and December 1958

	December 31, 1956	December 31, 1958
Agricultural cooperatives		
Number	1,617	2,755
Area (ha)	545,000	871,000
No. of members	96,000	139,000
Private farms (larger than .5 ha)		
Number	1,350,700	1,365,200
Agricultural area (ha)	4,594,000	4,532,000

Source: Donáth 1977: 162

payment of rent to cooperative members for the use of their land (Donáth 1977, 170–171; Berend 1983, 257–259). The extension of social insurance to cooperative members represented an inducement to elderly peasants, who had feared for their lives in old age once their land had been surrendered to the cooperative. Though on the books in the early 1950s, the payment of rent had been seen by many to be at odds with socialist forms of remunerating labor. As the party wished to offer the landed peasants a compromise, it encouraged rent payments. This policy ran into serious problems, particularly in cooperatives founded earlier by agricultural proletarians, who resented the privileges this offered to landowners.

Former landowners resented collectivization. They had been deprived of their land. Moreover, they were forced to work with people they considered less worthy, who had not worked as hard or held to the same values of property as they had during the difficult struggles in the 1950s. They were disillusioned further when the cooperative farm did not establish hierarchies of pay or status dependent upon one's property contribution.[3] So too, poorer villagers resented the attitudes wealthier peasants had about their abilities and work ethic, and toward their poverty in the past. These sentiments were not only expressed in Sárosd; they were quite widely held, as excerpts from a newspaper article illustrate:

> Economic differences from earlier patterns of stratification did not disappear and socialist consciousness did not develop immediately among those who when they joined the cooperative farm were from the working peasantry. That is why in the cooperative farm we find a way of thinking which developed on the basis of the old stratification. For example: "farmers (*gazda*) are still farmers," "they get income through rent without working," "how can I work together with him; I was his servant," "*kulák* blood flows in his veins, he's only behaving differently," "should I work with my former servant?" . . . The unity of the socialist peasant class is weakened . . . by keeping differentiations according to old patterns of stratification. This has pernicious effects on the life of cooperative farms. We must strengthen the view that we judge the membership of the farm—not according to the old patterns of stratification—but on the basis of work done at the farm, on one's conduct and relation to cooperative farm property. We must criticize and educate, while at the same time recognition is due to the cooperative farm member—wherever he may have belonged in the old stratification—who is disciplined in work, who with his every effort works to strengthen and enlarge the common farm, because he helps to improve the life of the entire cooperative community. (*Fejér Megyei Hírlap,* September 21, 1960)

Yet social discrimination clearly lived on. Cooperative farms were still organized territorially, allowing different social groups within the community to work with those who lived in their vicinity, in other words, who were their social peers as well as their neighbors. The attempt to gather the agrarian community into a happy collective of equals was clearly not effective. In fact, as Swain suggests, the party conceded to social prejudices in the final phases of collectivization.

> [F]ormer middle peasants were allowed to establish their own separate co-operative rather than join in with the

---

[3] Rent was paid to those who contributed land to cooperative farms during collectivization. But this was still far less remuneration than many landed peasants would have liked or expected for their contribution.

poor peasants who formed the core of the existing farms. This was beneficial because middle peasants were reluctant to join forces on equal terms with those who had formerly been their day labourers, while the former poor peasantry feared that, in an integrated farm, they would once again be relegated to the role of day labourers for middle peasant masters. (Swain 1985, 30)

Once smaller farms were consolidated into larger units, these divisions within the community would often be transferred into relations between management and labor, hence reinvigorating local hierarchies of power and prestige, though based no longer on property ownership, but on managerial control of cooperative production.

The most significant factor inducing peasants to join, however, was the campaign of intense propagandizing conducted in the countryside (Donáth 1977, 169). Workers from cities and other cadres swept down upon villages, joined in their effort by local party officials and the village intelligentsia. Agitators roamed the village, pounding on door after door to extol the value of cooperative production. When they were even allowed through the gate and into the house, their lectures often turned into marathon all-night sessions of heady debate. Administrative force, though not endorsed in public documents, was employed throughout the country. Psychological and physical abuse were not uncommon. In many cases the threat of physical pain or torture was enough to extract signatures to join the cooperative, since memories of the early 1950s were very vivid in people's minds (ibid., 170).[4]

The most successful tactic employed was to focus on local peasant elites (Juhász 1984; Swain 1985, 30). The traditional political influence enjoyed by these families was manipulated by agitators to convince the community of the value, and necessity, of cooperative production. As political elites were traditionally among the wealthier families, their capitulation to political pressures only underscored to poorer peasants the futility of trying to make it in the private sector. Once leading families had signed up, the rest of the community usually followed. Thus, the speed of collectivization—in some counties full collectivization took only two months to complete—may be attributed in large part to the success of policies targeting peasant elites. A recurring theme in propa-

---

[4] In fact, the strong association of private production with state policies of oppression and abuse was enough to prevent many villagers in the 1980s from even considering becoming private farmers again, even though economic policies permitted it on a limited scale.

ganda efforts during collectivization was to portray the cooperative farm as a large, extended family. This view of familial production would seem to be at odds with the common socialist representation of factory production built on worker solidarity and mechanized production. The predominance of factory production in socialism is usually tied to the push toward industrialization so crucial in the states of Eastern Europe and in the Soviet Union after Communist party ascendancy. However, Róna-Tas has argued that the centrality of factory production exceeded the industrializing project per se; it constituted the central organizing principle and image of the economy embraced by Lenin and his disciples (1990, 42–53). Lenin inherited from Marx and other early socialist thinkers an admiration for the technological organization, efficiency, and economies of scale factory production represented (ibid., 44–45). Although Marx was fully aware of the alienating character of capitalist factory production, he was far more optimistic about the wonders of factory production once society moved beyond capitalism. Buoyed by nineteenth-century optimism for industrial progress, Lenin and his followers fashioned a model of "one nation, one factory" (ibid.; see also Lewin 1974, 173) to structure the economy and facilitate economic planning. Yet the political implications of this model should not be overlooked, for central to this conception of national production was the denial of divergent interests within society. Hence, the exercise of power relations between different sectors of the economy and between groups of actors within the workplace or in society as a whole was excluded in principle, and severely constrained in practice.

In the final push for collectivization in 1959–1961, Hungarian party leaders chose to employ an alternative image, that of the extended family: an ideological concession by the party to the preference most landowners expressed for working with kin at home. Yet the image of cooperative farms as extended families must also be read in terms of social hierarchies of authority and power within the cooperative and at the family farm. Like the factory model, the family image of collective production denied power relations and divergent interests within the new organization. The family model of cooperative farming was not only a means of representing the new community in a favorable light, but was also intimately related to the party's strategy of enticing peasant elites into cooperative production. In other words, peasant elites were encouraged to join cooperative farms as the new "heads of the household," to lead their fellows as they would their family members into production. This strategy reinforced preexisting patterns of social stratification of local communities, ensuring that local political and economic elites would continue their privileged role as managers of the new farms, and poorer families would be relegated into subordinate positions within the cooperative hierarchy.[5]

Fears of alienation and subservience in collective production were also to be assuaged by the image of a welcoming family. For many fathers, the conviction of youth threatened their control of household affairs, a control they were afraid to lose in the new family of the cooperative farm. In one morality tale that was published in the county newspaper, this story of generational discord was told through the clash between a father and his daughter over her marriage plans. The young woman wished to marry the agronomist of the local cooperative farm, tying her fortunes to the future of socialist agriculture. Her father balked, fearing that her happiness would depend upon the cooperative, which he refused to join.

> And now his daughter wants to marry into the cooperative farm. In the old days it was the custom that if there was only a daughter in the house, then the son-in-law went there and farmed with his father-in-law. But now, now everything has been turned around, even accepted customs. New ones have come to replace the old and young people conform to these. (*Fejér Megyei Hírlap*, January 8, 1960)

The weight of outworn custom burdened the elderly, who could not be swayed by their forward-looking children willing to embrace new communities, ready to join new families. The rhetorical strategy of representing those resisting collectivization as a generation

---

[5] Evans offers a fascinating analysis of the manner in which the socialist state in Laos attempted without success to replace village forms of moral solidarity and economic action tied to Buddhist conceptions of merit-making and the redistribution of wealth. Though the particular religious conceptions differ markedly from those held in Hungary, the manner in which the socialist state targeted peasant families and failed to offer alternative ideologies of social community is quite similar. "Cooperatives also threw people together as individuals, expecting them to work as one big happy family. Many peasants complained about this because it robbed them of the ability to make choices about those with whom they would engage in reciprocal labor exchanges. The cooperatives therefore disrupted a crucial element in the incentive structure of the peasant economy, to replace it with, first, moral and political exhortation, and secondly, a work-point 'cash-nexus' incentive structure which had little point to it in an agrarian system in which no basic technological change had been introduced. Ironically, the cooperatives ended up reinforcing 'individualism' through their failure to construct a rationale for transcendent bonds. Hence the cooperatives rapidly collapsed back into the peasant economy as soon as state pressure was eased" (1993, 143–144).

out of step with the times was frequently employed by party writers, and was certainly prominent in socialist iconography full of young faces looking toward the future. However, an important aspect of this story is the manner in which the father, who eventually joins the cooperative farm, is portrayed as marrying the cooperative alongside his daughter: "Her father not only betrothed his daughter to the cooperative farm on that sleepless night, but himself as well" (ibid.). While this tale of recruitment may have been intended to be a hopeful one, it nonetheless captured the feelings many landowning peasants shared that they were repudiating timeworn customs of authority and prestige to subordinate themselves to stronger, more powerful members of the community, here embodied by the agronomist: son-in-law, head of the (cooperative) household, and professional of the future.

In articles promoting collectivization, private farmers were often represented as isolated individuals.

> Perhaps the most accurate way to characterize the pig herder of the Forward Cooperative Farm of Sárkeresztes would be that in the last two years he has learned to speak in the plural. If he mentions his own fate he says "we" instead of "I." And under the "we" he understands the large family, the cooperative farm, where, at his fifty-one or so years, he has found the meaning of his every day and the security of his future. This is all understandable since his prosperity depends on what the community—together—achieves year to year, to what degree common sense and diligence triumph. (*Fejér Megyei Hírlap*, January 17, 1960)

The pigherder had learned to speak in the plural, unlike in days past. Yet the opposition between private producers and cooperative farms was not one between individuals and groups, but one of smaller and larger groups. Private landowners included women and men, children, and grandparents. Nonetheless, in the party's campaign rhetoric, it was the male head of household who came to stand for the whole of the privately producing peasantry. This contrasted with earlier iconic images of the Stalinist period, when the peasantry was always represented as a woman, alongside the male intellectual and male factory worker. In that image, peasants were represented as backward, lagging behind the progressive politics and consciousness of the working class, hence their submissive, subordinated identity as women. In the revised image of the last collectivization campaign, the peasant being targeted for membership was the *male* head of household, the landowner, the manager, the *paterfamilias*. Hence, an additional consequence of reconstituting families in cooperative production was the institutionalization of patriarchal power within the farm. By bringing men into cooperative production as leaders and as managers, women became subordinated in production, often in ways unprecedented in peasant production of earlier years.

The restructuring of work and of the gender division of labor is conveyed in the following two articles, both of which boast about the improvements women experience with collectivization.

"Man and Woman with Equal Rights"

The cooperative truly liberates women from the old, great worries of woman's fate. . . . Because really the wife of the private farmer went out to the fields many times at dawn and only dark sent her home. Then—especially where there were many children—in the midnight hours she was forced to bake, wash, tidy up the house. Husbands found time occasionally to play cards or go to the pub, while we stayed home between four walls. In the cooperative we peasant women have become liberated people, equal in rights with men. We represent ourselves within management; we wish to be judged according to our work. (*Fejér Megyei Hírlap*, March 4, 1960)

"We, Girls and Women"

There is a great need for women's labor power in our cooperative, too. It has been my experience that women find their place, just like men, in crop cultivation and animal husbandry. To mention only a few examples, I would say that thinning crops, garden work, raising poultry is more women's work, than men's . . . I always find my work, I do it, but I still do not work as much as many private peasant women who participate in the most difficult work, in harvesting. And what do they earn? People never count that. Here at the farm the many thousands of forints prove that women also perform valuable work. (*Fejér Megyei Hírlap*, January 17, 1960)

I wish to draw attention to two issues in these articles. In the opposition between private farming and cooperative farming, both authors suggest that cooperative production would promise a lightened workload. Unlike their neighbors who continued in private production, women at the cooperative farm would be given reasonable tasks. Their jobs would complement those of men, soon to be their equals in production. Moreover, they would be remunerated for their labor at the farm. They would be paid money for work they had previously done within the nonmonetary economy of the household. It is worth stating that for many women, being recognized for their work, at least in the wage sphere of production, was seen to be a great benefit of socialism (Lampland 1989). This was true even though in the long run they continued to bear the heaviest burdens of housework and child-rearing; no promises were made to relieve women of these onerous tasks. The party did not institute policies to assist women in the household, the consequences of which were well illustrated in survey after survey on the discrepancy between men and women in the division of household tasks. The following numbers come from surveys conducted in the mid-1970s.

> Women working as manual workers in agriculture—employed or as members of cooperative farms—spend 5.2 hours of the total 11.5 hour work time in their average daily schedule on income activities. The larger half of their work time is devoted to their own farm, domestic chores, taking care of family members. . . . Women working as members of cooperatives spend twice as much time on domestic chores as women in industry or other employment (3.1 hours). (Kovácsné Orolin 1976, 216)

In addition to housework, women also carried much of the burden for gardening and other tasks in private farming activities which continued alongside cooperative production. However, in the minds of party officials, women's liberation would be found in wage labor, a formulation drawing in all its simplicity on Engels's famous treatise of a century past.

The other issue of note is the clear demarcation of male and female tasks in cooperative production. Private production was anathema not only because it was so difficult; it also required women to partake of tasks usually considered a man's job: to work in the fields, to harvest wheat. Cooperative production was represented as an appropriate division of labor by gender, permitting women to realize their equality by weeding seedlings, hoeing in the garden section, and feeding chickens. Restrictions on women's activities within the cooperative also extended to managerial functions. By restoring an appropriate division of labor, presumably based on traditional categories of male and female labor, women were deprived of their important role as manager in the peasant household. Hence restoration actually meant a significant shift in power and authority relations within production, sidetracking women in wage labor in ways not known in family production among subsistence farmers.[6] Needless to say, women's equality was not realized in management or in representative bodies of cooperative farms. Women were consistently underrepresented in both of these groups.

> In 1972, there were only 14 women directors among more than 3,000 cooperative farm directors throughout the nation. It is also clear that the level of women's training is lower; while 1,415 male agricultural employees possessed some kind of college degree in 1970, only 287 women had similar qualifications. The effects of a lower level of training are clearly felt in remunerations as well: while 28.2 percent of cooperative farm members earned below 20,000 forints a year, more than two-thirds of the women members, 67.2 percent, earned less than this figure. (Volgyes and Volgyes 1977, 31)

Overall, women fared far worse than their brethren in the wage structure and management categories. "[W]omen's earnings in agricultural management are 10–15% lower than those of men and . . . women with technical qualifications are persuaded to change careers and enter bookkeeping and administrative branches" (Swain 1985, 122).

The rhetoric of family production in which cooperative farms had been promoted should also be seen as directly tied to the revised policies of the Kádár regime regarding the politics of work and the home. As Goven has demonstrated (1993), there was a marked shift in the manner in which the state treated the family sphere after 1956. An important but often unremarked change in party politics following the 1956 revolution was the withdrawal of direct state intervention in the gender politics of the family, that is, the refusal to follow in the path of socialist emancipation projects so prominent in the Stalinist period. While the state may have intruded into the home by forcing peasants to give up their land and property to collective production, it withdrew from adjudicating issues of equity and social worth between men and women in the home. For example, during this period the state instituted legal provisions barring women from a variety of jobs that had

---

[6] This change in the gendered division of authority and labor would also come in second-economy activities, where women were increasingly demoted to the status of mere workers. Women carried out many of the basic tasks that sustained their husbands' second-economy commodity production; they no longer played a managerial role in these activities (see Répássy 1987).

once been seen as means to emancipating women in the workplace (Goven 1993).[7] And as the above quotations demonstrate, women's place was being redefined and restricted to certain kinds of jobs and certain kinds of labor in the wage sphere, while providing no means to liberate women from work in the household. The sanctity of the household and of the family, which became ever more important in social relations of embourgeoisement and the second economy during the 1980s, was established in these very early years of collectivization.

A final theme I wish to highlight in propaganda efforts promoting collectivization was the objective, self-evident character of numbers and calculations. Frequent references were made to the "sober mind" of the peasant who could calculate the advantages of cooperative production over private production. "People have common sense. They learned to count at one time, better than a professor, in the days of misery. But it's as if some of them have forgotten!" (*Fejér Megyei Hírlap*, January 28, 1960). In part this approach was designed to win over the middle peasantry, a group seen to be conscious of its own material and political interests; but the focus on numbers and reasoning was also central to party views of scientific socialism.

> The results of large-scale farming are already generally recognized facts and still here and there the rust is thick. Common sense and clever accounting do not set the pace, but some kind of undefined nostalgia for "freedom." Which turns out to have been mostly slavery, if the individual peasant grabs paper and pencil before deciding. . . . Life does not endure equivocation; it poses questions mercilessly,

---

[7] Although legal provisions were still in place during the 1970s to prevent women from filling certain jobs, it is interesting to note changes in explanations for the fact that women did not pursue certain professions, specifically the more technically demanding ones. In discussing the underrepresentation of women in technical fields, Kovácsné bemoans the constraints that women feel from housework. "[W]e must touch upon the apparent contradictory relation of women and agricultural machines. Namely, it is not so that women are intellectually incapable of operating complicated machinery. Indeed, we can agree considerably with the frequently quoted notion according to which it is more difficult to drive a car than the most modern agricultural machine, and thus the employment of women in plant cultivation systems has no obstacles. . . . However in practice we encounter other facts. A machine worker employed in the system works for 10 to 12 hours daily in constant readiness and nervous tension for many months. This is work which demands absolute accuracy and work discipline day after day, and which does not leave much energy for another 'shift' afterwards: taking care of housekeeping and farming around the house. Women see and know this" (Kovácsné 1976, 211).

to which sentiment cannot give the true answer, only intellect. (*Fejér Megyei Hírlap*, January 14, 1960)

There were means of determining the value of cooperative production to allay the fears of landowners, means that were not ideologized or driven by local party politics. Cool reason would win out over heated emotional attachments; the detached quality of calculations and scientific reasoning would demonstrate the irrationality of the past.

The party was nonetheless hard-pressed to explain that calm, sober calculations differed from the machinations of a scheming, clever actor out for individual profit and gain. As the following article illustrates, figuring one's preference for cooperative production did not entail finagling with markets, prices, or partners. Faced with a shortage of potatoes in 1960, the state faulted outmoded ways of thinking about production and marketing.

> The true reason is not [the potato beetle], but the neglect of state contracted production. This is a sort of perverted view, which is entirely foreign to the morals of socialist large firms and which feeds on the world of the small peasant—partly merchant, partly *robot* world—of living from day to day, on which our peasantry has turned its back. But this "commercial" spirit still hangs over the heads of most of the cooperative farms, if we pick up the planning pencil. . . . The people's economy takes considerable care that the chances of market competition not make marketing uncertain. Instead it ensures the realization of produce on actually calculated contractual prices. (*Fejér Megyei Hírlap*, March 11, 1960)[8]

State planning was designed to eliminate the unknown, the play and movement of markets, to be replaced with the sober calculations of planners and enlightened managers. Emotions and personal gain were maligned as antithetical to the new scientific world of socialist planning. Moreover, cooperative farms were not to have interests separate from the interests of national planning agencies. Such calls for quiet deliberation, often couched in terms of peasants figuring their advantages, demonstrated the degree to which the family farm was seen not as an enterprise placed within a capitalist market, but as a family concern working only to produce for the family of the nation. Such would be the goals of initial cooperative production set by the party.

[8] The term *robot* refers to feudal service, and evokes the connotations of unlimited toil and feudal backwardness.

Cooperatives in Sárosd

Progress Farm out at Jakabszállás weathered 1956 with a core of supporters. But the majority of members abandoned the cooperative, stripping the farm of tools and animals, initial contributions to collective production that they now could properly reclaim. The climate in early 1957 was one of renewed hope for private farmers; even outcast class enemies returned to the village to take up private farming again.

By 1958 there were three cooperatives in Sárosd, but only Progress conformed to the classic pre–1956 type of cooperative farm. In the village proper, there were two cooperative farms, representing different social strata from the village. Kalocsai cooperative, named after its president, was the farm for manorial workers in the village. Probably the descendant of a splinter cooperative formed in 1955 after a quarrel within Progress, the cooperative was better known under the name of "Everyone Stealing Whatever Is Handy" (Ki mit lop), as the leadership was known for drinking the proceeds of the farm at the pub whenever possible. Rákóczi Farm, manned by harvesters and day laborers from the village, was named after a famous eighteenth-century nationalist hero; its nickname was the forlorn "What Will Be, Will Be" (Lesz, ahogy lesz). Both Kalocsai and Rákóczi were farms of the second order, meaning they shared land and tools, but redistributed their proceeds differently from Progress, a third-order organizational form. In 1959, the peasants in the village established Dózsa Farm, recalling the sixteenth-century peasant rebel. Their nickname, "Take Everything Home" (Mindent haza) was a direct reference to their organizational form (first order), which stipulated that only tools need be shared jointly. The names for both the manorial workers' farm and the harvesters' cooperative were obviously disparaging references to the work habits attributed to them by former landed peasants. Cooperatives in the village were also given nicknames that depicted the relative wealth of their members: the Top Boots (peasant landowners), the Clod Hoppers (harvesters), and the Barefeet (day laborers and manorial workers).

Although cooperative farms were founded according to different organizational principles (first, second, or third order), the manner in which they conducted their business was nonetheless subject to supervision by local authorities. This was true even for Dózsa, the peasants' cooperative, which as a first-order cooperative shared only tools, not land. Finding their policy of *háztáji*, or privately farmed lands, contrary to statutes, the council president attempted to change their practices, though he met with strong resistance; he reported on these difficulties to the county newspaper:

> The council . . . pays attention to and assists [cooperative farms]: they must farm according to the articles of association. For example, in the Dózsa Cooperative Farm they misinterpreted the use of closed gardens, so that there were members who maintained a 2–3 *hold* garden under the title of a *háztáji* farm. The leaders of the council discussed this with the managers of the cooperative and with the membership as well. They requested that a general meeting be called. When they were reluctant to do so, the leaders of the executive committee of the village requested the agricultural department of the district council call a meeting as the authorized body. They have now held the meeting and settled *háztáji* farms *in accordance with the articles of association*. Not everyone accepted the activities of the council in the same way, but we must make them understand that we have to maintain state discipline, and its keeper is the local organ of state power, the village council. (*Fejér Megyei Hírlap*, May 12, 1960)

Thus despite their membership in a cooperative farm, peasants were still attempting to wrest a small parcel of land to work privately. But as the council president reminded them, the state's authority must be maintained, by ensuring that collective production predominate over private.

The propaganda campaign in the village was waged as elsewhere, heavy pressures being brought by outsiders and local officials alike on the peasantry. Some landowners chose to engage in debate, advocating consolidation of private holdings over cooperative ownership or production. Others shunned any contact whatsoever with the meddlers. Villagers were frequently called to the town council where they were forced to sit hours on end while the party secretary or other village personnel alternately coaxed and threatened them. Others were stopped on the street, or approached in their homes, annoyed and bothered whenever opportunity arose. One villager recalled that agitators explained to her husband that they wanted to free him from work. "Then what will we live on?" he asked. She could no longer stand the pressures, and they joined. To my knowledge, physical abuse was rarely practiced, though some claimed they were roughed up while at the town hall.[9] The simpler technique of threatening the

---

[9] A joke told at the time suggested that only people who had been slapped around joined the cooperative. A pun difficult to render in English, it played upon the

educational future of the villagers' children was sufficient for many families to join the cooperative.

But, as in communities across the country, it was the capitulation of a well-respected peasant which signaled the denouement of the propaganda campaign. Local party officials had targeted a well-known and well-respected member of the community, knowing that if he joined, others would follow. Married but without children, Uncle Pista lived well on his combined income from farming and rental cartage in the village. If Uncle Pista couldn't hold out against the pressures of the state, reasoned the villagers, then how could we, with children to feed and clothe? A series of meetings, at his home and at the town hall, were organized, attended not only by local officials, but also by agitators from Sárbogárd and county offices. After much discussion and persuasion, he joined. Learning of Uncle Pista's surrender, those remaining outside the cooperative appeared reluctantly at the town council to sign their membership forms. Needless to say, Uncle Pista became the president of the consolidated cooperative. Luckily, he proved a prime candidate, and worked tirelessly (though with obvious material reward) for twenty years until a quarrel with the party secretary deprived him of his office. No one abdicated the final decision by committing suicide in Sárosd, but several did so in the neighboring village, as did many nationwide.

Though resigned to the inevitable, many felt that they had been consigned to the pitiful state of servants, of manorial workers subject to the orders and dictates of a large, estatelike farm. Rumors abounded that collective production would surely result in collective sex, revealing interesting assumptions about the boundaries of property or analogies of productive acts with reproductive ones.[10] More than once I heard reminiscences of insufferable pain and heartbreak occasioned by the loss of property and the prospect of living in the dire poverty sure to follow cooperative production.

> When we joined the cooperative, we had the most beautiful rye and wheat in the village. They came and took it

---

similarities between phrases suggesting "willingly joined" and "such two blows": I joined the farm willingly, I was struck by such two blows (*önként álltam be, ó' két pofont kaptam*).

[10] A Hungarian proverb admonishes one not to lend his scythe, horse, or wife (*Kaszát, lovat, asszonyt ne add kölcsön*). When asked why, people explained to me that one never knows in what shape things that have been borrowed will be returned. As implied in the proverb, women are considered another tool or means of production, highlighting the concept of wives as property and their reproductive role as comparable to other constitutive or productive activities in the household.

away. I thought I would go mad. All that work down the drain. . . . I mourned that wheat for years. . . . We thought, why should we join with people who didn't bring anything? We worked so hard to have all we had, then others joined who barely brought anything at all, and then at first we didn't receive more pay or more money even though we had brought so much [to the farm]. . . . I thought I would go crazy. People didn't understand then why it was important [to join]. Now they do, but for so long I was sorry about that wheat.

For the first few years, villagers' fears of hardship and privation were realized, and social tensions between members festered.

It has long been held by many sociologists and other city dwellers in Hungary—as well as observers from abroad—that the final phase of collectivization was more benign than the policies and practices of the 1950s. To quote only one source, "Hungary's third collectivisation was altogether less coercive and punitive than the previous attempts had been. No quotas were set for the number of co-operatives to be formed in a given district . . . and a more positive approach was adopted toward the 'kulaks' and middle peasants" (Swain 1985, 30). One of the reasons that the final phase of collectivization has been considered less problematic is that it has been seen through the eyes of villagers whose consequent experiences in collective farming convinced them that cooperative production was a good idea. This is clear in the above quotation about losing one's wheat. Yet it took at least a decade, if not longer, to change people's minds about the wisdom of collectivization. Another reason the final collectivization process has been considered less painful is the awful memory many have of the 1950s. It is important to recognize the pains and privations of the 1950s: forced deliveries sweeping attics clean, leaving families with little if anything to eat; family members carted off to prison, others escaping to the city. However, I would contend that for landowners the final phase of collectivization was a much greater blow than the troubles of the Stalinist period. This was their final defeat. During earlier attempts at collectivization, people often survived on the land despite all the difficulties they encountered. After all, at the height of cooperative membership in the 1950s, only 20% of the agrarian population had joined cooperative farms (Ferge 1986, 48). Accordingly, 80% of the landowning peasantry resisted collectivization entirely. Having rejected collective production was a point of pride, as comments made by villagers above suggest. Resisting state agencies reinforced views about their own resiliency and pride in ownership, qualities they saw lacking in villagers who capitulated to cooperative production.

The issues of violence and the use of administrative measures during the final phase of collectivization have also been poorly understood. I would submit that their absence during the last collectivization campaign is not an indication of the degree of coercion employed to force people into cooperative farms. First of all, I do not think that threatening the future of one's children should be considered less coercive than sending someone to jail. Parents are often willing to take much greater pain upon themselves than to endanger their children's happiness. Also, it should be emphasized that collectivization was begun in 1959, during a period when severe measures were still being brought against those who had participated in the revolution of 1956. Moreover, three short years earlier the country had been invaded by a foreign army, to prevent its choosing another path toward national development. Hence in this environment, one's consciousness about flexibility and space for negotiation was quite reduced. All these considerations, taken together with the sad memories of villagers about the final collectivization, suggest that we must reevaluate our presentation of this quite difficult period in rural communities.

Three Farms Become One

In 1963, all cooperative farms in Sárosd were consolidated, donning the name Harmony (Egyetértés) to symbolize their new beginning as respectful colleagues in production. The consolidation of the farms was prompted by national and county officials who no longer countenanced the isolation of different social groups in production. However, the initial organizational form of the cooperative retained the territorial separation of productive centers based on three of the four farms. One center of production was located at Jakabszállás, one at the Puszta, and one at the peasant end of the village. Jakabszállás and the Puszta were better equipped with large-scale storage facilities and animal barns than the peasant center, having inherited granaries, workshops, livestock quarters, and even a small castle from their manorial predecessors. Because of the availability of sheds and barns at the Puszta and Jakabszállás, the animal husbandry units of the cooperative were based there; cows were based at the Puszta until larger quarters were built, and pigs at Jakabszállás. Despite the amalgamation of the three farms into one, the isolation of various social groups continued. Families still lived out at Jakabszállás, and in the village day laborers and peasants lived near to what had been their respective cooperative centers. The lack of a centralized transportation system for the farm at the time suggests that for several years at least, the social isolation of producers continued de facto.

The organizational structure of work performed at each of the three centers was fairly similar, despite some differences in equipment and facilities. Each productive center was managed by a brigade leader, who was responsible for allocating tasks on a day-to-day basis, and for calculating the daily work contribution of each member. He, in turn, had been informed of the day's tasks in discussions with the cooperative president, agronomist, and the other brigade leaders. For example, the brigade leader would charge several members with hoeing sugar beets. At the end of the day, the brigade leader would return to the field where the members had worked, and measure the acreage hoed by each individual. On the basis of these calculations, the members would be assigned the appropriate number of work units for the day. Thus, the managerial structure of the cooperative farm was relatively flat, comparable in this respect to the managerial structure of the manorial estate.

Before the economic reform, cooperative farms were constrained by strict national and county stipulations on production schedules, both in terms of the specific crops grown and the proportion of acreage to be sown. This has often led scholars to assume that the hierarchical command structure of planning and agricultural agencies was reproduced within the cooperative. This was not necessarily the case. Faced with very specific production quotas, which made few accommodations to local conditions, cooperative members were often drawn into long debates on how to meet the plan: where crops could be planted to produce the highest yield, what sequence of planting would bring the most out of the soil, and so forth. Since, prior to intense mechanization, work was still primarily labor-intensive and carried out in surroundings familiar to the workforce, members were fairly interchangeable in terms of knowledge and skills. Thus, during this period a level of shared understanding regarding agricultural practices existed which could form the basis of relatively democratic discussions. As Donáth observed, democratic decision-making was less in evidence if cooperative leaders were not from the village. During the 1960s it was not rare for local party officials to interfere with the selection of cooperative leaders, often bringing in outsiders. This led to greater alienation between management, however limited, and the general membership (Donáth 1977, 188). Harmony Farm was singularly fortunate in having consistently filled its managerial ranks with people from Sárosd (except for the agronomist and bookkeeper), which prior to organizational changes in the early 1970s eased relations between management and physical workers. The sanctification of technical expertise had not yet taken hold, leaving even the agronomist open to criticism by members.

In describing the managerial structure distinct to cooperative farms in the early 1960s, I have emphasized the relative absence of hierarchy and the potential for open decision-making processes. Villagers recall heady, contentious meetings lasting through the night, after which they had to trek out into the fields to start a day's work. Lengthy debates at the cooperative chronicled by informants testify to the opportunities given for each to voice his opinion. Yet the character of cooperative democracy was qualified by traditional ties, kin and otherwise, which constituted the fabric of the village community. Cooperative democracy did not extend to former manorial workers who were less apt to be accorded credence by peasant or day laborer members. Memories of greater ease in contributing to decision-making during the early years of the cooperative farm may be exaggerated to some extent, as they were contrasted with the practice during the 1980s of the near exclusion of nonmanagerial personnel from contributing significantly to cooperative policy. The degree of democratic involvement afforded different groups within cooperative farms thus varied with the degree of social harmony or animosity among groups. Novel organizational structures could not mitigate social prejudice.

Having lured peasant elites to cooperative farms during the propaganda campaign, the party then supported their disproportionate representation in leadership positions across the country, in recognition of their political and economic strengths (Juhász 1984). As political elites who garnered respect within the community, and as successful farmers who commanded extensive practical knowledge and experience, peasant elites were seen to be important actors in the development of cooperative agricultural production and in the building of cooperative social unity. In Sárosd a fragile truce was forged between former peasants, day laborers, and manorial servants, but it was a reconciliation of power wrought on the party's terms. Despite the membership's heavily manorial representation at Harmony Farm, former peasants and village dwellers nonetheless filled most of the leadership positions in Sárosd as well. The cooperative president, all three brigade leaders, the head of the cow barns and the man in charge of the granaries and storage facilities all were former peasants or day laborers, that is, village dwellers (from a total of nine managerial positions). Furthermore, from a membership of 500, the kin ties among cooperative leaders were surprisingly close: two of the brigade leaders were first cousins, and the head of the cow barn was first cousin to the cooperative president. The one crucial exception to the village monopoly of managerial positions was the party secretary, who had grown up on the manor (but in the privileged role as son of an overseer). Thus, in Sárosd as elsewhere, the appropriation of managerial power by peasant elites characterized the early years of cooperative production. Juhász observed that the concept of the cooperative farm as a "family" was tied directly to the close kinship and friendship bonds among the peasant elite. Even though cooperative farms had a relatively heterogeneous workforce from the very beginning, planners and county agencies conceived of cooperative membership in terms of a peasant elite "family," whom they had put in power (Juhász 1984, 2). This not only continued the fiction of the propaganda campaign for collectivization, it also sustained the political control of the managing elite over the workforce by denying the existence of different interests between management and labor.[11]

The commitment felt by members toward collective production was not uniform. Not all members shared a sense of responsibility for the farm, much less for decisions regarding specific productive activities. Despite their nominal membership in the cooperative, many chose not even to participate. Donáth has calculated that between 1961 and 1964, 20–25% of cooperative membership nationwide did not show up for work (although an increasing number of the defiant were of retirement age), and at least one half of all members did not do their share of collective work (1977, 200–201). Forms of resistance to collective production included the simple repudiation of peasant habits of diligence. One of Sárosd's venerable peasants, known for his hard working habits, became the most skilled loafer at the farm.[12] Others continued practices common at manorial estates: stealing grains and other products to take home. Morality tales published in the newspaper admonished such behavior, and encouraged members to see the farm not as a reconstituted manorial estate, but as a community of equals (*Fejér Megyei*

---

[11] This denial of divergent interest was characteristic of all economic enterprises. This is particularly visible in the policies of socialist trade unions, which represented both labor and management in one body, providing no alternative voice for labor against management or the party.

[12] Interestingly, the expression for not working hard or loafing was "to americanize" (*amerikázni*).

Hírlap, February 11, 1960). As the following numbers demonstrate, many families emphasized their subsistence activities on household plots over collective work.

	Percentage of income		
	1958	1961	1964
From collective work	56.9	45.6	47.7
From household work	43.1	54.4	52.3

Source: Donáth 1977, 202

As Donáth pointed out, these national figures obscured the differences in the degree of household production possible for former peasants and former agricultural proletarians. Since former peasants were well equipped with outbuildings and tools, they had a clear advantage.

The availability of labor for collective projects was constantly at odds with private work in family plots. During peak seasons this clash was heightened by the urgency of tasks outstanding.

> In these years it was a nationwide phenomenon that during work peaks the leaders went from house to house at dawn requesting the members to come work at the cooperative farm. Year after year . . . soldiers, students, officials dispatched from the cities harvested the produce or they completed other urgent tasks while a segment of the membership stayed away, or worked outside the cooperative. (Donáth 1977, 202)

These problems with adequate labor contributions seriously eroded yields and the quality of produce, thereby undermining the viability of cooperative production over all. Across the county, cooperative farms were experiencing difficulties with production, because people refused to show up to work. Party leaders were at pains to explain that diligence should not be associated only with private production.

> "Work is ennobling"—so goes the folk saying, and it means that only diligently working people can be thoroughly honest. This is great wisdom, verily a great truth and an eternal one. Because as long as there will be life on earth, this truth remains: only he can be an honorable man who works. And this applies to collective farms also. . . . That is why it is especially incomprehensible that in a good number of cooperative farms many people forget this basic truth and simply do not work. And at the same time they

say that large-scale farming is bad. . . . Those people who do not work, who within cooperative farms leave the cultivation of land and the huge farmwork to others, are saboteurs. (*Fejér Megyei Hírlap*, November 25, 1960)

Party officials were still battling to sever the association of honorable labor with owning private property. Not only would the party declare that diligence was an eternal quality; repudiating hard work was seen to be a repudiation of social community itself. As the party was forced to remind itself often in the early years of cooperative farming: "The change in people's thinking does not parallel the socialist transformation of economic structure, but falls somewhat behind" (ibid., February 14, 1960). This was only too evident, as people refused again and again to reap honor in cooperative production. Or as one journalist put it so succinctly, "We must struggle with the difficulties of getting started. But then there is no society where hunters shoot at roast pigeon" (ibid., January 14, 1960).

Resistance to cooperative work was not only a matter of principle. How members were paid, and how much, also influenced their commitment to cooperative production. The poor performance of cooperative farms, in part a result of insufficient labor power, meant meager pay. Nearly a third of working members earned less from their collective work than one-fourth the yearly salary of state farm members, and another 30% of cooperative members did not earn half the amount state farmworkers did (Donáth 1977, 201). Dissatisfaction with poor earnings was increased by the insecurity endemic to the system of payment for cooperative work, that is, remuneration on the basis of work units accumulated over the year. The value of work units assigned to any particular task was constant, although several revisions in the relative value of units did occur during the 1960s. The problem, however, was that the monetary value of a work unit was not determined until the end of the year. Poor crop yields, the result of insufficient labor, inadequate tools and machines, bad management, or simply problems with the weather, all contributed to the final calculation. Furthermore, numerous investments and other costs, so crucial in the developmental stages of cooperative farms, took priority over distributing profits in the form of the work unit. Cooperative members had no way of anticipating their yearly income, and were virtually without money most months of the year. As was true in the 1950s, families often diversified, sending the young off to urban industry and nearby state farms to ensure the household a regular source of money, keeping the elderly at the cooperative farm to maintain their household plot and access to grains, and to continue receiving rent payments from the land they had contributed to the farm.

The First Compromise

The chaos of cooperative production in the early years fed its underdevelopment: the vicious cycle of poor performance and inadequate labor was further aggravated by the need for extensive state investments that were not forthcoming. Widespread estrangement from collective work, coupled with a strong commitment to household production, forced a rethinking of cooperative policy. The solution reached, though hotly contested, was to allow members and their families to sign sharecropping contracts with the cooperative farm (Donáth 1977, 205–207). Certain factions among the peasantry had long been advocating the policy, especially for cultivating corn and potatoes and mowing hay. The party had explicitly forbidden the practice in the 1950s; it had been considered a step backward, reminiscent of capitalist forms of exploitation, a move away from collective, socialist relations of production. In the renewed debate, these concerns were increased by the prospect that sharecropping would strengthen household-based production, regarded by hard-line critics as the undesirable continuation of the private, nonsocialist sector.

> It seemed an absurdity that the collective farm would be fortified if cooperative work organization and work performance were replaced with familial work organization and individual work completion, even if only temporarily until the appearance of technology that would replace careful handwork. Even if only for a short time, the retreat seemed unacceptable: household production being strengthened and individual cultivation gaining ground at the cooperative farm. (Ibid., 206)

Advocates of the policy considered the goal of increased agricultural production to outweigh these problems, and they succeeded in implementing the program against strong objections.

After the surrender of land to cooperative production, household-based agricultural activities became more devoted to raising livestock: chickens, pigs, cows. Yet access to feed was limited: the plot of land allocated each family by the farm was less than half a hectare and there were no retail feed stores serving the private sector. The opportunity to increase the amount of acreage under cultivation by sharecropping made the contracts very desirable. The lack of rigorous work patterns at the cooperative, owing to rampant absenteeism and disillusionment with collective work, had sown discord among members. Thus, the ability to work within the domestic division of labor, so familiar and clearly preferred to collective forms, also contributed to the popularity of sharecropping contracts. Finally, collectivization signified the appropriation of control from the individual head of household to the cooperative membership, a forced alienation of one's labor to the dictates of the group. As with peasant work before the war, sharecropping offered a context where the supreme value of control or mastery over the process of working could be recaptured and realized. Yet it bears emphasis that mastery over work was still a male privilege. Women constituted the large majority of family members assisting in private production, but they earned no direct wage for this. "Between 1962–1967 . . . the numbers of family members helping out grew enormously. (The proportion of women in the age group above twenty years of age [who worked in private production] was approximately 90% everywhere.) Yet they did not have an independent income. Their work increased the income of the member, under whose name the land had been issued for family sharecropping" (Kovácsné Orolin 1976, 223). Throughout the 1960s sharecropping was very popular and constituted a significant proportion of labor invested in grain production. In 1966 a full one-third of the work done in cooperative grain production was conducted under sharecropping contracts; in 1970 it still represented 27% of work time in grains (Donáth 1977, 207). (For a discussion of sharecropping policies in Romania, see Kideckel 1993, 111.)

Employing family members to supplement the cooperative farm workforce was not restricted solely to sharecropping contracts. At the height of the agricultural season, family members and others were hired to assist in the big projects of the summer.

> The widespread participation of family members appears to be the most expedient means of alleviating seasonal labor shortage. According to last year's statistics 28,388 cooperative members participated in collective work; outside family members surpassed five thousand. Moreover, last year nearly 19,000 seasonal workers were employed by cooperative farms, and the demand for labor appearing at peak labor times verifies this best. These were family members, by and large. (*Fejér Megyei Hírlap*, June 6, 1968)

The compromise struck by planners and peasants was generally successful. It led to the increased quality and quantity of produce, while also making better use of labor resources within the village community. Sharecropping also played a very significant role as a labor-intensive activity when mechanization was in its embryonic stages in cooperative farms.

Mechanization of Agriculture

Investments toward mechanizing agriculture were very slow in coming; in the 1950s state investments were focused primarily on heavy industry. Many cooperatives were without adequate tools for cultivation. In 1957–1958 only 6.5% of cooperative farms had five of the basic tools needed—plow, harrow, roller, horse-hoe and sowing-machine—and 47% had none of these. Thirty-five percent of smaller farms did not own any kind of implements whatsoever. Among private producers, many former agricultural proletarians still did not have adequate equipment for production at the end of the 1950s (Kulcsár and Szijjártó 1980, 200). Machines such as existed were concentrated in state farms, and at tractor stations. During the final push for collectivization the number of tractors tripled, but since the amount of land in cooperative cultivation increased so dramatically in this period, the level of mechanization actually decreased (Donáth 1977, 192). At Harmony Farm scythes were used to harvest wheat at least until 1964.

In the 1950s tractor or machine stations were set up to serve a number of communities. Tractors and other relatively rudimentary machines were to be made available on request to cooperative and state farms. Machine stations clearly were established to concentrate the limited capital investments being made in agriculture in this period, as well as to reward those people willing to produce collectively. Tractor station personnel were also engaged in the propaganda campaign to convince the peasantry of the superiority of mechanized agriculture. State farms were gradually allowed to build up their own machinery base, and by 1968 tractor stations were effectively disbanded to oblige cooperative farms to bear the financial burden of mechanization.

State investment policies continued to favor industry over agriculture in the 1960s; on a yearly average 18% of all national investments between 1961 and 1965 were made in agriculture, although this did represent a 50% increase over state investments made between 1958 and 1960 (Donáth 1977, 191). Despite the increase, the push to mechanize was hampered by the policy embraced by the party of collectivizing both grains and animal husbandry. A large portion of the investments made prior to 1968 were devoted to construction, primarily of barns and stables for livestock. The need for quarters was urgent, so initially makeshift barns and stables were constructed, which then had to be replaced with more substantial buildings (see table 4.4). Furthermore, until the abolition of machine stations and the rethinking of credit policies in agriculture in the mid-1960s, investments in machinery by cooperative farms had been restricted by state and county agencies (see table 4.5). Once the reform redefined the transitional character of cooperative production, the state was less reluctant to allow a concentrated effort to mechanize cooperative farms, as they had allowed in state farms.

The conception of large-scale agriculture held by the party and planners assumed its mechanization. Policies of intensified industrialization required that good numbers of workers abandon agriculture to feed the burgeoning industrial sector. Their loss would have to be compensated by machines. In fact, between 1959 and 1965 one-fourth of the agricultural workforce (450,000 people) left agriculture for industry, heightening the need for mechanized supports to the remaining labor force (Kulcsár and Szijjártó 1980, 81). Table 4.6 chronicles the increasing mechanization of different agricultural tasks.[13]

Planners and agricultural experts welcomed mechanization; villagers did not. Employees of machine stations, and later cooperative farm personnel, recounted stories of strong resistance to mechanizing agriculture. People were simply skeptical of the ability of machines to replace the special combination of human and animal power in agriculture.[14] Many were concerned about provisions for fertilization. Stores of manure to spread on the fields always accumulated at home, as well as being provided by animals during work in the fields. If draught animals were to be replaced by machines, how would lands be fertilized? This concern was well-founded, as the availability of chemical fertilizers was limited. Another problem raised by peasants was the effect of machines on yields. How could a machine replace the care taken by a careful reaper to capture all the grains? Wouldn't the loss of grains increase dramatically? The poor quality of machines available in the early years of mechanization merely substantiated the fears of peasants. With time, however, cooperative members realized the value of machines to supplant their own physical labor. The free-

---

[13] Corn harvesting was not as rapidly mechanized as wheat. This may be explained, at least in part, by the predilection for members to cultivate by hand the plots of corn they were assigned in cooperative lands, which gave them much higher yields than if the cooperative worked them with machines.

[14] Many of the same concerns were voiced in the nineteenth century, when machines were initially introduced into agriculture.

TABLE 4.4. Investments in agriculture according to financial-technical compositions and social sector (1960–1970)

	1960–1964	1965–1970	1960–1970
Construction			
State	3,907	6,880	10,787
Cooperative	9,553	25,713	35,266
Domestic machinery			
State	2,390	2,259	4,649
Cooperative	2,717	5,551	8,268
Imported machinery			
State	2,335	2,932	5,267
Cooperative	3,690	5,721	9,411
Other			
State	2,440	3,271	5,711
Cooperative	3,450	6,085	9,535
Totals			
State	11,072	15,342	26,414
Cooperative	19,410	43,070	62,480

Source: Pető and Szakács 1985: 483
Note: In millions of forints at current prices.

TABLE 4.5. Distribution of tractors in Hungary according to social sector

		Percentage of total[1]		
Year	Total	State farm	Tractor station	Cooperative
1956	26,200	30.8	61.8	1.1
1960	47,900	25.4	54.7	14.3
1964	83,800	18.3	24.8	51.6
1968	93,800	19.3	5.0	75.4

Source: Donáth 1977: 192
[1] Percentages do not total 100; the remaining tractors were presumably owned privately, by peasants.

ing of labor for other tasks and the reduction of toil in work have been major reasons for the strong support of cooperative production among workers during the socialist period.

By the 1970s mechanization began to have a serious impact on the social relations of cooperative production. The introduction of mechanized methods of cultivating and harvesting produce required cooperative personnel to command new kinds of knowledge or skills. It was no longer considered sufficient to have extensive practical experience. Training in new techniques and methods, in

TABLE 4.6. Percentage of mechanization of major agricultural tasks, 1960–1970

	Agriculture[1] total 1960	1965	1970	State farms 1960	1965	1970	Cooperative farms 1960	1965	1970
Wheat Harvest	42.6	78.9	93.4	97.4	99.2	100.0	53.3	83.1	98.7
Potatoes Harvest	12.2	17.7	24.0	83.7	89.4	95.2	27.3	34.4	64.3
Picking, loading	—	—	5.4	—	—	38.1	—	—	13.9
Sugar beets Harvest	35.8	49.1	72.0	74.7	89.7	91.8	33.0	46.5	70.8
Loading	—	—	15.7	—	—	64.5	—	—	12.3
Corn harvest	—	2.6	23.8	—	24.0	89.0	—	0.9	40.0

Source: Petö and Szakács 1985: 489

[1]The work of machines operated by state farms, cooperative farms, and machine stations calculated for the area of the country.

the use of new machines, was encouraged, as the following editorial argues.

> Leaders, parents, young people all recognize that old knowledge inherited from the past is no longer enough. It is not enough to drag [people] into agriculture, casually, for want of better opportunities, assuming that "he can [learn to] hoe anyway." The true interest of communities requires that young people get professional training, to cultivate on a higher level the knowledge gathered by their parents and grandparents—at the price of cold dawns, bruised palms, and deadly fatigue—so that there be new kinds of "hoers and scythers" who combine the strength of their arms and love of the land with intelligence and knowledge. (*Fejér Megyei Hírlap*, June 4, 1968)

Promoting careers in skilled and semiskilled work within agriculture was an important new trend, but it did not solve the immediate problem of supplying cooperative farms with skilled personnel. Cooperatives began to send members to special short-term training courses. The other solution was to allow cooperatives to hire skilled workers, who maintained a pure wage-labor relation to the farm. They enjoyed none of the privileges of cooperative membership, such as a share in yearly profits or rent payments, or a possible role in decision-making (Kulcsár and Szijjártó 1980, 86). This new policy represented the first step in reevaluating a central principle of cooperative farm organization, that is, those who worked at the cooperative owned the lands collectively, since they had all once contributed their property to the farm. The relation between cooperative farm employment and collective property ownership no longer held, now that people were being hired solely as wage laborers.

The shift toward increasingly mechanized agriculture thus brought changes in the constitution of the labor force. This phenomenon was at odds with official representations of the cooperative peasantry. Cooperative farms were less and less a collection of aging peasants, and increasingly characterized by a young and frequently changing workforce (Juhász 1979). New theories about the worker/peasant were born (Márkus 1973). In the early years of this shift, between 1965 and 1970, the average age of the cooperative workforce (not counting retired members) decreased from 47.1 to 43.6 (Kulcsár and Szijjártó 1980, 88). The establishment of repair stations and technical support groups for the cooperative's machine park, and later the development of industrial workshops allowed by the reform at cooperative farms drew many away from industry and back into agriculture. The increasing popularity of the agricultural sector for skilled workers came to be seen as a serious threat to the industrial sector. In 1972, the industrial working-class faction within the party and among planners staged an antireform drive to ensure that agriculture not undermine the supply of adequate numbers of skilled workers to industry. Many workers still chose to return to agricultural cooperatives, if only to have better access to resources necessary for participation in second-economy activities.[15] Moreover, working at cooperative farms often meant giving up the arduous commute to industrial jobs in the nearby county seat or capital city.[16] Nonetheless, working in agriculture, even in skilled occupations, was still looked down upon by many in the late 1970s. "Many studies of skilled worker trainees who have

---

[15] Many who sought employment at cooperative farms during this period were *returning* to agriculture, as Juhász has shown in his survey (1976). It should be emphasized, however, that a good number (approximately one-third in Juhász's survey) had never worked in agriculture before. Such numbers bolstered the fears some party officials voiced about skilled workers abandoning industry for the benefits of cooperative farm employment.

[16] In Sárosd, young people often cited commuting as a reason for leaving other jobs for those at the cooperative farm. The county seat was only a half hour away, not a far distance to commute. However, this attitude demonstrated that workers in Sárosd had not often participated in long-term commuting for jobs, during the socialist period or earlier. This contrasts strongly with other regions in Hungary where male workers would commute for several hours daily, or sometimes weekly, to keep a job in the city. Not incidentally, these regions were areas that had been populated by migrant laborers before the war, in other words, areas where workers had been forced by economic circumstances to engage regularly in long-distance commuting to find work.

completed their studies have shown that over 50% of them leave agricultural employment within four to five years of graduation ... and in a study of 700 students in an agricultural training college ... 50% of all students, and 60% of the girls, admitted that they did not want to get a job in agriculture" (Swain 1985, 89).

Mechanization also altered the values accorded different kinds of work within the cooperative. The classic 1950s image of "the new socialist man on his tractor" became a reality, and in time working with machines became a truly valued occupation, as did other skilled work in agriculture (Kulcsár and Szijjártó 1980, 86–87). Table 4.7 shows early changes in proportions of workers in different branches of agricultural production between 1967 and 1972; table 4.8 shows the distribution according to years of experience. By the late 1970s, the numbers employed in a variety of tasks not considered traditionally agricultural had grown substantially. In a survey of thirty cooperative farms conducted by the Cooperative Research Institute in 1977, "32.4% of the membership worked in crop production, 4.4% worked in market gardening, 17.5% worked in animal husbandry and 21.5% worked in machine shops or in other forms of industrial-type occupation" (Swain 1985, 88). Those in skilled jobs and those in mechanized occupations were increasingly paid higher wages, as were those who worked in animal husbandry. Those employed in crop production fell behind the others, in salary wages and in prestige (see table 4.9).[17]

The revaluation of occupations accompanying mechanization also had consequences for the distribution of social groups within the farm. Having appropriated managerial and other leadership positions within the cooperative, peasants had isolated themselves from agricultural workers, who did the more menial tasks of the farm. But the social separation of different strata in work was further solidified by divergent patterns of career choice among the children of former peasants and agricultural proletarians. Peasants encouraged their children to acquire advanced training in agriculture. Former agricultural proletarians tended to send their children off to trade schools after they graduated from elementary school (Juhász 1984, 1–2). Table 4.10 shows the family histories of work-

---

[17] Working in the construction brigade of the cooperative farm, for example, may have not been as lucrative as a job in the city. But workers made up for the loss on weekends by engaging in well-paid private construction jobs. The tempo and pace of work at the cooperative farm was seen to be less demanding than at jobs in the city, and even if the farm had demanded a more strenuous output, the construction brigade would have resisted, seeing their primary responsibility at the farm to be to rest up for their private work on the weekend.

TABLE 4.7. Changes in numbers and proportions in types of work groups, nationwide

	1967		1972			Increase
	No.	%	No.	%	Diff.	%
General nonskilled workers	264	7.7	577	18.8	+313	+118.6
Plant cultivators	1,655	48.5	993	32.4	–662	–40.0
Workers in animal husbandry	709	20.7	537	17.5	–172	–24.3
Machine, chemical workers, and industrial workers	428	12.5	659	21.5	+231	+54.0
Administration and commercial workers	101	2.9	143	4.7	+42	+41.6
Management	251	7.3	157	5.0	–94	–37.5
Unknown	13	0.4	—	—	–13	—
Total	3,421	100.0	3,066	100.0	–355	–10.5

Source: Juhász 1976: 252

Note: Figures are based on a representative sample survey conducted by the Cooperative Research Institute in 1967 and again in 1972.

ers in different branches of production, providing a picture of the family backgrounds of a sample of cooperative workers and their occupations. The consequences of these choices became evident in the shifting domain of cooperative occupations and their value.

> [T]hose who were bound to private farming, and their offspring, "lagged behind" relatively on the acquisition of skilled training. In contrast, former agrarian proletarians—almost "thinking ahead"—realized their trade aspirations in greater proportion; they became skilled workers or participated in training for semiskilled occupations. Thus for understandable reasons those strata acquired a larger role who had been the poorest, those who, if they hadn't even brought land or tools to the cooperative, by joining had augmented the investable resources of cooperatives with the most necessary "capital," *with their training, their industrial experiences.* (Kulcsár and Szijjártó 1980, 93; emphasis in original)

Thus, mechanization began to undermine the peasant elite's political control of cooperative production, as their practical knowledge of farming had lost its value in production. Hence the class divisions of the prewar village which had lived on in early cooperative farms were slowly being eliminated. The once hallowed relation between property ownership and competent farming no longer held, shunted aside by a demand for the skills required in mechanized production, skills frequently sought by descendants of the agricultural proletariat. The position of former peasants was further eroded by the authority vested in educated managers of the technocratic elite who dominated the cooperative farm by the mid-1970s.

Reform of the Economic Mechanism

The first series of major economic reforms instituted by the state, known as the Reform of the Economic Mechanism, is generally associated with the fateful year of 1968, but its influence on agricultural policy began in early 1966. Considered a watershed in Hungarian economic policy, the reform introduced changes whose primary effects were intended to influence the industrial sector, and did so.[18] Nonetheless, several shifts in agricultural policy, specifically in relation to cooperative production, do warrant brief mention.

The reform provoked a reinterpretation of the status of cooperative property and a redefinition of the transitional role of cooperative production. Formerly, cooperative property had been distinguished from other forms of collective ownership, such as state-owned enterprises or state farms. As group property owned by a limited number of individuals, it was considered less socialist than enterprises owned by the state, and so by extension, by all citizens of the country. With the reform, cooperatives were accorded full socialist status, and as "legitimate" enterprises, expected to play a greater role in the national economy (Erdei 1972, 74–75). The following passage illustrates how awkward these definitional debates were, and demonstrates that the change in status was to have immediate economic consequences.

> Co-operative property is the joint property of the members—it is collective property. This property is more limited than public property but *it represents socialist property of the same rank*. So co-operative property is *socialist social property*, one of its forms of manifestation. To be quite exact in wording it: it is socialist *co-operative group property*. It serves the interests of the socialist state that the group-ownership—as a socialist one, the collective ownership of co-operatives—should get stronger. It naturally implies that in our economic policy the planned direction of economy must be founded upon the co-ordinated and simultane-

---

[18] The Reform of the Economic Mechanism in Hungary has been quite extensively studied. Among others, a general overview is offered in recent works by Berend 1990; Kornai 1992; Nee and Stark 1989; Swain 1992.

TABLE 4.8. Distribution of 1972 sample according to type of work group and length of membership in the cooperative

Years in cooperative	Without special skills	Grain production, gardening	Animal husbandry, equipage	Machine, chemical, industrial	Admin. and sales	Managers of subordinate units	Branch and upper management	Totals
0–3 years								
Numbers	177	230	97	318	64	34	36	956
Percentages	30.7	23.2	18.1	48.3	44.8	43.6	45.6	31.2
3–6 years								
Numbers	58	122	81	97	10	10	16	394
Percentages	10.1	12.3	15.1	14.7	7.0	12.8	20.3	12.9
Over 6 years								
Numbers	342	641	359	244	69	34	27	1,716
Percentages	59.2	64.5	66.8	37.0	48.2	45.6	34.1	56.0
Totals								
Numbers	577	993	537	659	134	78	79	3,066
Percentages	100.0	100.0	100.0	100.0	100.0	100.0	100.0	100.0

Source: Juhász 1976: 253

Note: Figures are based on a representative sample survey conducted by the Cooperative Research Institute in 1967 and again in 1972.

TABLE 4.9. Annual earnings in May 1st Cooperative, 1976

	Earnings (Ft)
Mechanics for fodder works	48,286
Full-time manager, independent leaders, and administration workers	44,384
Pigherds	41,897
Drivers and loaders	41,635
Sheep and cattle slaughterers	40,685
Mechanized crop production workers	39,872
Warehousemen	39,615
Cattlemen (dairy)	36,816
Workers in grass production	35,869
Calf rearers	34,541
Mechanics in general workshop	33,909
Construction workers	30,104
Traditional crop production workers	22,000

Source: Swain 1985: 90

Note: These aggregate figures do not reflect the very high wage rates that combine-drivers, and those involved with the mechanized side of harvesting, can earn at harvest time.

ous development of state and collective ownership representing two forms of social property. (Fehér 1972, 16; emphasis in the original)

The elevation of cooperatives to full socialist status necessarily altered, or at least qualified, the degree to which cooperative property was considered transitional within the economy. Though never totally repudiated, the idea of cooperatives as transitional institutional forms was downplayed.[19]

In simple terms, the reform represented a reinterpretation of the role of markets and the character and degree of economic planning. Devolution of decision-making to the level of the enterprise and a greater reliance upon the market for setting prices and production goals were the prominent features of the policy. The prevailing explanation for the need for markets was that socialism must live within the constraints of human nature. The introduction of incentives in the workplace, both in agriculture and industry, was also justified by a specifically utilitarian understanding of action. The

---

[19] In the final analysis, the legal distinctions made by party theorists and planning agencies between cooperatives and state farms—that cooperatives were group property and state farms were state property—simply obscured the state's imposition of collective ownership on the workforce.

TABLE 4.10. Original family social status of interviewees

Original status[1]	Without special skills No.	%	Grain production No.	%	Animal husbandry, equipage No.	%	Machine, chemical No.	%	Admin. and sales No.	%	Management No.	%	Totals No.	%
Unknown														
No.	40	6.9	54	5.4	27	5.0	87	13.2	17	11.9	36	21.9	261	8.5
%		15.3		20.7		10.3		33.3		6.5		13.8		100.0
Agrarian worker or semiproletarian														
No.	131	22.7	216	21.8	135	25.1	69	10.5	13	9.1	17	10.8	581	18.9
%		22.5		37.2		23.2		11.9		2.2		2.9		100.0
Small, middle, or wealthy peasant														
No.	216	37.4	472	47.5	284	52.9	174	26.4	42	29.4	43	27.4	1,231	40.2
%		17.5		38.3		23.1		14.1		3.4		3.5		100.0
Crafts, commerce, or agricultural worker														
No.	22	3.8	50	5.0	15	2.8	63	9.6	7	4.9	9	5.7	166	5.4
%		13.3		30.1		9.0		38.0		4.2		5.4		100.0
Industrial worker														
No.	127	22.0	154	15.5	60	11.2	231	35.1	31	21.7	18	11.5	621	20.3
%		20.5		24.8		9.7		37.2		5.0		2.9		100.0

	Without special skills		Grain production		Animal husbandry, equipage		Machine, chemical		Admin. and sales		Management		Totals	
Original status[1]	No.	%	No.	%	No.	%	No.	%	No.	%	No.	%	No.	%
White-collar employee														
No.	40	6.9	45	4.5	15	2.8	33	5.0	26	18.2	11	7.0	170	100.0
%	23.5		26.5		8.8		19.4		15.3		6.4			5.5
Professional/intellectual														
No.	1	0.2	2	0.2	1	0.2	2	0.3	7	4.9	23	14.6	36	100.0
%	2.8		5.6		2.8		5.6		19.4		63.9			1.2
Total														
No.	577	100.0	993	100.0	537	100.0	659	100.0	143	100.0	157	100.0		
%	18.8		32.4		17.5		21.5		4.7		5.0			

Source: Juhász 1976: 260–261

Note: Figures are based on a representative sample survey conducted by the Cooperative Research Institute in 1967 and again in 1972.

[1]Family social status before the establishment of the cooperative farm (head of family).

term "economic mechanism," suggesting that economic forces are mechanically interrelated, derived from the same logic, one premised on the preoccupation with incentive among human agents, a fascination with means, a fixation on utility. No longer would the interests of an undifferentiated national economy hold forth, as the following article suggests.

The Twilight of an Incantation

> At one time the interest of the people's economy was an incantation and was heard as an absolute argument in debates. . . . At one time we believed that the interests of the people's economy were expressed in a direct and absolute fashion. On this assumption rested the system of the centralized command economy, just as did the frequent and direct intervention in the economic activities of enterprises. It is already evident that we can consider the assertion of interest of the people's economy only as the result of a process, a tendency: the clash of enterprise interests. . . . It is not necessary to praise the social values of local decisions and "ideologize" their significance to the people's economy. However it is essential that the analysis of the comprehensive, wide-ranging enterprise interests precede decisions, that making quick and easy profit should not be typical. Rather, long-term interest and winning and extending markets should be. Instead of apparent and alleged interests, we must pay attention to the real interests of enterprises. (*Fejér Megyei Hírlap*, June 23, 1968)

Unfortunately, the competitive pressures assumed to grow from a more active market were essentially nonexistent in the industrial sector, since nearly all Hungarian enterprises were monopolies. The cooperative agricultural sector, on the other hand, was large and very diverse, and differences in the character of production, be they due to local conditions or to distinct organizational patterns, became even more salient. Ironically, then, policies that were designed primarily to improve industrial production became more influential at the outset within the cooperative agricultural sector. Indeed, it is possible to view cooperative farms in some respects as the experimental stations for many policies later instituted in various stages of the reform. Dependent upon their own resources, bearing the full costs of success and failure, cooperatives were allowed to introduce all sorts of interesting organizational forms, incentive systems, and wage policies that, once judged efficient, were embraced by the industrial sector as well. The transition, however, was not always easy. First and foremost, cooperative presidents had to learn new ways of managing, of working in an altered economic environment. As one cooperative president explained, when discussing the influence of the new planning system: "Today we cannot compete with state commerce and other forms which had developed for years. We are aware that we need competitors, but we don't know how to do this. This is what we must learn" (*Fejér Megyei Hírlap*, December 18, 1968). Specialized knowledge about markets, about competition, about products had to be acquired. So too, managers had to acquire the independence of thought and action required by the decentralization of authority and control.

> An economic attitude is beginning to prevail and be strengthened day by day. It boils down to this: in our society the old habitual attitude that it is sufficient to accept plans coming from above and to fulfill them more or less is gradually disappearing. In ever wider circles the more realistic approach is carrying the day, in which every economic organization and collective must take responsibility for itself, for what it sets before itself, and for how profitably it does its work. (*Fejér Megyei Hírlap*, December 29, 1968)

Local decision-making could also increase the arbitrary use of power on the local level. The party needed to remind its officials of the lines between responsible, independent thinking and authoritarianism. As the secretary of the Central Committee stated quite clearly: "A manager should consider power to be like medicine, excessive use of which can act like a ravaging poison" (*Fejér Megyei Hírlap*, November 17, 1974). These cautionary remarks were important, coming as they did after years of abuse of privilege and power in the name of central authorities.

Organizational innovations at cooperative farms took essentially two forms: the introduction of nonagricultural productive activities into the cooperative, and the strengthening of household-based agricultural production. With the reform the restriction on agricultural cooperatives engaging in nonagricultural activities was rescinded. Initially the emphasis was placed on repair and servicing shops for machinery, to replace the work of tractor stations that had effectively been dismantled. However, with time, other kinds of craft or industrial shops were established, activities ranging from highly skilled work in electronics to simple, unskilled work in, for example, assembling products on contract for a factory in Budapest (Rupp 1983). These subsidiary industrial units (*melléküzemág*) helped to subsidize the primary agricultural activities of cooperatives. Not bound by the seasonal fluctuations in cash flow which so plagued agricultural production, industrial units lessened the dependence of cooperatives on outside credit sources. Furthermore, they provided employment for members during the slow winter months when agricultural activities were reduced. Industrial units also helped to strengthen cooperative farms that were burdened with poor lands, providing alternative sources of income for the membership. The big-industry faction of the party looked askance at these new activities, and tried frequently to restrict their operation.

> Sixty-four cooperative farms are engaged in supplementary sideline production activities. They ease their worries with meshing wire, making reed flooring, soda water, repairing scales and all sorts of other activities. However, two things are conspicuous in these activities. . . . These activities do not occur in the vicinity of cooperative farms, but in far away places, where obviously only employees [i.e., not cooperative members] are able to work. The second is that some consider these activities primarily as a way to increase their money! . . . The authorities have already pointed out, and not just once, that auxiliary enterprises should assist or complement agricultural activities first and foremost, should serve the purpose of complementing products and offer cooperative members opportunities for jobs in the first place. They should not be ways of increasing the numbers of cooperative employees. (*Fejér Megyei Hírlap*, June 6, 1968)

The dangers of drawing workers away from industry and into the agricultural sector were unsettling to party officials committed to the priority of industry over all else. Agriculture should be supported through these activities; workers should not benefit by higher wages or increased job opportunities. Others welcomed the introduction of small-scale industry into rural communities, offsetting the historical tendency to concentrate industrial employment and infrastructural developments in a few large cities (Konrád and Szelényi 1976).

Household-based production, though long permitted on a restricted scale, came to play a much larger role in agricultural production with the reform. The Hungarian term I am translating as household-based production, *háztáji*, literally means "in the vicinity of the house." It should be made clear, however, that household-based production covers not only work in stables and gardens behind the family house, but also any farming done on lands allotted by the cooperative farm to members (e.g., cornfields, potato and alfalfa plots), and on any additional properties owned or rented privately in the village. Akin to the redefinition of socialist property relations, household production was elevated to a new status. No longer considered the simple survival of private production, and so antagonistic to the cooperative sector, household production was now considered a supplementary or complementary realm to large-scale farming. Cooperative farms and consumer cooperative agencies set up more favorable contracts for purchasing produce and livestock. Campaigns were mounted, for example, by the Women's Council, to encourage women to increase their production of eggs and poultry. Women were praised for their diligence, even though this was demonstrated within the household.

> The kilograms of eggs and the tons of poultry delivered to the buyers all speak to the housewives' diligence and desire to help out.... The results of last year's competition was nearly 21 million eggs and 48.5 freight cars of poultry meat. This is a huge quantity of goods, for which the housewives receive in return many millions of forints. (*Fejér Megyei Hírlap*, April 21, 1968)

Incentives were offered if families agreed to produce specific quantities, thereby incorporating household-based production more directly into the planning schedules of contracting farms and agencies. Farms were encouraged to improve access to grazing fields and to make more feed available, both in terms of membership contracts and for purchase at large within the village.

The contribution of household-based agricultural activities to the national economy soared. Estimates of the percentages of produce and livestock sold by the noncooperative sector vary, but the dimensions are truly significant.[20] Depending upon the specific product, for instance, pigs, chickens, or vegetables, the figures range from 30% to 80% of total production. Furthermore, participation in the agricultural second economy was not restricted to employees or members of cooperatives and state farms. Fifty percent of the nation was engaging in private farming by the mid-1970s, whereas only 20% of the workforce was employed in agriculture. According to the 1981 survey on small-scale agricultural production compiled by the Hungarian Central Statistical Office, 42% of the Hungarian population were engaged in some sort of domestic agri-

---

[20] Even fairly reliable statistics on household-based production could be misleading, for most agencies did not attempt to estimate how much of what was being produced was actually withheld from the market for family consumption. Those who did attempt to make calculations could only approximate possible figures.

cultural production. Within this group, 31.2% belonged to the working class, 11.2% were classified as members of the cooperative peasantry, and 26.9% were retirees (Central Statistical Office 1982b, 20). The second economy offered excellent sources of supplementary income, allowed wives to remain at home with the children, and engaged elderly members of the family in productive activities.

In some eyes, it also increased the autonomy of villagers and challenged the legitimacy of the socialist sector. Now that private activities were gaining respectability once more, workers were taking liberties in freeing up time to work elsewhere. In an article entitled "Order and Public Property Must Be Protected," the bookkeeper of Harmony Farm described to the county newspaper new measures taken by the farm to prevent abuses by workers.

> We passed a resolution at the annual meeting that the cooperative would have a inspector keep an eye on those who go on sick leave, so that no one can cheat the community by not fulfilling his daily work. Among those on sick pay it has happened that someone took on a separate job, naturally for additional pay. We have eliminated this possibility by hiring a sick leave inspector for the cooperative farm, who regularly visits those concerned and ascertains whether they are at home and resting, so that they can return to work healthy as soon as possible. There is no question that keeping a separate inspector for the cooperative means additional costs, but we need one at the farm.
> (*Fejér Megyei Hírlap*, May 5, 1968)

The everyday problems of keeping workers at their jobs clearly preoccupied managers and party leaders. Others considered these problems to be much more fundamental. Hard-liners within the party and ministries continued to perceive domestic production as the unfortunate remnant of capitalist agriculture, and worked consistently for its abolition. By 1973 a fight was led against the reforms in general, and the agrarian sector in particular, gaining support within the party from the faction most closely allied with industrial interests. Editorial commentary appeared explaining conclusions drawn by the Central Committee on "the accepted directives on strengthening the leading role of the working class and the further improvement of its position" (*Fejér Megyei Hírlap*, August 18, 1974). Having won the upper hand in policy-making forums, this group succeeded in reducing monetary and institutional supports for household agricultural production. Learning of the change in policies, villagers across the country butchered thousands of sows in the summer of 1974, choosing to protest by depriving the country of renewable resources in pork meat (Róna-Tas 1989, 26–33). As a result of both state policies toward agriculture and private sabotage, major food shortages hit the country in 1974 and 1975, creating an economic and political disaster. For the first time in recent memory, basic staples like potatoes had to be purchased on the international market. Party and state agencies had sown disfavor not only among rural agricultural producers, but also among urban dwellers, who were outraged at the scarcity of foodstuffs and the prices demanded for produce actually available. The "food fiasco" marked a turning point in the support of the agricultural second economy, which for the rest of the socialist period was perceived and portrayed in official forums as an integral component of the socialist sector.

Of course, the implementation of reformist legislation on the local level was ultimately dependent upon the good will of local officials. Their support, or their interference, influenced the outcome. During the campaign to raise the level of egg production, participants voiced their reservations about local support for their efforts.

> One such comment heard in several different places found fault with how the eggs were received in kilograms. Others asserted that the abundance of goods in eggs and poultry meat has pampered the buyers and the workers of the enterprise. In some cases the buyers and workers are indifferent to their work and even allow themselves to speak gruffly. Several criticized the leaders of the town councils because they did not provide pasture for geese, without which breeding is difficult. People have requested that tour days be held punctually and the buying scales be accurately calibrated. (*Fejér Megyei Hírlap*, April 21, 1968)

As the role of efficiency and productivity was to increase in the economy, the concerns of individual producers would become more important to officials, who needed their good will and enthusiasm to guarantee higher contributions from the household. Slovenly behavior, indifference, or deliberate barriers to individual effort could hamper efforts to increase contributions from the household. Though the party and state authorities were not always able to reverse these habits or prevail upon intransigent officials, they could certainly encourage individual producers to persevere, through newspaper articles or other media.

The policies encouraging decentralization and a greater reliance on market forces in planning were intended to increase the overall efficiency of the economy. The greater responsibility of enterprise managers for decision-making was matched by a more considered policy of capital investments by the state. The previous pattern of providing unlimited funds to factories, state farms, and to a lesser degree, cooperative farms, regardless of their profitability (or actual losses), was changed; good money would no longer be thrown after bad. Credit and other forms of state investments were now to be distributed to enterprises deemed successful. The shift in policy meant unequal access to state supports, thereby possibly fostering economic inequalities. The case of cooperative farms was particularly problematic, as the success of cooperative production relied upon a complex set of independent variables, such as the quality of land and weather conditions, as much as on the shrewd policies of management and the hard work of the membership (Donáth 1976). Editorials with titles like "Egalitarianism versus Differentiation," reminiscent of Stalin's attack on the bourgeois principle of egalitarianism, dismissed these concerns and argued for tying wages more directly to performance. The embrace of monetary incentive as the motive principle in economic behavior was explicit.

> Interest, which is the main moving spring of human behavior, is generally concrete and immediate. The reform placed the increase in profit at the center of enterprise work, and it is natural that various decisions and activities tend in this direction. The materialism which derives from profit interestedness is not contradictory to our socialist principles, since the individual works not only from simple selfish, egotistic motives but is obliged to work subordinated to the interests of the larger enterprise community. (*Fejér Megyei Hírlap*, June 23, 1968)

No longer was honor in diligence a principle of labor uninhibited by other factors or considerations. Now interest and materialism were legitimate considerations in the workplace, although these qualities were subordinated to collective production. Human nature had now been rehabilitated, defined in "concrete" and "direct" terms of personal interest and monetary incentive.

Money

An important change that coincided with the reform platform of the 1960s was the shift from paying cooperative farm members in work units to calculating wages in terms of hourly wages, piece rates, or monthly salaries. The shift to a monetary payment also corresponded with changes in the structuring of the workplace, as tasks were differentiated within the organization in terms of the kind of wage paid and the level of monetary reward given. It also corresponds, I would contend, with a move away from the model of a naturalized economy of socialism to an increasingly market-driven, commoditized universe.

Work units had been the primary form of remunerating cooperative farm members from the inception of collectivized production in the late 1940s until the reforms of the late 1960s. Each kind of job or task had its own abstract work unit value, which specified the amount of activity to be completed over the course of a day, and which also included calculations of physical difficulty and skill. Work unit measures had been calculated for every single conceivable task in farming, from shoveling manure to the artificial insemination of stock. Table 4.11 gives a few examples of different tasks, and their relative values. Each cooperative farm member would have his or her daily tasks recorded by the brigade leader, an elaborate project that required cautious calculations of the minute details, for example, of the quality of soils cultivated, acreage covered, and time expended. There was considerable attention paid, then, to the particular character of work, but without immediate reference to wages or monetary calculation. At the end of the year, once the profits of the cooperative farm had been calculated, the monetary value of work units were determined. Cooperative farm members would then be paid their share of the cooperative's earnings according to their accumulated work units.[21]

Work units are often seen as preeminent instances of the baroque calculations characteristic of socialist scientific planning, elaborate calculations required to avoid the evils of commodification, exploitation, and the alienation of wage salaries. Yet the work unit system (*munkaegységrendszer*) was adopted in its entirety from a system developed for manorial estates, inspired by scientific management theories common in Germany and the United States during the 1930s. At the time, the work unit system probably represented the first piece-rate system of remuneration in Hungarian agriculture. This manner of quantifying the labor process thus presaged the wholesale quantification of production in the plan. Taylorism and scientific management were important influences in

---

[21] In addition to their work unit payments, members would also be paid, for example, rental fees or premiums awarded to excellent workers or certain managerial personnel. In addition to the work unit, cooperative farm members were given allotments of feed (usually grains like wheat and barley), which were distributed four times during the year. These constituted important sources of income for cooperative farm members, aiding household production.

TABLE 4.11. Examples of agricultural tasks and their assigned work unit values

Task	Fulfillment value	Daily assignment	Work unit
Fertilization with horse-powered machine			
	50 kg/kh	6–8 kh	1.50
	100 kg/kh	5–7 kh	1.50
	200 kg/kh	4–6 kh	1.50
Plowing: single furrow plow, at a depth of 10–15 cm			
Loose and medium soil		1,000–1,300 sq. fathom	1.00
Hard soil		700–1,000 sq. fathom	1.00
Sowing: horse-drawn seeder			
12–14 row seeder		3.5–5 kh	1.25
15–17 row seeder		4–6 kh	1.25
18–21 row seeder		6–8 kh	1.25
22–24 row seeder		7.5–10 kh	1.25
Row hoeing potatoes with a hand hoe			
Loose soil		650–750 sq. fathoms	1.00
Medium soil		600–700 sq. fathoms	1.00
Hard soil		500–600 sq. fathoms	1.00
Picking sugar beets[1]			
Under 100 quintal[2]/cadastral acre			
Loose and medium soil		110–120 sq. fathoms	1.25
Hard soil		100–110 sq. fathoms	1.25
100–200 q/c. acre			
Loose and medium soil		100–110 sq. fathoms	1.25
Hard soil		90–100 sq. fathoms	1.25
Over 200 q/c. acre			
Loose and medium soil		90–100 sq. fathoms	1.25
Hard soil		80–90 sq. fathoms	1.25
Picking fruit fallen to the ground		300–500 kg	0.75
Leading cattle onto a truck			
Without a ramp		25–40 animals	1.25
With a good ramp or on an incline		60–100 animals	1.25

(Continued)

(Table 4.11 continued)

Task	Fulfillment value	Daily assignment	Work unit
Successful conception of livestock			
Natural fertilization		1 cow	2.50
Artificial insemination		1 cow	3.00
Healthy piglets on the second day of the litter		10 piglets	3.50–4.00
Hog		100 kg weight gain	1.30–1.40
Ram		100 kg wool at delivery	14–16
Blacksmith: removing a used horseshoe, adjusting it, and reshoeing it		18–22 pieces	1.75
Plastering, whitewashing on a brick surface			
Master		20–35 sq. meters	1.75
Assistant		20–35 sq. meters	1.25

[1]Complete process: lifting out with beet digger, cutting off the top of the beet, cleaning it, piling it up, and covering it with leaves.
[2]Quintal = 100 kg
Source: Munkaegységkönyv [Work Unit Book] 1960, 5th ed.

early Soviet theories and policies on wages and industrial organization. Thus the work unit system was easily adapted in the socialist period, bearing many similarities to Soviet policies, despite its capitalist origins.

With the reform, cooperative farms successively abandoned the work unit system of payment, replacing it with monthly salaries for some jobs, hourly wages for others, and piece-rate scales for still others. Monthly salaries were usually given to those in management positions, and hourly wages or piece rates to workers. The distinction between piece rate and hourly wage was seen to lie in the ability to apportion the task to units or parts. Some work was continuous, and hence not seen as amenable to division. This was the general explanation, but workers often saw the use of hourly wages or piece rates as means to reward some groups and reduce the pay of others. For example, I witnessed a debate between women working in the garden brigade and their boss over the possibility of being paid in piece rate when picking tomatoes. The women would have preferred to have been paid on a piece-rate basis, since they could easily have made their daily wage by noon, returning home to attend to the always present tasks of housekeeping. However, the manager of the garden section refused their request, since this way he would have had to pay them a higher overall salary. In the women's eyes, the boss was saving money that he could use to pay their male colleagues higher wages. Tractor drivers, for example, were frequently paid in piece rates, calculated on the number of rows they plowed or planted per day. Unfortunately, this means of remunerating workers did not always ensure quality plowing, which one could often see when rows of corn meandered across the field in interesting patterns.

The shift from work unit to wage finalizes the equation of labor with money, rendering each phase of work and each unit of time simultaneously a monetary expression. Not everyone welcomed the wage system. When discussing the poor quality of work performance at the cooperative farm—the universally bemoaned work ethic of socialist workers—one villager commented: "The work ethic deteriorated when we changed to money." This critique of monetary wages addressed two issues of wage policy. The first was the alien character of money, the pernicious quality of monetary indices that interfere with the natural quality of work on its own terms. Even though the work unit was a condensed calculation of skill, time, and effort, the character of work was evaluated in terms of its own immediate properties and requirements. It was defined in ways specific to the project itself, not with reference to any outside or external qualities. Hence, for cooperative farm members, the work unit fit very closely with their notion of work as a process with its own integrity; work units suited their concept of *dolog*. Moreover, work units were also consistent with the naturalized economy of socialism, where work had an immediate and self-evident value, divorced as it was from property relations and the strange convolutions of markets and prices.

But the pernicious quality that some villagers attributed to money alone is insufficient to explain people's trepidations about the shift to a monetary wage system at the cooperative farm. After all, there were many uses to which money could be put in the growing consumer economy of Hungary's reform period.[22] Implicit in

---

[22] First and foremost, villagers could expand their second-economy activities by buying more feed and investing in new stables, barns, and livestock. But the post–1968 period was also the time when many basic amenities—such as refrigerators, televisions, decorative fencing, indoor plumbing, central heating, cars, new homes—could be financed by state-sector monetary wages and private income, improving the quality of village life and bringing it up to the standards of many city dwellers for the first time in memory (Róna-Tas, n.d.).

the criticism of monetary wages was the means by which these wages reinforced, or recalibrated, the system of hierarchies within the organization. Workers found the coincidence of restructuring the organization with a shift to monetary wages significant. Now the hierarchy of the cooperative farm was expressed in how different grades or ranks of workers were paid, and how much say they had in their salaries. The abuse of premiums or differential wages and even the decision not to pay piece rates were frequently commented upon by workers, since they saw their fortunes manipulated by a corrupt elite management team. The perniciousness of money, of privilege, in the new cooperative farm organization drew people away from a pride in work, in the process of work itself. That is why it was so easy to malign the quality of work being done on piece-rate or hourly wages; no one took pride in their work. This pride in labor was also intimately related to controlling one's work within the organization, an experience that was denied workers in the reorganization of cooperative farms with the reforms and with the rise of a new technocratic elite.

Rise of Technocracy

The reform prompted organizational changes, changes in wage policies, and changes in attitudes toward skill and knowledge. More than ever before the value of education and skill was acclaimed. This was true for the man on the shop floor, just as it was central to the increased importance of managerial personnel. Technocracy was now on the rise as a socially validated form of social stratification. This did not occur without resistance. The working-class faction of the party decried growing social differentiation. Tensions born of increasing inequality between managers and workers led to a major debate within the party in the early 1970s. The reform faction argued that the reform's success depended upon rewarding planners, managers, and other skilled personnel, whereas the opposing faction warned against abandoning central socialist principles of social equality. Although in the short term compromises were made to lessen income differentials (Sabel and Stark 1982, 470), in the long term the technocratic conception of the reforms prevailed, marking the final watershed in the rise of technocratic power in Hungary (Konrád and Szelényi 1979; Bauman 1974).

The political coup of planners and managerial interests bore significance beyond differential wage scales alone. Knowledge and qualifications were now to be valued for their own sake. Many—

both party leaders and workers—were suspicious of the value of book knowledge or specialization per se, believing it fell far from the important abilities one gained with practical experience. The party wished to alter attitudes toward specialization, representing it as practical knowledge, as the means to reach goals. In an article in the county newspaper, cooperative farm presidents were asked about the numbers of skilled personnel they employed, in response to the Minister of Finance's call for improvements in agricultural production. The minister argued that to reach European standards in yields and livestock, specialists were indispensable. "In long-range planning every collective farm has considered its possibilities and requirements concerning professionals. One cannot dispense with these people in creating a modern technical-material basis, people who not only know how to plan, but also know what to do so that their conceptions can be realized" (*Fejér Megyei Hírlap*, January 10, 1974). Once the party sided with planners and managers, local control of managers and skilled personnel was strengthened, and the line separating physical labor from executive activities was hardened (Braverman 1974).

> This final conception of the farm president and other managers as a distinct group of professionals is encapsulated in the regulation concerning personnel work passed in 1974 . . . While rejecting "Western bourgeois notions" which had emerged amongst some leaders and introducing the "triple requirement" for management posts of political suitability, necessary academic/political qualifications and management ability, it strongly reinforced the notion of a separate management endowed with specific political and educational qualifications not shared by the ordinary membership. By the early 1970s, the conception of cooperative farm management that had emerged was one of a separate body of carefully chosen professional individuals whose duty it was to organise the efficient running of the farm. (Swain 1985, 117)

Decentralization of decision-making and responsibility carried the potential of increased power to be wielded by managerial groups over their subordinates in production.

Clashes over the exercise of power and control of local affairs could result in the abuse of privilege, in extreme cases even bringing the party to expel members for their abuses. In one case in Fejér County, a renowned cooperative farm president went too far, spending his time hunting, going to horse races, mounting elaborate banquets and drinking fests, no longer fulfilling his duties to the community or to the cooperative. The party's criticism of his personal dictatorship says a great deal about the shift to an informed and enlightened party elite.

> The students in basic-level [party] seminars know by heart what are the three basic requirements that we must demand of every leader: political qualification (paired with the capacity and desire to execute policies), professional knowledge, and the capacity and suitability to be a leader. Although it would be wrong to simplify any one of these conditions, the first two are criteria that can be relatively easily assessed: educational level, experience, . . . the ability to solve larger tasks now and then, . . . in our socialist society a person taking on any kind of office should know that the task entrusted to him is a service, and is not the opportunity to use (or abuse) power. . . . as Comrade Biszku Béla, secretary of the Central Committee, has said . . . with regard to this question at the Tenth Congress: *"power does not belong to persons, but to class. . . . To confuse the power of the working class with one's own power is reprehensible."* (*Fejér Megyei Hírlap*, November 17, 1974; emphasis in the original)

The manner in which power was to be wielded in the new, technocratic world was not to be so crude, so obvious, so glaring. Finer methods of persuasion were to be employed.

Corresponding in time to the debate over technocratic power, new organizational structures were encouraged by the party which introduced much greater complexity and hierarchy into cooperative farms. The spatial organization of productive branches, as characterized Harmony Farm, was replaced with an organization whereby branches were devoted to specific tasks within the two broad categories of grain cultivation and animal husbandry. Several tiers of "managers" (e.g., brigade leaders, foremen) were instituted and the activities of workers assigned to each branch were restricted to the simple tasks of that narrowly defined group (see figs. 4.1 and 4.2). Workers selected as foremen or other bosses were sent to special night classes, where they were taught the skills presumed necessary for their positions (and where they acquired the nominal justification for their power, a diploma). The diversity of tasks formerly performed by workers assigned to a specific production center no longer characterized cooperative work.

The transformation of cooperative farms into increasingly specialized, hierarchical organizations substantially altered the perception cooperative farm members held of themselves and their work. Agricultural work was seen to be quite different from what peasants had once done. This attitude was particularly prevalent among younger workers, but was also characteristic of older workers who themselves had once been peasants. When asked whether she was a peasant (*gazda*), a young worker responded:

> "Farmer?" Go on . . . My father isn't one and neither am I." Of course this appellation is not right for a smiling girl of seventeen. "I am a gardener, not a farmer." Not even every member of the older generation declares himself to be a peasant or a farmer; instead, they are stonemasons, truck drivers, mechanics, cowherds, loaders at the cooperative. . . . Young people expect the cooperative farm to be their employer, to ensure them a steady job and to take care of their professional training and advancement. They expect to be able to perform a specific job, one aspect of a whole process as gardeners, poultry raisers or even loaders, but in no way to do everything, as jack-of-all-trades. (*Fejér Megyei Hírlap*, May 23, 1968)

New attitudes toward work were being formulated. But the distinction that held in the minds of cooperative workers between an identity as a skilled worker and one as a peasant also contained a crucial, but unspoken element: landownership. Peasants own their own land, and in the best of all worlds, cultivate it without any outside interference. Even if workers had been able to participate in all aspects of production, which was no longer possible, they had been denied a much more central role: that of landowner.[23] The absence of a relation to the land is made explicit in an article published in 1974: "The main point for this stratum [young people and the middle generation] is no longer being tied to the earth, but the work itself, its circumstances and income" (*Fejér Megyei Hírlap*, July 14, 1974). This description clearly captured attitudes among the young. But it should also be noted that the party had worked to sever the relation between production and property, and reinforced this view as often as possible in the media and in the workplace. While they may have achieved this goal in coopera-

---

[23] When I asked, nearly everyone at the cooperative farm denied that they were peasants. They identified themselves rather as cooperative farmworkers (*termelöszövetkezeti munkás*). When pressed, people explained that they did not own their own land, and so could not be peasants. The one exception to this response in my survey was the young party secretary, who proudly claimed his peasant identity. This is perhaps not surprising. His family was very proud of its peasant identity (the Pál family, see chapter 6). As a manager and party secretary, this man was also more inclined to identify with party statements that portrayed the cooperative farm as a collective family of farming peasants.

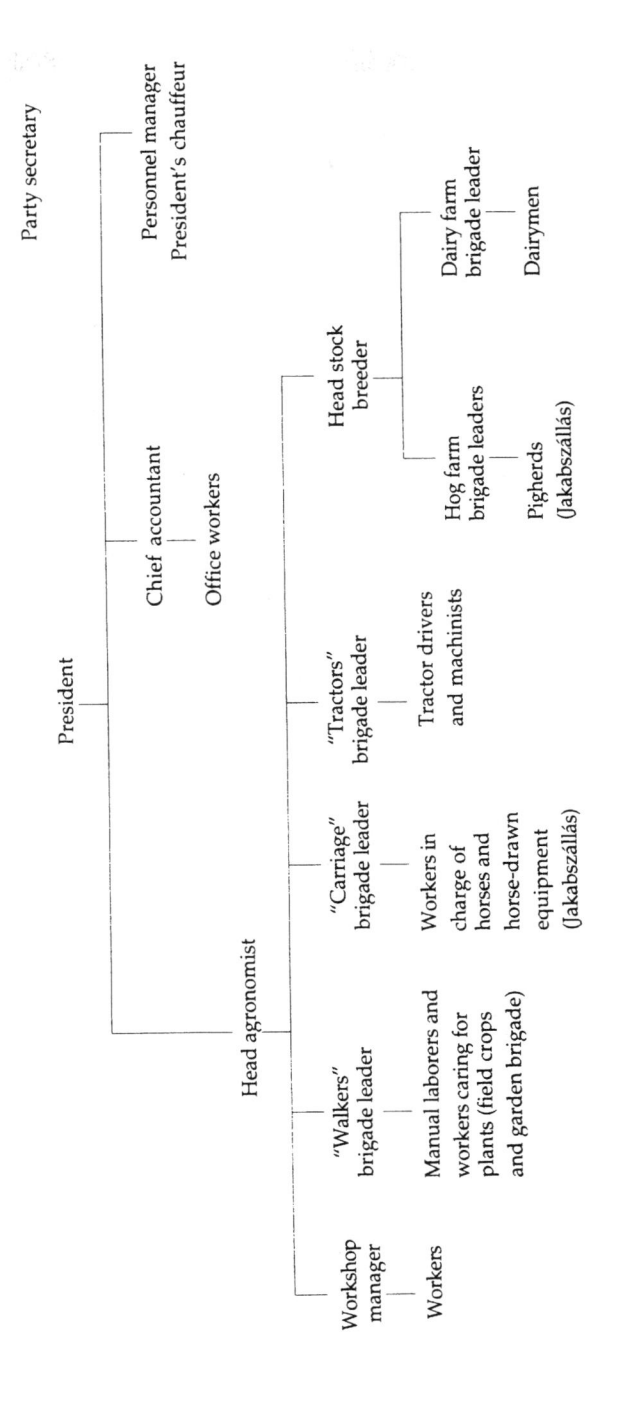

FIG. 4.1. Organizational structure of Harmony Farm (1960s)

```
                                      Party secretary

                                         President
                                            |
        ┌──────────────────┬────────────────┼────────────────┬──────────────────┐
   Head agronomist                                      Chief accountant   Personnel manager
        |                                               Office workers    President's chauffeur
        |
   ┌────┼─────────┬──────────────┬──────────────┐
Workshop    "Walkers"       "Carriage"      "Tractors"    Head stock
manager     brigade leader  brigade leader  brigade leader  breeder
   |             |               |              |             |
Workers    Manual laborers  Workers in      Tractor drivers  ┌────┬────────┐
           and workers      charge of       and machinists   Hog farm   Dairy farm
           caring for       horses and                       brigade    brigade
           plants (field    horse-drawn                      leaders    leader
           crops and        equipment                           |          |
           garden brigade)  (Jakabszállás)                   Pigherds   Dairymen
                                                             (Jakabszállás)
```

FIG. 4.2. Organizational structure of Harmony Farm (1980s)

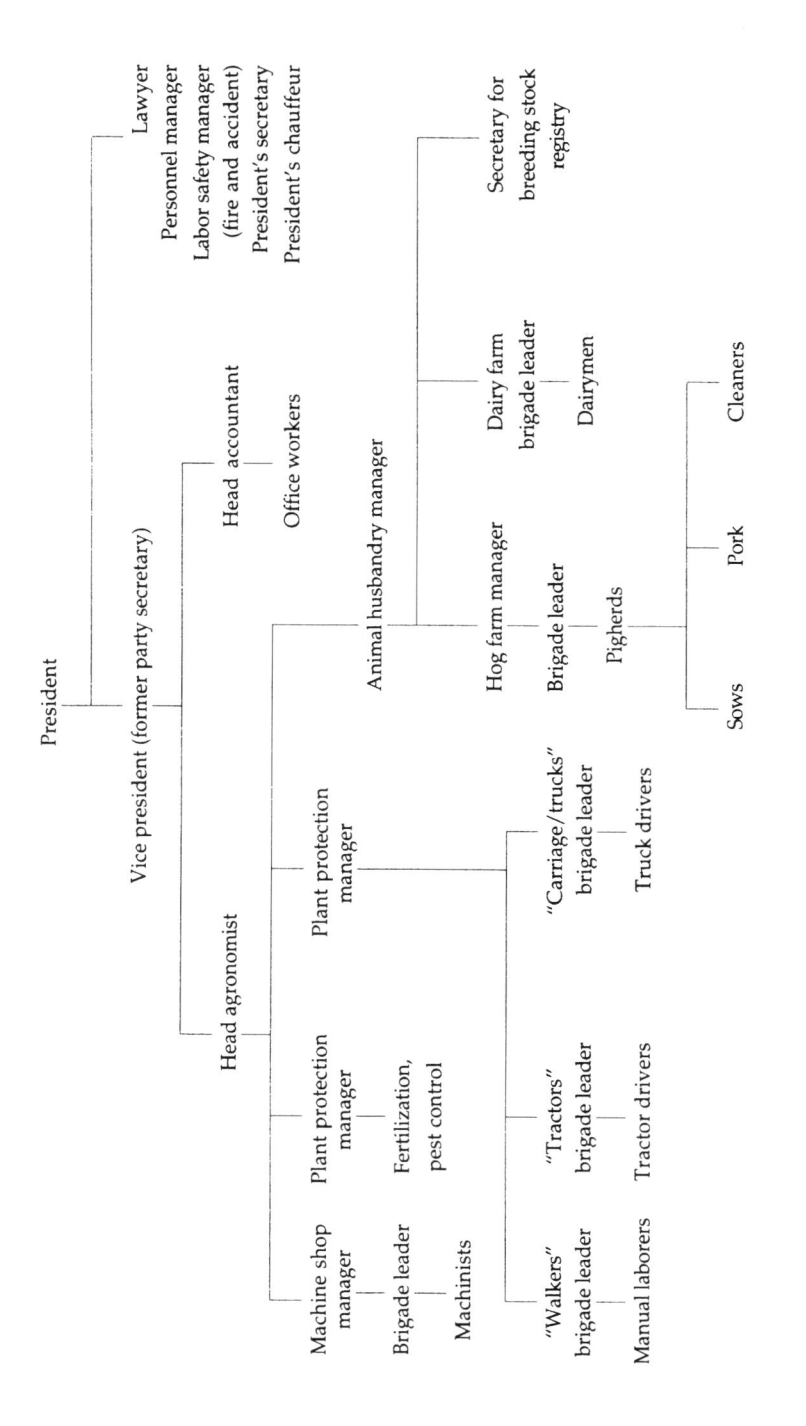

```
President
│
├── Lawyer
├── Personnel manager
├── Labor safety manager (fire and accident)
├── President's secretary
├── President's chauffeur
│
└── Vice president (former party secretary)
    │
    ├── Head agronomist
    │   ├── Machine shop manager
    │   │   ├── Brigade leader
    │   │   └── Machinists
    │   ├── Plant protection manager
    │   │   └── Fertilization, pest control
    │   ├── Plant protection manager
    │   ├── "Walkers" brigade leader
    │   │   └── Manual laborers
    │   ├── "Tractors" brigade leader
    │   │   └── Tractor drivers
    │   └── "Carriage/trucks" brigade leader
    │       └── Truck drivers
    │
    ├── Head accountant
    │   └── Office workers
    │
    └── Animal husbandry manager
        ├── Secretary for breeding stock registry
        ├── Hog farm manager
        │   └── Brigade leader
        │       └── Pigherds
        │           ├── Sows
        │           ├── Pork
        │           └── Cleaners
        └── Dairy farm brigade leader
            └── Dairymen
```

tive farms, they were to be confronted with the renaissance of private ownership and a pride in work in second-economy activities.

The transformation of management into a skilled elite achieved several goals simultaneously. It rehabilitated book knowledge, conferred greater powers on management leaders, and excluded larger and larger numbers of workers from decision-making. On more than one occasion managers justified the exclusion of workers from decision-making by saying that they were unschooled in modern agricultural techniques. When I asked why members would have to demonstrate an arcane knowledge of fertilizers to speak to problems of income differentials between management and workers, managers responded by explaining that workers' opinions were based on a limited understanding of farm management. Ironically, workers were criticized for their exclusion from general cooperative affairs, the explicit agenda of the technocratic elite.

The process of reorganizing cooperative management was also an opportunity for some managers to be excluded, but perhaps for other reasons. The story of one villager comes to mind. From 1960 until 1974, he had been a hard-working brigade leader at one of the cooperative centers in the village. He was respected and well liked, and extremely good at his job, able to figure the complex calculations of work units in his head, not needing pen and paper. He was known to be fair, and stood up for cooperative members when he thought they were being ill treated.[24] During the reorganization of the cooperative farm, the brigade leader was told he would have to take a test, to prove that he knew what he was doing in his job. He found this offensive, and suggested that twelve years of service to the cooperative farm should be enough to demonstrate his abilities. He fought this requirement for two years, continually refusing while the cooperative president consistently pleaded with him to take the test. He finally conceded, but flunked the test, having answered a question on the basis of coop-

---

[24] In fact, he chose to testify for several women in the village when they took the cooperative farm to court for not having paid their retirement benefits when they had sharecropping contracts. Not only could he testify to their having worked, he also remembered the yields of every crop they had produced over several years of sharecropping contracts. Refusing to pay for benefits for part-time members was a common tactic employed by cooperatives across the country in the 1960s. Luckily, however, many of those affected were able to recover their benefits, if they took the cooperative to court. Unfortunately, not everyone had the temerity to confront local leaders in this way. It was also courageous for the brigade leader to challenge the cooperative; indeed, some have suggested that the reason the brigade leader was removed from managerial ranks was that he had sided with the women in this suit against the farm.

erative farm practice rather than according to state regulations. To all observers, it looked as if the test had been rigged to ensure that the brigade leader would be ousted from his job. Infuriated, he resigned to become a pigherder, where he proceeded to earn twice as much money for half the work. The cooperative succeeded in excluding a loyal and hardworking manager, whose politics did not ally him with the new technocratic elite. Perhaps this was all to the good, since the brigade leader, one of the brightest people in the community, believed that knowledge had to be demonstrated in work, not in management offices or in the cozy confines of a hunting lodge. The brigade leader was not alone in finding the managerial ranks uncomfortable. Another villager was approached to become a brigade leader at the cooperative farm, but it would have required his attending a technical school. He decided against it, both because at his age (thirty-five) he would have found it strange to go to school, but also because he had plenty of work at home with his second-economy projects. His attitudes about managers surely influenced his decision as well. "According to the old *cseléd* talk, a good worker will not be a good boss, but a bad worker will become a good boss, because he will tell people what to do so he won't have to work."

The radical separation of groups in production, as a result of the new branch system (*ágazati rendszer*), essentially eliminated the institutional structures for democratic decision-making among cooperative members (Donáth 1977, 269), and membership involvement in planning was further eroded by the new ideology of managerial expertise. The difficulties of organizational change were captured in a newspaper article reporting on the annual meeting of Harmony Farm in 1974. The article was entitled "Complete Agreement in Sárosd," but the final comments disputed this claim.

> After a difficult year, however, many tasks still need to be accomplished. We must concern ourselves more with the socialist competition movement, must rectify work discipline and the quality and quantity of work. Managers in special areas must feel greater responsibility. The 1974 year plan of the cooperative of Sárosd wishes to correct work organization. The relations between managers and the membership, and last but not least the relations among managers will be the key to development. (*Fejér Megyei Hírlap*, February 13, 1974)

Changes in cooperative farm organization and managerial hierarchy were viewed with concern by many villagers, but there was little they could do to halt the rapid process of their disenfranchisement. In discussing this change, one villager commented: "When we first started up collective farming, it was difficult, but since we had no choice, we felt we might as well make a go of it. We had a sense that the farm was our responsibility. Now they can no longer make us believe that the farm belongs to us."

Conclusion

This chapter chronicles the complex shift in socialist policy and politics which took place between 1957 and the early 1980s. Collectivization, mechanization of agriculture, economic reform, the growth of the second economy, changes in wage policy, and the rise of technocracy significantly altered the character of agricultural production and village life. All of these processes required a serious reformulation of socialist policy and entailed a substantial restructuring of social institutions.

Agricultural production was restructured with the final phase of collectivization. The state effectively appropriated virtually all agricultural landed property, tools, and livestock, reducing those who remained in agricultural production to the status of laborers bereft of property. Though cooperative farms maintained the legal fiction of collective ownership, the agrarian workforce had essentially been fully proletarianized. The party's success in achieving collectivization was not matched with success in fashioning a willing and able workforce. Massive resistance among agricultural workers forced a significant concession to family farming practices with the institution of sharecropping contracts.

The eventual success of mechanizing production restructured work, contributing to a significant shift in the composition of the workforce. No longer characterized by brute physical effort and hardship, cooperative farmwork came to be far more attractive than in previous years, certainly than in the early years of cooperative production, and in some ways more attractive than in the arduous years of private farming. The new technical skills required of the workforce in dealing with machinery and the new chemical revolution in fertilizers and pesticides undermined the former control of peasant elites who had managed production from the inception of most cooperative farms. Their practical knowledge in farming was dismissed as less and less relevant to the images of modernized production promoted by the state. Thus the class divisions of the prewar period—divisions that were based on the assumption that greater knowledge of farming was held by those who owned property—were no longer the primary means of discriminating among members of the cooperative workforce. Far more salient were the social relations structured by skill and managerial expertise, that is, new kinds of knowledge and novel practices tied to the modernization of agricultural production which would differentiate villagers within the political and economic hierarchy of the cooperative farm.

The increasing importance of mechanization and modernized production methods in cooperative farming is often associated with the organizational shifts in cooperative farm production introducing greater hierarchy and complexity to the enterprise. However, the rise of technocracy and the ascendancy of a managerial elite within the cooperative farm cannot be attributed simply to the demands of technology. The rise of a technocratic elite constituted a substantial restructuring of knowledge and power intimately associated with the economic reforms of the party and the development of the intelligentsia as a class in socialist Hungary (Konrád and Szelényi 1979). New notions of expertise and new habits of management fundamentally restructured the political landscape of the cooperative farm.

The practices of evaluating and remunerating labor were restructured as well. Wages were redesigned at the cooperative farm, bringing in hourly wages, piece rates, and monthly salaries. Markets—in labor and in commodities—were strengthened. Labor, time, money, and objects were increasingly tied conceptually and in practice, as equivalent and interchangeable qualities. While formerly disparate qualities and quantities were increasingly equated within the complex relations of markets all across the body politic, so too the discrete integrity of different domains became increasingly important politically. Producing in the private sphere of home and family became increasingly viable, promoted by state policies it also became increasingly preferable, as it reinvigorated a private, personal economy against the state-controlled sector. While the site of family production was seen to exist outside the oppressive constructs of state production, it nonetheless increasingly became commoditized, demonstrating features of the markets in commodities visible in the primary economy. Gender relations within the household also shifted, as the commodified character of second-economy activity contributed to the demotion of women's labor in managing household affairs. The rise of household-based production, encouraged by the reform, gained even greater impulse from the political disenfranchisement of the cooperative membership at the farm, and so the significance of private activity increased in direct proportion to shifts in the cooperative sector.

GALLERY

1. Wheat harvest at a manorial estate in Ercsi (Fejér County) in the 1930s. The man cutting the wheat with a scythe is followed by a woman who gathers the wheat with a sickle (swath-layer) and a woman who binds the wheat into sheaves. Courtesy Néprajzi Múzeum, Budapest.

2. A woman carrying a sheaf of wheat, 1930s manorial harvest. Courtesy Néprajzi Múzeum, Budapest.

3. Wheat harvest in the 1930s. Wheat is stacked high on a wagon drawn by oxen. Harvesters are seen leaving the field, surveyed by a wealthy woman seated in an elegant carriage. Courtesy Néprajzi Múzeum, Budapest.

4. The poster, dating from 1954, exhorts people to submit their wheat to the state directly after threshing. Advantages are offered for quick delivery, including monetary bonuses. Failure to fulfill delivery obligations brings sanctions, increasing one's obligations and levying additional duties. Courtesy Országos Széchenyi Könyvtár, Budapest.

5. The party/state, personified as an agrarian worker, uses a strong arm to keep wealthy peasants from joining the cooperative farm. The text reads: "The agricultural cooperative belongs to working peasants, women, and youth" (1950). Courtesy Országos Széchenyi Könyvtár, Budapest.

6. Evening classes are being advertised to teach "working peasants" better cultivation techniques and to train people in new socialist thinking. Courtesy Országos Széchenyi Könyvtár, Budapest.

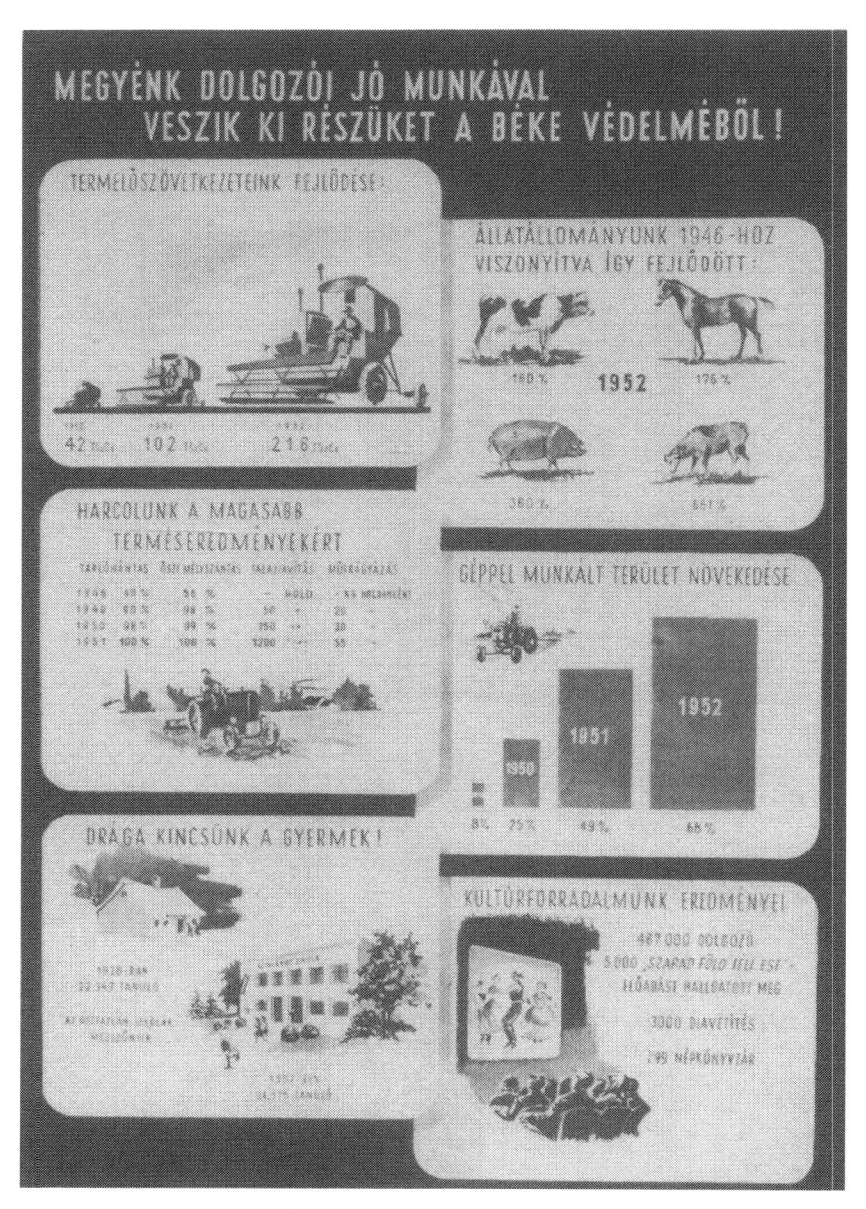

7. A common genre of socialist iconography is the depiction of numerical increase in factory production, national services, and cultural activities. Entitled "The Workers of Our County Do Their Share in Protecting Peace with Their Good Work" (1953), the poster boasts growth in a variety of domains: an increase in the number of cooperative farms, better cultivation of the land, more students in school, an increase in the size of animal herds, growth in the acreage worked by machines, and advances in socialist cultural services. Courtesy Országos Széchenyi Könyvtár, Budapest.

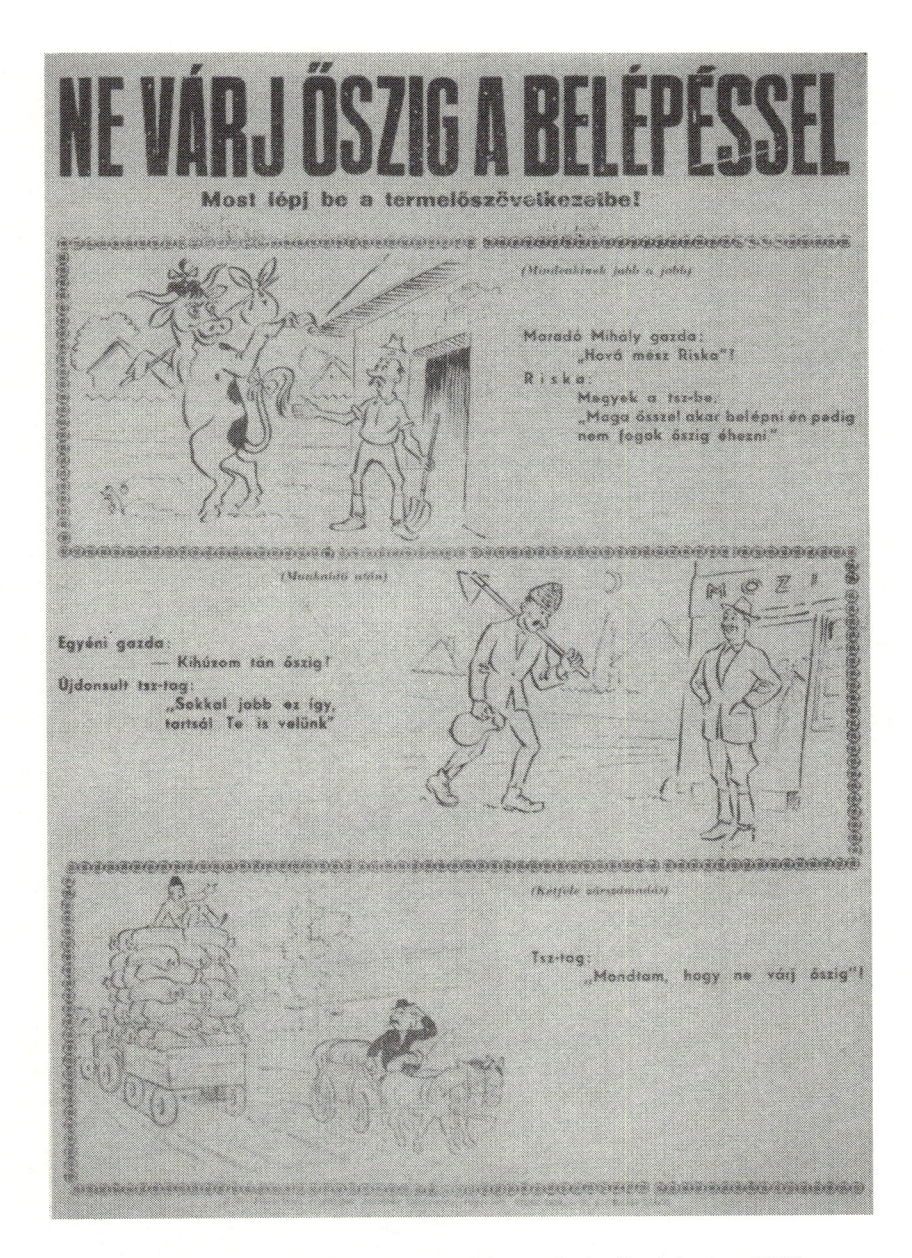

8. This poster, which dates from the second period of collectivization (1959), uses humorous stories to convince peasants of the value of joining the cooperative farm, depicting cooperative members doing less work and earning higher yields. The text reads: "Don't wait until fall. Join the cooperative farm now!" In the first segment, entitled "The better is better for everyone," Farmer Mike Stay-Behind asks his cow where she is going. "I'm going to the cooperative farm. You want to join in the fall, but I will not starve until then." Courtesy Országos Széchenyi Könyvtár, Budapest.

9. Pigs being weighed at the cooperative farm in preparation for their transport to the stockyards. Photo by author.

10. Women in the garden brigade sorting potatoes at the cooperative farm. Photo by author.

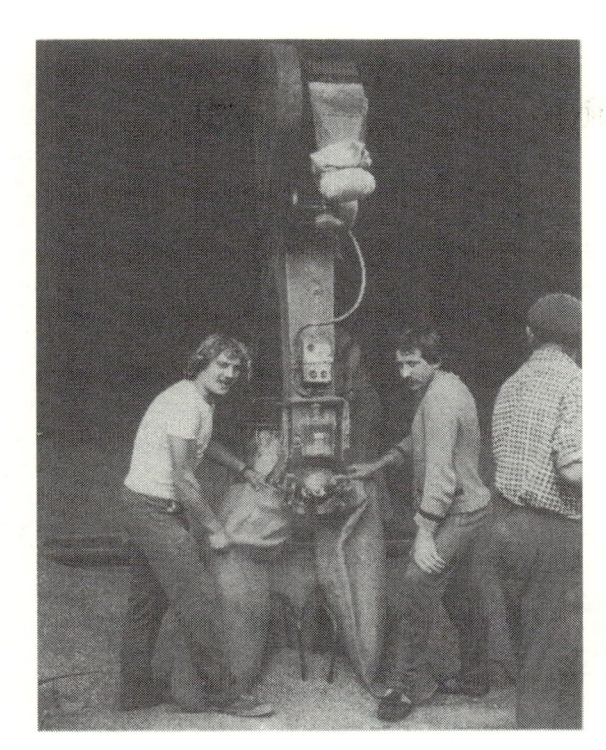

11. Two cooperative farm workers holding sacks while a large mechanical sieve sorts dry peas. Photo by author.

12. Pigs being herded from one barn to another by cooperative farm members. Photo by author.

13. Women in the garden brigade cutting broomcorn to make brooms for the cooperative farm. Photo by author.

14. Cooperative farm members guiding sheep through a special disinfectant (sheep dip) to clean their spring coats. Photo by author.

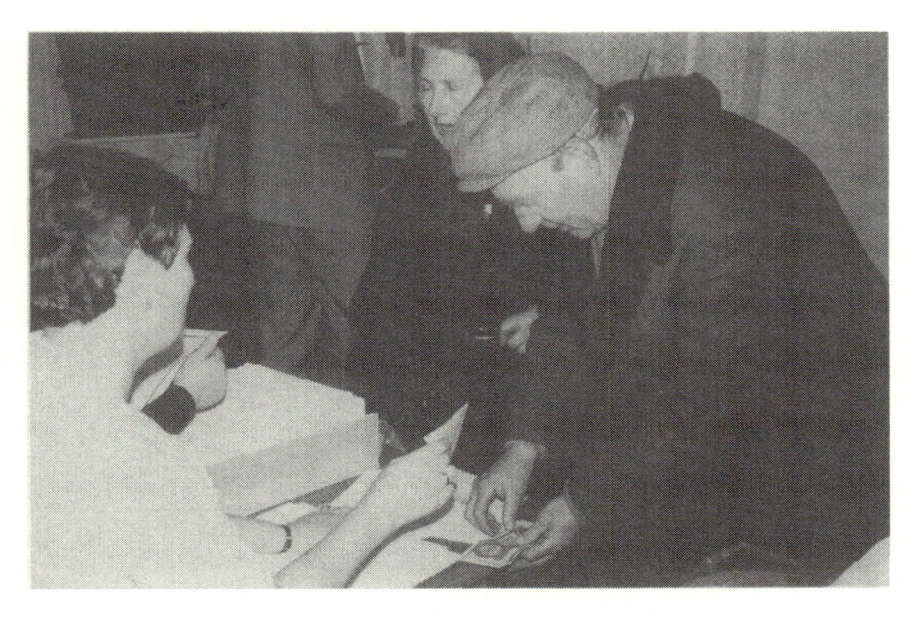

15. Cooperative farm members at the central office collecting their bonus payments, rent, and other benefits paid annually. Photo by author.

16. Annual meeting held at the local restaurant. The lack of enthusiasm on the faces of the cooperative members expresses their sentiments toward voting, and the annual meeting generally. Photo by author.

17. Women in the garden brigade of the cooperative farm hoeing row crops in early summer. Photo by author.

18. A couple hoeing their backyard garden in springtime. Photo by author.

19. Cleaning out a pigsty in the family's backyard. Photo by author.

20. Shoveling corn into the family's corncrib to dry during the fall months. Photo by author.

21. Harvesting beans grown in between rows of corn on a private family plot. Photo by author.

22. Processing paprika in the family kitchen. Photo by author.

23. A family helps haul bricks for building a private home. Photo by author.

24. Butchering chickens for a wedding feast. Photo by author.

25. The local blacksmith sees to his private customers. Photo by author.

26. Tending a private vineyard. Photo by author.

FIVE

Planning as Science, Planning as Art

> My unforgivable sin had been to exchange the proceeds of string-pulling for money and thus to disregard the advantages of the "planned" economy over those of the market.
>
> Kenedi 1981, 13

In Sárosd the terrain of politics was uneven. The political landscape was marked by rifts and fissures: rifts over the stated goals and true meaning of actions taken, and fissures between groups within the community according to their relative control over public definitions and private pursuits. The rugged contours of political life were thus shaped by circles of power and influence, fragments of opposition, and stretches of disaffection.

The character of local politics is the subject of this chapter. I define politics broadly: it is the process whereby meaning is contested, reality constructed, and power negotiated. In this chapter, however, my concerns are narrower: I will examine, in particular, the political bases of economics and material relations. Specifically, I will consider how socialist planning has been conceived, that is, the assumptions that have fueled its particular character. I examine the social scientism of Marxism-Leninism, a factor that has far-reaching implications for how the economic goals of socialism have been constructed and realized.

The scientific project of historical materialism entailed a commitment to quantification, that is, to the numerical representation of social goods and the calculation of social progress. The abstract character of a quantified universe helped to obscure the inherently political character of planning and economic decision-making. However, the actual practice of fulfilling plans laid bare to most villagers the economic and political advantages held by elites. The process of developing abstract social goals to be achieved through planning actually facilitated schemes for the personal aggrandizement of those in charge: party bureaucrats, local officials, and farm managers. The representation of planning as serving the long-term interests of the people's economy contrasted starkly with the everyday project of enhancing one's personal fortunes and reaping significant individual gains among managers and party officials.

It was quite clear to villagers that the representation of planning and its everyday practice were far apart. Few had time for the public posturing of officials; rather, they dismissed planning as a scheme to reward some and punish others. Hence planning as realized in the workplace also strengthened views of human nature as ruled by the calculation of utility and the pride of individualism. The ironic consequence of planning in village life has been to intensify sentiments regarding the inherent value of immediate gain, of the very ephemeral character of economic transactions, and the necessity of cynicism.

Socialist Planning

Socialist planning has been the subject of much mystification. Traditionally, Western observers have caricatured it as intrusive, stifling, and unnatural, interfering with the superior mechanisms of market forces. To quote one famous advocate of market freedoms, "Wherever the state undertakes to control in detail the economic activities of its citizens, wherever, that is, detailed central economic planning reigns, there ordinary citizens are in political fetters, have a low standard of living, and have little power to control their own destiny" (Friedman and Friedman 1979, 55). Rather than being compared with the national budgetary process in Western European countries and the United States, to which it bears similarities, socialist planning has often been compared with an ideal notion of how the market and rational choice function under capitalism.[1] It is equally problematic, however, to view socialist planning within the Soviet bloc as simply a more complex form of the budgetary process characteristic of nation states, or of capitalist enterprises, in the twentieth century. In socialist societies, the plan acquired added significance as a tool, the scientific instrument of a state that portrayed itself as a vast apparatus executing scientific socialism.

It is curious that the avowed social science realized in the institution of Marxist-Leninist states has generally gone unremarked in

[1] In the social science literature, the tendency to compare socialist institutions (or institutions of other, non-Western, countries) with ideal types rather than with Western empirical cases is rampant, and interferes with any serious comparative work (Burawoy and Lukács 1985; Pakulski 1985).

the literature on contemporary Eastern European societies.[2] This is all the more surprising when one recalls the ubiquity, in socialist societies, of reference in newspapers, television shows, radio programs, and movies to the *scientific* principles of Marxism-Leninism, and to the unique role of socialist societies in the development of world history. Surely many contest the scientific validity of Marxism-Leninism, but this does not justify the consistent neglect accorded the question of science and politics posed by Communist ideologues and practitioners. The tendency to dismiss those aspects of socialist society deemed irrational or indefensible has contributed to a consistent misunderstanding of the practices of these regimes and the social consequences of their policies. The centrality of scientific principles in Marxist-Leninist ideology, and in everyday practice, has had a profound effect on the folk model of socialist politics. For with the transition to socialism in Hungary, politics became science.

In socialist Hungary, the manifesto, as social plan, was replaced by the dictionary. The reigning mental image of the angry Communist Manifesto, so closely associated in Western minds with socialist politics, became obsolete. In the scientific rationalism of Marxism-Leninism, moral suasion was superseded with precise definitions and assertions of universal laws. Party secretaries and propagandists did not mine the rich prose of Marx and Engels's stalwart proclamation; rather, they consulted handy desktop dictionaries, such as the fourth, revised and expanded edition of the *Political Pocket Dictionary* (1980) or the *Pocket Dictionary of Party Life* (1984) or the second, unrevised edition of the *Pocket Dictionary of Political Economy* (1984).[3]

This genre of instructional handbook had a very respectable pedigree in the socialist world, dating to the volume written by N. Bukharin and E. Preobraschensky in 1919 entitled *A.B.C. of Communism*. In the preface they explained the reason why they wrote the manual.

> The daily experience of propagandists and agitators has shown that a work of this kind has become a pressing ne-

---

[2] Barney Cohn first alerted me to the peculiar "social scientism" that characterized the socialism of Marxist-Leninist states.

[3] Another handbook was the "What Must Be Known" series, which featured editions such as *What Must Be Known about the 1977 Plan of the People's Economy?* or *What Must Be Known about Hungarian-Soviet Economic Cooperation?* or *What Must Be Known about the World's Unions?* The shift from the question "What Must Be Done" to "What Must Be Known" is quite telling, indicating a shift in the party role from social critic to omniscient leader.

cessity. . . . The old Marxist literature—the "Erfurt Program," for example—is obviously no longer serviceable; . . . We look upon the "A.B.C." as an elementary course which will be used in the party schools. (1921, 5–6)

Bukharin and Preobraschensky designed the *A.B.C.* to serve a broader audience, beyond the cadre in party schools, writing in a form accessible to all workers or peasants who wished to acquaint themselves with the "task and goal of Communism" (1921, 6). Throughout the text crucial phrases and sentences were highlighted to allow the reader to skip over lengthy discussions, to reach the axioms of the argument itself. The dictionary form—a compendium of axioms—followed from the character of argumentation and assertion characteristic of the *A.B.C.*

The advance of the dictionary over the treatise exemplified the advance of Marxism-Leninism over earlier socialist programs. The surety of scientific accuracy was coupled with a knowledge of historical destiny. Advance itself became axiomatic. Witness the following definition of Marxism-Leninism:

> Marxism-Leninism: a scientific *theory* addressing the most general laws of the development of nature and of society; it concerns the revolution of oppressed and exploited masses, above all the working class, and the victory of socialism and the building of Communist society. It is the scientific *worldview* of the *working class*, the *ideology* of the revolutionary *workers' movement* and the theoretical basis of *Communism*. . . . The three constitutive parts of Marxism-Leninism are: 1) *dialectical and historical materialism*; 2) political economy; and 3) the theory of *scientific socialism*. (Dús 1984, 82–85; italicized terms appear as separate entries in the dictionary)

Marxism-Leninism, then, was a theory both of nature and of society, and also the worldview of a social class. But it was above all a science. Moreover, it was the ultimate, demystified worldview, the true science, clearly expressed in the definition of dialectical and historical materialism:

> [D]ialectical and historical materialism: the philosophy of Marxism; the unified, consistently materialist and dialectical philosophical study revealing the most general laws of nature, society, and thought; the *worldview* of the revolutionary *working class*, of Marxist-Leninist parties. Dialectical and historical materialism is not solely the scientific, true reflection of the most general relations of reality, but—by virtue of this—it is also the method and

tool of the revolutionary transformation of reality. (Dús 1984, 23)

The scientific program of dialectical and historical materialism was founded on the relation between knowledge and action, between consciousness and revolutionary practice, an absolutely central tenet of the Marxist tradition. The genre of dictionary and handbook was significant, then, not only as an index of the presumed exactitude of socialist principles, but also as it was understood to be a tool for the transformation of consciousness, for the enlightenment of socialist citizens. Repudiating the idealism of other political traditions, Communist party officials understood these exercises in definition, debate, and deliberation to have immediate political and social consequences. Commenting upon the language of domination characteristic of Marxist-Leninist states, Fehér, Heller, and Márkus write:

> A strange feature of the language of Bolshevism could perhaps be best described as perverted *Lebensphilosophie*. The Party raises a claim not only to the exclusive interpretation of the theoretical heritage, but also to the expression of "life" as well. The Party embodies life in its real, profound sense. (1983, 198)

Language and life were intimately related; the proper use of revolutionary vocabularies would transform social experience. The ubiquity of texts, banners, and formulaic sayings in socialist rituals and everyday practice were not merely trappings of the regime—finery displayed to confirm party rule—but integral to the politics of transformative science. The *Lebensphilosophie* of the party was embodied and enacted in linguistic artifacts that altered consciousness.[4]

The dictionary and handbook gained further significance in a Leninist state, where it was assumed that an elite, trained cadre would lead the masses to socialism and communism. Bukharin and Preobraschensky made this point very clearly.

---

[4] Verdery argues that language occupies as central a role in socialist cultural production, as do intellectual products such as dictionaries, encyclopedias, and national histories. "For a Party bent on transforming consciousness, control over language is one of the most vital requirements. Gross captures an aspect of this when he writes that communist rule changes language so it no longer reflects or represents reality; metaphor becomes more important than prosaic discourse, and magical words replace descriptive and logical ones (Gross 1988, 236–238). But whereas Gross sees this as an aspect of the destruction of language, I see it as the retooling of language *qua* means of ideological production" (1991, 89).

> Our program must be known in all its details by every member of the party. . . . Only those who "recognize" the program (that is, those who believe it to be right) can be members of the party. It can be considered to be right only by those who understand it. . . . Without a knowledge of the program no one can be a real Communist-Bolshevik. (1921, 12)

Introductory statements that appeared in Hungarian editions of *The Pocket Dictionary of Party Life* and *Political Pocket Dictionary*, though far less strident in their tone, nevertheless stressed their educational value to party members, who should keep abreast of new and revised concepts, of "technical terms and categories suddenly emerging at every moment in everyday party life" (Dús 1984, 5).

Thus, the construction of socialism—the building of new economic forms, new political organizations, new modes of thought—was a scientific activity, based upon the laws of Marxism-Leninism. As scientific activity, it was presumed to exhibit regularities and consistencies in form. Its practitioners aspired to rational calculation, yet they were to be open all the while to incorporating the "suddenly emergent" knowledge of new experiences. The central arena for constructing these explicitly political goals was, of course, planning: in the planning of the national economy, and through the plans developed by individual enterprises to guide their activities.

Planned economy is defined in the fourth, expanded edition of the *Political Pocket Dictionary* (1980) as:

> One of the basic characteristics of the socialist economic system. . . . The [state] plan stipulates the overarching goals of economic development and the economic rules subordinate to these goals. The goal of a planned economy is to ensure the planned, proportional development of the people's economy. Meanwhile, it strives with optimal balance to help the most effective economic growth, coordinate interests of the people's economy, enterprises, and individuals, ensure the realization of economic-political goals, and by all this, guarantee the growth of the country's economy and national welfare. (Fencsik 1980, 337)

The tone of this description, with phrases like "optimal balance" and "economic growth," may suggest why some observers view planning as comparable to the budgeting process in capitalist societies. *The Pocket Dictionary of Party Life* characterizes the process of planning thus:

> [T]hat phase of *management*, in which they work out the goals of the organization, as well as—taking into consideration the powers, tools, methods available—determine the solvable tasks in the interest of reaching the goal (*goal and task*). In the course of planning they examine and indicate the different paths of the achievement of the goal. They fashion the harmonious working of the elements of the system. Planning is in close relation to *decision*, concretely an organic part of the process of *preparing decisions*. (Dús 1984, 156)

The pragmatics of calculating politics lies in decision-making, in managing tasks. "The goal is one, the tasks are multiple, and these must be solved simultaneously or one after the other" (Dús 1984, 20). The construction of socialism became the careful tooling of means, for the end is one. What else is the Reform of the Economic Mechanism?

The pragmatics of calculus was reflected as clearly in changes in planning strategies as in its classic forms. From the mid-1950s on, reform packages were crafted one after another, with successive modifications in structure and emphasis. As Kornai points out, the attempts to qualify economic projects were not meant to question the basic system of economic practice the party embraced, but only to alter various components.

> The initiators of [the "perfection" of control] . . . realize the classical system has encountered some serious inherent contradictions. They see that something has to be done to remedy the ills. Yet they remain convinced of the correctness of the classical system's basic principles and of its superiority. So they argue that all the difficulties are solely (or at least mainly) caused by a failure to apply the correct principles consistently enough. The system would operate substantially better if a few secondary principles, legal measures, and institutions were replaced by more effective ones, and resolutions applied more consistently, while the primary principles were kept unchanged. (1992, 396)

In the early 1980s, new legal measures were introduced, secondary principles reevaluated, and new institutions established, all in the name of socialist development. The degree of accommodation and adjustment was great, though judging by the constant theme of crisis in newspaper articles and essays, fully warranted. Nonetheless, the approach taken by party officials was to perfect varying components of the economic equation. Old saws about diligence and hard work were reiterated.

240   *Chapter Five*

Well Balanced

We must adapt more quickly and more resiliently to changing conditions, and we must make the economy of higher quality by taking advantage of our possibilities and building on our endowments. And of course we must work. Not from sunrise to sunset, but in an organized fashion, and diligently. More disciplined and in better coordination. (*Fejér Megyei Hírlap*, December 4, 1982)

But more sophisticated strategies were also employed, targeting organizational structures, modifying markets, and promoting limited forms of interest coalitions.

Resilient Adaptation

In the second half of the 1970s it became more and more obvious that the building and functioning of the organizational system of enterprises which existed until then did not create favorable conditions for attaining our economic political goals—since in the meantime the circumstances of the economy changed significantly. . . . It is evident that one must not consider the modified enterprise organizational system as a final state of affairs. We must carry on with the promotion of rational economic competition, eliminate monopoly, or at least significantly reduce it, since these all encourage the increase of the efficiency of the economy. Of these, one of the most important tasks is the modernization of the internal management system of enterprises. A key issue is making internal interest relations more open, bringing to light the success of various isolated units, and developing an internal incentive system proportionate to performance. (*Fejér Megyei Hírlap*, January 9, 1983)

All of these suggestions revealed a desperate hope in the redeemability of socialist economics.

Various components of the economy were to be perfected. But economic rationality was finally expressed in the process of quantification; the plan was an elaborate document filled with numbers and more numbers, which ensured that it had been rationally designed and constructed. The rational or scientific character of socialist planning rested firmly on the quantitative calculation of the goals of production and distribution. Plan indicators were numerical figures; goals were represented as percentage of increase, as tonnage produced. The counting of objects was also paired with the counting of time. The plan itself was always constructed in terms of a number of years: one-year, five-year, ten-year plan. The marking of time as the march of numbers contrasted sharply with the seasonal conception of time among villagers, a cyclical, circular movement foreign to the forward progress of the socialist economic miracle. The language of numbers derived from specific models of economic rationality and utility, in which the scientific calculation of action necessitates its quantification (Hacking 1990; Porter 1986).

The numerical calculation of action and value exemplified by planning had its precursors, moments in the construction of modern economics and politics. Certainly the cadastral survey of Franz Josef in the mid-1850s represented an initial attempt to calculate value, anticipate revenues, and restructure politics in the Empire (Lampland 1992). During the 1920s and 1930s, grand debates flourished among economists in Central Europe considering the relative advantages of planning versus markets, debates that continued to play a role in economic policies and especially in reform platforms throughout Eastern Europe during the socialist period (Kornai 1992, 479). Yet in the course of the debates no one questioned the basic premise that quantification was a necessary component of rationalizing economic practice. Indeed, the extent to which numerical calculation and scientific rationality were seen to be valuable tools for economic practice across ideological barriers is best illustrated by Lenin's fascination with Taylorism, piece rates (e.g., work units), and statistics as important components of industrial progress and economic development. Lenin appropriated those forms of calculation, which he saw to be progressive, modern, technologically innovative. His fondness for Taylorism is well known. Lenin believed strongly in the power of numbers to capture social reality, a belief reflected in his approach to centralized control and planning. His attitude toward statistical compilation was quoted in the Hungarian law on state statistics passed in 1953: "No kind of constructive work, no sort of planning can be conceived of without proper accounting. And accounting is inconceivable without statistics. Accounting without statistics cannot move forward. Statistics . . . is one of the most powerful weapons to come to know society" (in Balogh 1986, 473). In other words, the modern character of scientific economic practice was to be perfected under a socialist regime, taking the original contributions of capitalism one step further.[5]

---

[5] The progressive mentality of socialist history is often ridiculed in jokes, such as the following: "Capitalism is standing on the brink of destruction. In socialism we are one step ahead." On a more serious note, I might add, the hypermodernism of socialism is worthy of further analysis. (I am indebted to Barry Carr for first bringing this issue to my

The curious result of rendering the plan an infallible tool for economic calculation was that it *always* had to be fulfilled. Whether for the nation or for the lowliest of enterprises, the plan existed to be fulfilled. The predominant and significant tendency among planners and management was to proclaim the full realization of the plan, despite how frequently this goal may actually have been thwarted, and their proclamations of success were echoed in newspapers, magazines, and television shows. The necessity of always fulfilling the plan, then, seriously qualified the manner in which the plans were constructed in the first place, how they were fulfilled, and how they were handled as a central ritual in political life. The plan was reified, mystified as teleology; it was a process in which the means of state socialism became an end in itself. The relation between ideology and reality may have been tautological, but it was fundamental to the political culture of Hungary; it, finally, was what gave practice the illusion of praxis.

All efforts, then, were directed toward substantiating the true science of Marxism-Leninism. Planning entailed a constant process of bargaining and negotiation among agencies and institutions throughout society (Kornai 1959; 1980b) to ensure that the stated goals of the plan, of all plans, would appear to be fulfilled. The necessity that planning succeed resulted in widespread falsification and exaggeration of figures, which divorced the quantified, material expression of the plan from the substance of the economic behavior it was designed to coordinate (for a discussion of comparable problems in the Soviet Union, see Lewin 1974, 144–147). The quantification integral to planning required that supervisory agencies declare numbers that enterprises should realize, numbers that, once submitted, formed the basis of future plans. However, enterprises consistently attempted to resist fulfilling plan targets as allocated. In a context of rampant uncertainty—generated as much by changes in planning targets as by the difficulties of a shortage economy—it was in the interest of the firm to reduce expectations of their performance, to resist the "ratchet principle" of year-to-year planning (Sabel and Stark 1982, 447).[6] Since the goal of ful-

---

attention.) While many observers have considered the relation between early Soviet socialist art and modernist styles of the West, the more mundane consequences of a heightened modernist consciousness have not been sufficiently studied. For two exceptions to this rule, see Bauman's recent (all too brief) discussion of socialism and modernity (1990), and Fehérváry's (1995) analysis of Sztálinváros as a modernist city.

[6] "[For managers] to survive they must defend themselves against the unforeseen, and the most unforeseeable part of their job is the ability to meet plan targets assigned

filling the plan was sacrosanct, enterprise strategies of accommodating central expectations to their own conditions frequently entailed falsifying plan figures (see Lewin 1974, 144). Strange though it may seem, then, all those participating in the process of writing and fulfilling the plan actively and consciously separated appearance from essence, in a society where ideology—understood to be the expression of the separation of appearance and essence—was proclaimed no longer to exist. Thus, the reification of planning and its teleological mystification pervaded Marxist-Leninist ideological discourse and the politics of economics alike.

In socialism, planning constituted the quantification of social good; it was the calculus of political practice and so its reification.[7] The plan became an actor, a historical subject, that could "strive . . . ensure . . . guarantee" (Fencsik 1980, 337). Moreover, the logic of utility and rationality came to be seen as necessary structuring principles for actions taken by the state and by individuals, rather than as ideological assumptions that were taken from positivist economics and the scientism of politics. During this century, quantitative techniques had become the dominant means of economic calculation (McCloskey 1985), so it is not surprising that Marxist-Leninist planners employed these techniques as well. Perhaps more surprising was the degree to which economists came to identify planning as a technique that could be improved independently of the political process of which it was so centrally a part. This confusion could arise only if one assumed that the technique of calculus was a transparent, neutral medium for achieving one's goals, divorced from the broader social and cultural project of politics and power in which it was embedded. Kornai recalls the problems researchers confronted when attempting to devise sophisticated mathematical techniques and computerized programs for the planning process.

> The pioneers of mathematical planning hoped their methods would be welcomed. Their disillusion was all the

---

to each enterprise by the central ministries. The better they are at fulfilling their obligations one year, the more likely they are to face a stiffer one the next. Their natural inclination, like piece-rate workers on an assembly line, is to hold back production and understate their capacity. But this is only a stopgap measure, since the planners can anticipate this defense and impose higher standards" (see Sabel and Stark 1982, 447–448).

[7] Vajda's discussion of the congruence of Husserl's phenomenological and Lukács's Marxist critique of science has been helpful in structuring this argument (1983), as has Lukács's seminal criticism of orthodox Marxism and positivist science in *History and Class Consciousness* (1971).

greater when the planning bureaucracy, which often used rough guesswork, rejected their refined, lucid range of devices as an alien body. If one considers the actual sociology of the planning operation, not its declared principles, the resistance is understandable. The political leadership has no wish to state openly its real political goals in the form of "welfare functions" and "planner's preferences." All members of the planning apparatus strengthen their negotiating position in both vertical bargaining and horizontal reconciliation by keeping some items of information back and distorting others, upward or downward, according to their own interest. Complete candor would make their own lives harder. The aim of their calculations is not to serve any absolute "social interest" but to support their own viewpoint. That can only be hindered by careful outside checks and collation of their figures before they are fed into a computer; their manipulations are upset by the inexorable logic of a system of mathematical equations. (Kornai 1992, 404)

Sorely lacking in the discussion of planning, or the science of economics more generally, is attention to the political consequences of calculation, an attention to the final consequences of means, of practice, of action. The construction of knowledge is a social project: the "inexorable logic of a system" of politics and epistemology at the center of planning.

The embrace of science and economic calculation as the path toward rationality rendered the political project of planning invisible. It removed planning from the ongoing debate over social values and placed it in the realm of scientific experts. It was no longer amenable to critique by lay people. Yet to take the plan as solely or even primarily economic calculus was to confuse its representation with its goal: "the preferences of the economic policy are dictated not by considerations of profitability, but by the criterion of how far the apparatus of power retains a direct control over the means invested" (Fehér, Heller, and Márkus 1983, 67). The control of the party over decision-making was bolstered over time with an increasingly sophisticated notion of intellectual knowledge (Konrád and Szelényi 1979; Szelényi 1982). Konrád and Szelényi's critique of socialist economics as "rational redistribution" identified the crucial role intellectuals played in setting societal goals and distributing surplus. However, their analysis did not directly address the techniques of calculation used to enhance the position of intellectuals as a new class. I believe it is important to examine not only how intellectuals structured the process of rational redistribution under socialism, but also to consider the epistemological bases of the class power of intellectuals. If we consider the epistemological question of calculus and politics, then in this respect intellectuals involved in the economic sciences share many attributes across the socialist-capitalist divide.

The reification of planning and economic rationality found its expression in the near universal dismissal by villager and city dweller alike of ideology and politics. Opinion was essentially unanimous: there was no role for socialist ideology in Hungarian society. I was repeatedly told to abandon my search for the part played by socialist principles in everyday life, and to look to economics, the state's true and only concern.[8] The construction of intention and purpose in terms of economic utility was perceived to be natural, indeed *human* nature, and thus was seen as the basis of individual and institutional actions. Everyone simply denied that politics had any role in Hungarian society, quite an ironic comment considering its Marxist-Leninist pedigree. Yet by the 1980s, everyone had come to see economic strategies as dominant, identifying the state's preoccupation with economic fine-tuning as the means of socialism, overshadowing politics entirely. When I challenged people on this point, they would concede that certain basic principles of socialist politics still held sway: the leading role of the party, limitations on private property, the absence of a profit-driven market. Despite this recognition, however, they found these constraints to be relatively minor and uninteresting. Yet this very refusal to consider politics significant played into the means of legitimation employed by the Kádár regime: politics was denied, as if the party were now engaged only in innocuous games of economic calculus.

Villagers pointed to the manner in which the plan was fulfilled at the cooperative farm and other local enterprises, seeing very little in the way of socialist conviction and much in terms of personal aggrandizement. Yet economists in Budapest also ridiculed my concern with the hows and whys of planning, assuring me that plans were not written to be fulfilled. This response would seem to fly in the face of everyday socialist practice. How does one reconcile this attitude with Kornai's description of the planning process?

---

[8] Rofel describes a similar turn in China during the past decade. "State discourse in China on the Four Modernizations . . . has inscribed its version of the universal efficacy of science, progress, and rationality. There was a rigorous attempt to create a 'factual' separation of economics and politics, as the state urged the populace to leave off with ideology and 'seek truth from facts'" (1992, 95).

[O]ne can state that elaboration of the plan is a monumental piece of bureaucratic coordination aimed at prior reconciliation of the processes of the economy. Thousands upon thousands of functionaries in the party apparatus, the state administration, the firm and cooperative managements, and the mass organizations negotiate, calculate, renegotiate, and recalculate before the millions of planning commands finally emerge at all levels. . . . The starting point for discussing implementation of the plan is [that] . . . implementation is compulsory. This principle is so important that some authors have referred to it as the most characteristic feature of the system by attaching such names to the system as "directive planning," "imperative planning," or "command economy." (1992, 114)

At all levels of planning projects, the command to fulfill the plan was crucial. For economists and so many others involved in the everyday project of planning, however, finagling, accommodation, and adjustment clearly dominated their experience, obscuring the public representation of plan fulfillment.

All throughout the socialist period in Hungary, a stark opposition was drawn by Hungarians between public and private domains, between what was seen as the posturing of vacuous politics and the real, substantial, truthful site of the home and hearth. Thus, although virtually every adult worked in the public domain, and every child attended a public school, everyday practices in the public sphere were dismissed, seen to be devoid of substance or significance. What happened at home, behind closed doors or in one's back yard, was what was truly significant and formative of one's identity. Everyone still engaged in public activities: wrote and fulfilled plans, built socialism in the factory, and learned lessons about party heroes in school. Once they returned home, however, Hungarians ridiculed the plan, criticized their bosses, worked fervently in their private gardens, and forgot as much of the Communist moralizing preached in school as possible. Adamant about the contrast between public and private worlds, few recognized that the assumptions one held about human nature, personal utility, and individual interest permeated the entire social fabric. The economistic thinking of the planning agency and factory director also characterized the scheming of the entrepreneur in the second economy and the villager out to wrest a favor from local officials. "You scratch my back, I'll scratch yours" (*Kéz kezet mos*) permitted everyone to manipulate public affairs for private gain. Yet while they actively attempted to subvert the system, Hungarians also maintained the inequalities the system structured, thus reproducing the system itself. People felt powerless to change the system, unable to withdraw entirely, and thus regularly participated in its reproduction. Forced to follow directives they did not believe in and politics they found alien, Hungarians kept a singular distance from their own actions day in and day out. Meanwhile, Western markets and capitalist societies represented havens of normality and reality. Ironically, for Hungarians, their own actions at work and at home veered closer and closer to the cultural universe and social pragmatics of capitalism, while they continued to feel straitjacketed by the empty politics and productive economics of socialism.

The widespread skepticism Hungarians demonstrated toward planning and politics is best explained if we examine how planning occurred in actual practice. It is incumbent upon us to examine more directly how a project so universally ridiculed was regularly practiced and hence contributed to the reproduction of cynicism, irony, and despair.

Planning at Home

How was the plan actually constructed and realized at the cooperative farm? How did the concepts of rationality and utility influence cooperative affairs and private activities? What was the nature of local-level politics? The construction of the yearly production and marketing plan for the cooperative farm was undertaken by farm management, discussed by the directorate council (an elected body of representatives from the wider membership), and finally approved by the membership during the annual meeting. However, the members of the directorate council, and the broader membership, had no real say in planning decisions. With rare exceptions, senior cooperative management made all decisions and the forum of directorate committees and annual meetings was staged merely to accord with the legal requirements of cooperative "socialist democracy."

The cooperative farm scheduled a regular series of meetings at the end of the fiscal year, to be held at different levels of the organization, in order to address questions of the membership who worked in various branches and units of the farm and to acquaint farm members with the results of the year's efforts. Meetings were also held for retired members, to keep them informed of current policy and changes in farm operations. Serious effort was expended to arrange these meetings, but no measures were taken to ensure participation by cooperative farm members in wider decision-making. Management followed the formal rules of enterprise democracy, but no one among the physical labor force took these meetings seriously. Never once did I hear anyone question the plan—in detail or overall structure—and only on one occasion did any worker raise a question about cooperative farm policy. As I discussed problems of enterprise democracy with one of the more enlightened managers of the farm, he explained that members "had been weaned of the habit of speaking up." When I suggested that workers could be encouraged to participate more actively by management, he resisted, explaining that the average worker at the cooperative farm was not competent to judge the complex problems of farm organization, the scientific problems of chemical fertilizers, or other innovations in agricultural economics. Planning and farm management had to be left to the technocratic elite, whose broader, more comprehensive understanding of economic and scientific affairs made them better qualified to run the farm than the physical workforce.

Among workers, debate about farm affairs was lively. They did not find technocratic ideology convincing, since they were continually having to deal with the negative consequences of poor management, bad organization, and inadequate training. The strong antiintellectual bias of villagers was reinforced by practices at the cooperative farm which encouraged managers to pursue advanced training, but did not require them to demonstrate new knowledge or altered practices at the workplace. Managers were regularly enrolled in courses issuing diplomas and certificates, which were used as justifications for their positions within the hierarchy, positions they had held long before acquiring further education. Although these managers had the requisite diploma, their judgment and abilities were rarely informed by the training they received, since they had been enrolled only to justify the position they already held. And even if managers had acquired training in legitimate ways, villagers rarely believed that book knowledge could supplant or excel that found in practical, everyday experience. In the opinion of one villager, describing a new pig barn that had numerous structural and design problems: "Those who designed the fattening barn were well-educated, that's for sure, but it's also absolutely certain that they didn't know what they were doing." When discrepancies in pay scales between management and physical workers were criticized by the membership, I was told that cooperative leaders "dangled diplomas in [their] faces."

Although workers did not speak up in official meetings, they were keenly aware of the complex strategies employed by management to represent the cooperative as solvent and successful. At one annual meeting, the president was reading off a litany of fulfilled plan goals: wheat yields, corn yields, prices paid for various crops. Overwhelmed by the monotony of numbers flowing through the air, and not very schooled in spotting inconsistencies, I sat quietly and listened. As the president droned on, one cooperative member sitting next to me leaned over to point out an apparent discrepancy between wheat yields and sales. "Isn't it strange that the goal for reaching yields in wheat was met, but not for sales? How could it be, that if we fulfilled the goal for wheat production, we fell short of the plan in sales?"

Since there was a fixed price for wheat, it was clear that the discrepancy between plan goals and year-end results had been snuck in under the rubric of sales, not production.[9] This would fit conveniently with the overall attention to brute increase in plan fulfillment, and socialist ideology generally, which paired a fascination with bulk with a disregard for the subtleties of markets and pricing. Such strategies, however, did not bolster the confidence of cooperative farm members, who were reinforced in their belief that management merely produced numbers for the sake of show, but not to reflect the actual practices of managing collective resources. Moreover, since cooperative management never took the time to explain reasons for discrepancies, the attitudes of physical workers were never shaken.[10] It was hard to sympathize with managers and local party officials when they complained about the difficulties of organizing and motivating villagers. The indifference felt by so many had been fostered by patterns of political practice that consistently excluded villagers from a say in their own affairs.[11]

---

[9] This was not the only occasion when a villager generously called my attention to discrepancies between representation and reality. At a village meeting to elect the new members of the water board, nominations were solicited from the floor, giving every indication of a democratic process. The selection committee retired to a back room, and several minutes later returned with a typed ballot. The cooperative lawyer, sitting next to me, whispered in my ear, "Did you hear a typewriter? I didn't."

[10] Conceivably problems with the quality of wheat or with drying it sufficiently may have influenced the price, but these fine points were never explained to the membership. This furthered their conspiracy mentality, and taught them nothing about the vagaries of markets and prices.

[11] The Romanian public meetings described by Kideckel bear a strong resemblance to those staged in Sárosd. "[A] general assembly of a collective farm or a consumer cooperative was fraught with inter- and intra-class tensions and was a raw exercise in symbol manipulation by state authorities. . . . Speeches by the local elite and visiting dignitaries were replete with household metaphors suggesting an identity of interest

Long-standing attitudes about the dangers of speaking up on one's own behalf strengthened management's ability to exclude the physical labor force from participating in decision-making or from influencing issues of organization or tasks at the farm. Many Hungarian proverbs warn against attempting to rectify wrongs or speaking up against injustice, perhaps the most frequently quoted being "If my mouth doesn't open, then my head won't hurt" (*Ne szólj szám, nem fáj fejem*). Reviewing recent Hungarian history confirms the sad wisdom of this adage. But some at the farm found this reticence quite stifling. Describing the crazy organizational innovations his boss had initiated at the farm, one villager bemoaned the silence of his fellow workers. "Everyone keeps still, but then [the management] can trample them even better. Since so many of the workers are in their jobs because of friends or kin, they prefer not to endanger their good jobs. So they don't say a word." Another worker, renowned for making his opinions known to fellow workers and bosses alike—regardless of the consequences—explained very simply: "One man alone is never right. The others don't say anything, so if I say anything, it's in vain." When people have been convinced of their vulnerability for so long, the politics of exclusion is easily achieved.

Although the yearly plan consisted of goals for production and marketing set for the upcoming agricultural season, it also incorporated decisions stemming from longer-term considerations, such as the sequencing of grain cultivation dependent upon crop rotation, the amortization of machinery and credit obligations, or shifts in the remuneration and taxation of the labor force. Cooperative management would spend four to six weeks compiling figures, calculating costs, and balancing competing considerations in production, credit forms, and investment. A hallmark of the economic reform was its emphasis on decentralized decision-making. Increased independence of control over production goals and investment decisions, intended to improve the profitability of enterprises, was considered to have been most successfully realized in cooperative farms. However, county and state agencies continued to influence decision-making in several crucial ways, and, in so doing, interfered with attempts to improve economic planning and profitability.

State agencies were directly involved in setting policies concerning wages and income taxes at the farm, and also in facilitating

---

between them and the villagers. The split between them, however, was all too visible" (1993, 193).

access to funding, such as credit sources. Social security and retirement provisions were determined by national agencies. Policies on the remuneration of the workforce were also stipulated by national legislation, and enforced by steep taxation levels on percentages in wage increases at the farm. Shifts in policy allowed cooperative farms some leeway in the composite payment schedules of the farm. For example, for a brief period in the late 1970s, taxation was levied only on the average pay of cooperative membership. In response, many factories and cooperative farms, including Harmony Farm, increased the pay scale of certain skilled (primarily male) workers, and hired many unskilled (primarily female) workers at very low pay, offsetting total labor costs. Legislation was later changed, shifting attention from average pay to individual incomes.

Managers frequently complained about the difficulties that labor taxation could cause for profitability. They felt unable to provide adequate monetary incentives to the labor force, and so were forced to hire more people at low pay when it would have been more effective to hire fewer people and pay them a decent wage. Efficiency would have improved, as would have morale. Wage scales were significantly influenced by state policies, policies that tied inflation to consumer spending, and so led to the suppression of pay increases. Wage increases were discouraged by introducing a progressive tax that was scaled to increments of increase. Nonetheless, during one period of restricted wage increases, top management voted themselves a substantial raise, and so incurred an increased tax burden for the farm—all the while explaining to the rest of the labor force that the wages of the physical labor force could not be increased owing to strict penalties in the legislation. This sort of behavior reinforced the workers' attitude that regulations were consistently manipulated for the benefit of top management.

The issue of credit sources highlights an additional and important aspect of outside intervention in decision-making and planning. Credit was distributed by county agricultural agencies. Although the rise of subsidiary industrial units at cooperative farms reduced their dependence on state loans, credit still remained an important resource for the development of new branches of production, for the improvement of machinery, and for other investment possibilities seen to enhance the profitability of production. Control over this resource, therefore, was a crucial form of control over individual farms or enterprises. As a result, county agricultural offices continued to have a significant role in planning. Within the county, cooperative farms vied with one another for attention, hoping to gain the support of county personnel. County officers' pet projects, and even styles of management, were soon emulated in the hope of financial benefit.[12] Thus, the choice of farm projects was strongly influenced by considerations other than profitability, or the specific needs of the enterprise itself. Needless to say, this kind of influence also extended to crucial decisions regarding the hiring and promotion of personnel within the farm, a point to which I shall return.

County offices intervened in a variety of issues at the local level. Beyond specific regulatory functions, county agencies also influenced the pace and character of reform legislation. Whatever the intention behind recent reform legislation encouraging decentralization of decision-making, considerable scope in interpretation was exercised by county officials. Even if it were beyond the formal authority of the county to influence decisions at the level of the farm, its political presence in agricultural affairs meant that cooperative managers acceded a large measure of control to its offices. County offices differed markedly in the degree to which they encouraged or discouraged the implementation of reform policies drafted by national ministries and planning agencies. The profile of agricultural production in different counties demonstrated these discrepancies.

Village officials could also refuse to implement directives from the center. Despite a clear mandate from central authorities to encourage higher levels of milk production, the head of the animal husbandry branch of the cooperative farm in Sárosd thought cows should be held collectively, not in private hands. He maintained artificially high fees for pasturage, thus discouraging villagers from keeping cows. All the while, one heard constant pleas from state agencies to increase milk production and encourage private initiatives in stabling cattle. This constituted another level of the "small king syndrome" (kis király), a term that referred to the disproportionate control local officials exercised over the implementation of national policies.

The very nature of the interference from outside agencies contributed to the problem of constructing a plan whose utility and rationality actually translated into improved production. The fol-

---

[12] I was most struck by this when I incurred the wrath of several managers during the ceremonial dinner at the annual meeting. Not realizing the importance of creating the proper image for county officers, I recalled some of my experiences at Harmony Farm, experiences that farm managers were sure did not reflect well on the cooperative or themselves.

lowing examples should demonstrate this. In March 1983, a party representative from the county offices participated in a meeting of the directorate council, during which several aspects of the cooperative plan were to be discussed. It had been proposed that sugar beet production be phased out, as the special machinery used for planting and harvesting was deteriorating rapidly and probably could not be used another year. Cooperative management was reluctant to replace the machinery, a decision made on the basis of their investment strategies—they had recently acquired two very expensive West German combines—and on the relative profitability of sugar beet production. Tractor drivers were also strong advocates of abandoning sugar beets. The old and cumbersome machinery was increasingly difficult to use, especially during the late harvesting season (late October to early November), when the fields were muddy.

Serious consideration was also being given to eliminating the garden branch of the farm. For several years, it had been a drain on the budget. This branch was small and production was limited, making it difficult to establish lucrative contracts with canning firms or other food industries nearby. The cooperative farm's store, which sold garden produce in the village, could not compete in price or selection with either the state-run grocery, or a newly opened private shop. Their business was also hampered by the fact that nearly all the villagers had their own vegetable and fruit gardens. Management had been reluctant to abandon garden production altogether, in part on the general principle that the farm should cultivate a variety of crops, and also because it offered employment to ten women who had long been members of the farm.[13] But, as productivity continued to decline, and there was a possibility of establishing a sewing shop to provide the women an alternative job, the leadership was strongly in favor of dismantling the garden branch once and for all.

Both proposals were based on a careful process of weighing the social and economic costs to the cooperative farm and its workforce. Neither proposal, however, was finally implemented that year. After some debate among members of the directorate committee, all in strong support of the proposals, the representative from the county rose to voice the party's concern that these decisions represented too radical a shift in farm production. The party view

[13] It was ironic, though not surprising, that women in the garden section engaged in the most arduous (and, needless to say, poorest paying) work at the farm, yet were excluded by national legislation from more remunerative and less difficult jobs, such as tractor driving, in the interests of their health.

held sway, since managers did not want to alienate party officials at the county level. This experience, reproduced time and time again, demonstrated to local management the extent to which they were in control of the fortunes of their enterprise. At another meeting of the leadership committee, the cooperative farm lawyer discussed new regulations issued by the county agricultural office concerning vacations, enterprise lunch policies, and compilation of information about physical workers and farm personnel on maternity leave. Rather than even debate the merits of these new regulations, the cooperative president simply said: "We cannot debate these provisions, so let's just vote them in." Thus, despite reforms to delegate greater authority and control over local affairs, many local leaders felt constrained in the exercise of their duties. The experience of local leaders simply being channels for the execution of central party directives was made particularly clear to me when I congratulated an acquaintance on his successful bid to be elected to the local town council. "What does a council member do?" he asked me with a twinkle in his eye. "Nod in agreement."

County agricultural offices also made recommendations in such obscure areas as the preferred piglet mortality rate and the piglet/sow ratio. In Fejér County authorities at the county agricultural office strongly advised that no more than 8% of piglets die before weaning. Such recommendations apparently were made to encourage greater vigilance among the pig herding staffs in cooperative production. No one at the farm knew how this particular figure had been reached (someone suggested that the average mortality rate was 25%), but it nonetheless became a figure to reckon with. The farm therefore never counted the number of piglets actually farrowed until after weaning, so that their figures more closely approximated those specified by the county. The piglet mortality ratio illustrates the particular features of the politics of science. Central agencies, whether they be county agricultural offices, ministerial bodies, or party organs, stipulated a wide range of indices for production. These directives, however, were not developed in consultation with those involved in the process of production. In a society devoted to the development of a new "praxis," political agencies consistently neglected practical input from the productive process itself. Such directives exemplified the centralized, top-down character of the science of planning. No longer content merely to lead the working class to socialist utopia, the party also now had to interfere in the most mundane affairs, demonstrating time and again its enlightened superiority in every aspect of social life. Now the party had designated itself as final arbiter in such arcane matters as the piglet mortality rate. The structure of planning, of party politics, of local authority structures excluded local knowledge from influencing decision-making, locally or nationally. Hence Marxism-Leninism, though declared a science by its practitioners, would be better described as a politics of scientism. Lacking any empirical basis, it could not aspire to a science of knowledge grown from experiment, conversation, or disagreement.

Figures concerning mortality rates did not directly affect planning calculations and production per se at the farm. But other county recommendations did. A further twist at the pig herding branch concerned the piglet per sow ratio. The county suggested that the breed of hog at Harmony Farm should average eight to twelve piglets per litter. This number was expected of the cooperative, regardless of intervening circumstances. To meet expectations, the cooperative started to keep 280 sows, but to list only 220 on the ledgers. This made it possible to maintain a high piglet/sow ratio, but it also required the falsification of other figures in the plan. For example, the amount of feed purchased by the animal husbandry branch, normally acquired from the grain cultivation section, had to be manipulated, in order to maintain the fiction. The figure given for the quantity of grain sold also had to be altered to cover the discrepancies between internal and external distribution, and so on.

Discrepancies between figures submitted and figures withheld complicated the ongoing production of numerical indicators at the county level. Yet it was precisely this process—the production of numbers divorced from actual practice—which fueled the fiction. The scientistic quantification entailed in planning required that the county declare numbers that farms should realize, numbers that once submitted formed the basis of future plans. County offices thus declared and fulfilled, demanded and substantiated numbers as objective indicators of production levels known to be otherwise. Kornai (1959, 1980a, 1980b, 1992) has called this process of accommodation, of negotiation, and falsification, "plan bargaining" (cf. Kalecki 1964; Antal 1979, 1981). The constant demand for increases in production, the predominant concern of higher agencies, was met by local enterprises through manipulated statistics and schedules. In offering figures that underestimated production, the actual amount of increase required or desired by supervising offices could be reduced. Ironically, then, in an economy that portrayed development as dependent primarily on brute growth, this process of bargaining contributed to problems of endemic shortage.

This kind of elaborate accommodation of production figures to expectations explains why it took management nearly twice as long to determine whether, and how, they had fulfilled the plan than it did to write the plan in the first place. In the words of one manager, discussing the wheat harvest, "The harvest will reach 50 [quintals per hectare] only if we really play with the numbers!" His language—"The harvest will be 50"—captured the numbers-centric rhetoric of socialist planning. This dry, abstract rendering of the most sacred event in the agricultural calendar contrasted sharply with the more common attitude toward harvesting wheat. Riding in a combine during the harvest, the driver told me as we mowed down the wheat in his path, "There is no more beautiful job than providing the nation with bread." When I first learned of these elaborate accommodations in fulfilling the plan, I asked a friend how the cooperative farm was able to prosper this way. "The question," he said, "is not how the farm survives, it's how the country survives."

The subtleties of influence and negotiation in planning were masked in public forums by constant reference to the identity of interests across all sectors, the absolute unity of goals at all levels of economic decision-making. Concessions to divergent interests were voiced on occasion, especially when reforms were initiated to prompt greater responsibility at lower levels of management and control. Nonetheless, no one publicly repudiated the basic fiction that planning, carried out by party officials and bureaucrats, achieved the goals of an abstract social will, independent of class politics or narrow social interests.

> But to presuppose that there is a logic of planning beyond . . . material and social determinants . . . is an extreme case of fetishistic thinking revealed by Marx: a transformation of Smith's "hidden hand" into a directly mystical-divine (because devoid of any institutional correlate) force of social reason per se. But this is precisely the official apologetic ideology of Soviet-type societies—the objective existence of a rationality of development, independent of the conscious needs and will of the population, of a logic of the future, the requirements of which are realized through the practice of central bureaucratic planning, as the actualization of general social interests. (Fehér, Heller, and Márkus 1983, 12)

In this way differing interests in society were denied, as were the political and social interests of elites to maintain their control over resources and power. Lower-level managers had to resist these measures consistently, finagling to protect their own interests. While describing the success he was having with a new chemical fertilizer, one manager openly discussed his partisan approach with me. The fertilizer was considered to be good only for three years, and yet the fields were flourishing far beyond that time. When I asked him whether he was going to inform the appropriate authorities, he laughed, saying he had absolutely no intention of doing so. At this point he was being given the fertilizer for free; if he told them, he might have to pay for it. Furthermore, he explained, the people who manufactured the stuff were not the same people who marketed it, so it didn't matter. When I asked him directly whether his interest was the same as the interest of the state, he said flatly, "No. For them it's just a drop in the sea, [the money] they could save knowing this. For me, on the other hand, it means a great deal if I can get this fertilizer for free."

The Reform of the Economic Mechanism in 1968, and subsequent reforms, introduced a new economic vocabulary, as the complexity and sophistication of enterprise planning grew. Terms such as "productivity," "profitability," and "efficiency" graced magazine and newspaper articles, and embellished the speech of local county officials, management personnel, and party members. Increased independence of enterprise decision-making with the reforms, and especially the removal of automatic monetary supports for cooperative farms in the mid-1960s, also forced cooperative farms to become more and more concerned with the profitability of their activities, as they bore the costs of inexpedient measures directly. However, at the cooperative farm of Sárosd, the primary and overarching goal continued to be the simple increase of gross production levels. Thus, the predominant concern in the construction of plan goals—as elsewhere in rural Hungary—was the increase in output and improvement in yields.[14]

While the sophistication of planning mechanisms grew with the reforms, the inability of party officials to improve overall

[14] I was struck by the contrast with American farmers, for whom a bountiful harvest bodes ill. An abundance of goods drives the price of their commodities downward, whereas only scarcity will increase the price. A good harvest in Hungary was an unproblematic good, for both management and membership. Clearly this attitude stemmed in part from the joys associated with bounty, and the symbolic centrality of the wheat harvest to notions of family and national welfare. But the relation of cooperative farms to marketing agencies run by the state also buffered them from the concerns of the market and price fluctuation, since one institution was responsible for production, the other agency for marketing. Since the transition, villagers can no longer rejoice in their harvests, as markets—local and international—have become an increasingly menacing factor in their finances.

economic performance became more and more glaring. Though a new vocabulary was created to convey the sophistication of market reforms and new means of achieving efficiency and productivity, it often fell on deaf ears. Years of poor performance, heralded in print as the triumphant achievement of socialism, only convinced everyone of the emptiness of these phrases. In an interesting editorial in the *Fejér Megyei Hírlap*, the writer encouraged party members to drop empty phrases and demonstrate their true intent by changing their behavior.

> I am angry with the president of an activists' meeting, who closed the conference [by saying]: "Let us work with active commitment for the realization of our further goals!" . . . Let us strip away the cover of phrases more bravely than before. . . . Today our socialist system expects of everyone that social will, consistency, service, endurance, diligence, steadfastness of purpose, strictness, and patience prevail in their actions. Everyone should work hard on his own front for our prosperity. (January 7, 1983)

Action spoke louder than words. Villagers also found so many of these new terms to be empty buzzwords. "Efficiency. There are better Hungarian words for that: to work economically, produce better."

Workers regularly discussed problems with work organization, suggesting better solutions to one another casually. But their input was not solicited by management. When problems arose at the farm, leaders were seen to change regulations to their own benefit, rather than deal with the structural problems affecting production. Manipulation of regulations for private gain constituted one more example of management choosing to comply with legal stipulations for appearances' sake, but no more. Witnessing such behavior, workers were convinced that officials—be they party leaders or enterprise management—had no interest in doing more than keep up appearances to further their own ends, rather than address serious problems of organization and production. Since management was paid more than the average worker, workers considered problems with waste and inefficiency in farm practices to be of less concern to the management. They could surely absorb the loss in their paychecks, unlike physical workers, whose salaries were more directly affected by low productivity. Differences in status, pay scales, and perks were explained by workers in terms of management personnel taking care of their own. "How is pay distributed at the farm?" one villager asked me. "First me, then me again, then my buddies, then the workers, if anything is left over."

Calculating costs of waste and inefficiency could also pit management against workers. To illustrate this point one villager recalled the disagreement he witnessed in an extension course between a cooperative president, who argued for higher production values, and a bookkeeper, who explained that the increase in yields could become very expensive, if one calculated the costs of fertilizers and so forth. But since cooperative presidents were rewarded for higher yields, they could opt for that choice, undercutting the profit of the farm to the disadvantage of workers. This exchange between the bookkeeper and cooperative president rang true to the villager's experience back home, where management frequently chose options that enriched them but impoverished the farm's coffers.

Villagers did not refer to the process of planning in the specialized jargon of late reform economics. They spoke of it as irrational and artificial. Cases like the piglet/sow ratio only convinced them that irrelevant and illegitimate concerns interfered in decision-making. While planning always seemed problematic to them, it was considered especially so in agricultural production, which was ultimately subject to the whims of nature, not human will. The true problem, however, was not the individual instances of contradictory practice—as for example, when, in order to maintain its bus for transporting workers during the oil crisis, the cooperative had to drive the vehicle round and round the village without any passengers to chalk up enough mileage to comply with new restrictions on enterprise transportation facilities. It was simply against human nature, they said, to elevate abstract social goals over personal utility.

One of the most common sayings one heard in Sárosd was that "Even saints line their own pockets first" (*Minden szentnek maga felé hajlik a keze*). In his book on the practices entailed in building a house through private channels in Hungary, *Do It Yourself*, Kenedi János describes this mentality well. In the following excerpt he explains how officials in charge of assessing properties work.

> The state could dispense with valuers by setting up a complete land register. One could then simply look up the price of such and such a property. However, instead of such a gigantic ledger, the council employs official valuers, who have all the data in their heads. A valuer can follow the book or he can behave like a human being. A valuer following the book is hopeless. If he behaves like a human

being he will use his head—meaning he will accept a bribe. (Kenedi 1981, 19)[15]

Moreover, not only did villagers expect personal utility to take precedence over all other considerations; one was strongly admonished if one did not keep personal reward one's first concern. This is expressed in the saying, "If you don't steal from the state, then you're stealing from your family" (*Ha nem lopsz az államtól, akkor a családodtól lopsz*).[16] As one villager explained, this way everyone is taken care of. The presumption that one should look out for one's own, coupled with the increasing prosperity garnered from expanding possibilities in the second economy, fostered a classically Horatio Alger–type bootstrap mentality.[17] More than once I heard villagers say that anyone who isn't well off in Hungarian society has no one to blame but him- or herself. This created a very hostile environment for the development of social welfare policies designed to assist Roma and other socially and economically deprived communities.

Thus, *pace* Marxist-Leninist rhetoric, planning epitomized irrationality: the futility of demanding that people not act according to their nature, and the superfluousness of constructing abstract social goals when the pursuit of personal utility was more effective

---

[15] People commonly assumed that it was impossible to prevent employees from appropriating public resources for their own personal ends. In fact, people believed that the quality of work and commitment of employees would have deteriorated if officials had been vigilant in policing abuses. This was illustrated to me once when the local veterinarian visited the pig section of the cooperative farm. He had been called out to castrate a few pigs who had been overlooked when the rest of the boars had been castrated a few weeks earlier. (Boars must be castrated if they are to fatten up properly.) Half jokingly, some workers suggested he butcher the lot, and distribute the meat among them, rather than try to save them for fattening. Hoping to impress a visiting ethnographer, he asked the workers to describe this mentality with a proverb. He was fishing for the classic expression, "One doesn't cover the mouth of a threshing horse" (*Nyomtató lónak nem kötik be a száját*). Another villager merely quipped, "In his own trade, every Hungarian is a crook" (*Minden magyar ember saját mesterségében gazember*).

[16] The prewar equivalent was, "If you don't steal from the manor, you are stealing from your family" (*Ha nem lopsz az uradalomtól, akkor a családodtól lopsz*). (See above, p. 86.) This attitude does not condone stealing from one's fellows; the emphasis is on taking advantage of powerful (and paternalistic) institutions.

[17] Various illegal schemes were openly discussed to evaluate their feasibility and rationality. I once witnessed a discussion among several cooperative farmworkers and the local policeman on the value of stealing sheep. One man suggested to his fellow workers that since it was common knowledge that the shepherd always stole sheep, if one or two broke loose from the herd, it would be a good idea to steal them. The policeman discouraged him from this plan, saying simply: "It isn't worth it to steal only one or two."

and proper. Beliefs in the higher good of collective action or social improvement were universally scorned. Although by the 1980s nearly every member of the cooperative farm had come to believe that it was better to have a job at the farm than to be farming as a private peasant, nonetheless few believed that collective efforts had any inherent benefits. One often heard the saying "A shared horse has a dirty back" (*A közös lónak túrós a háta*). In other words, if one owns something in common with others, then it often goes to ruin. When I overheard several young office workers cite this proverb at the cooperative offices, I countered with another proverb, one I had seen in a proverb dictionary but never heard used: "There is much success in cooperative work" (*Közös munka sok siker*). All three of the office workers looked at me in amazement, and one asked, "Where is that true?" The folly of collective efforts, unmitigated by personal reward, was demonstrated in the poor work performance of cooperative members. Without a decent salary, no one gave a hoot; if their salaries were not tied to performance, then little could motivate them to change their slovenly ways, especially when they could supplement their earnings at home in the second economy. Discussing the poor performance of the Hungarian national soccer team, one villager commented: "Well, here life is guaranteed. If someone tells you that life is difficult here, then he only has himself to blame. Once someone joins the national team, then he no longer has to play well, because he already has his salary. Perhaps they should pay soccer players according to their performance, and then they would play better." Hence personal advantage was always seen to prevail over wider social needs. Does this mean that villagers considered corruption, the diversion of social goods for personal ends, to be rational and just?

The ability to divert resources, that is, the right of privileged access to scarce goods and services that certain members of the community enjoyed, depended upon having a personal relationship either with senior personnel in local economic institutions, such as the cooperative farm, hardware store, or glove factory, or with local political figures. The degree of personal intimacy was important in this regard: kin or sexual ties were of greater value in gaining access to privilege than were qualifications or position. In other words, the kin of the party secretary were in a much better position than were party members to benefit from village resources—just as the mistress of the former head of the hardware store profited more from his embezzlement than did some of his accomplices.

That some people could take advantage of opportunities denied to the rest of the village did not necessarily entail strict illegalities. For example, the extensive network of acquaintances whom the head of the marketing cooperative could marshal within the county made it much easier for him to buy bricks for building when they were a scarce commodity, or to ensure that his children were accepted to high schools where enrollment was highly selective. The mobilization of social ties to facilitate one's personal needs is called "protection" (*protekció*), and is presumed by all Hungarians to be necessary and natural for bureaucratic and official negotiations.[18] Hankiss's discussion of corruption begins with a lengthy list of words for finagling, protection, and corruption. He suggests that the number of terms for these practices underscores their ubiquity and pervasiveness in Hungarian society, just as the Eskimos are reputed to have an extensive linguistic repertoire for snow (1979). Mocking claims by the party that socialism had changed the character of local politics, one villager stated: "*Protekció* has been eliminated in socialism; now one has 'socialist connections.'" This was confirmed by another villager's comments. "[*Protekció*] makes the world go round. As long as the world is the world, this is the way things will be. That'll *never* change!"

The line between *protekció*, privilege, and corruption was difficult to draw. The right of cooperative management to set their year-end premiums at 40% of their annual salaries while stipulating that members receive 4% of their (much smaller) salaries in profit-sharing was granted to them by changes in legislation following the reforms. Viewed from the eyes of the physical workers, such legislation merely reinforced management privileges. Management received their premium for fulfilling the plan, a plan they had written to be fulfilled and then calculated at year's end to appear to have been fulfilled. That their job was to write and fulfill the plan, for which they received a higher salary in the first place, apparently was irrelevant to policy makers. Thus, although it further distanced the economic fortunes of management and membership, this practice was fully legal and promoted by national policy.

The "sweet-corn episode" offers another example of the privilege of connections. Five members of the cooperative management— the president, vice president, legal counsel, head of the tractor

---

[18] Hungarians are not alone in identifying the importance of personal connections to negotiating within a socialist economy. For a discussion of networks, personal ties, and socialist planning in Romania, see Kideckel 1993, 67–68, 166.

brigade, and section head in charge of fertilizers and insecticides—struck a private deal with a nearby canning factory to deliver sweet corn. The canning factory stipulated that it had to be picked by hand, and that it should not be subjected to the usual barrage of weed killers. The group of five devoted their allotments of private corn lands within the cooperative farm to plant sweet corn, which in itself was certainly legitimate. But they employed cooperative resources and personnel to do all the work cultivating the corn. Only when it came time to harvest did the respective "owners" of the corn need to act on their own behalf. With the exception of the section head in charge of fertilizers and insecticides, all those involved paid for the labor involved in harvesting the corn. (The section head, who also was in charge of the garden branch, instructed his workers to harvest the corn for him; they were not given any additional pay for this rather unusual task.)

Many viewed the sweet-corn episode to be a totally inappropriate diversion of collective resources for private gain, and a blatant abuse of managerial position. The success of the project, one conducted with extensive monetary and service support of cooperative resources, was immense, close to the average annual salary of unskilled laborers at the cooperative farm. All five participants pocketed the full proceeds with no qualms. In recognition of the rather unseemly and very public manner in which they pursued their private ends, however, the following year the management allowed other members of the cooperative to join the sweet-corn deal. Proceeds were not as great for the newcomers, though, because the canning factory chose to favor other producers, with whom they had had longer commercial relations. This delayed transport for so long that by the time it was delivered, the value of the corn had decreased markedly.

Although many villagers were critical of this project, they would have been hard-pressed to resist. The seduction of easy money affected even upright villagers. I was standing in front of the young party secretary's house when we learned of the huge profits the five principals were pocketing. He had long been growing tobacco privately, and was respected for his private diligence. He was also respected for his public behavior, which was rigorously honest and untouched by corruption or scandal. Despite its shady trappings, he found the canning corn project quite attractive, easily pursued in one's private time at home. "It would be so nice not to have to work, just grow corn and tobacco." When I suggested this sounded very much like private production, and not the sort of thing one expected to hear from a party secretary, his father laughed, saying: "See, 'he preaches water but drinks wine'" (*Vizet prédikál, és bort iszik*). This proverb, usually a comment on the hypocrisy of the priesthood, was easily transposed to political leaders. The similarities in temperament demanded of those preaching for the church and for the party had already been pointed out to me by the party secretary's father, who had considered it a toss-up whether his son would become a priest or party leader.

Leaders of economic institutions relied heavily upon personal relations in order to circumvent the difficulties of a shortage economy. Examples from the Soviet Union—such as the exchange between factories of raw materials in return for a month's labor force—were striking, and seemed especially baroque to Hungarians, whose markets were much more developed. Yet despite, or perhaps because of, the rudimentary development of markets and noncentralized means of control during the 1980s, managers often had to become quite astute in manipulating regulations and personal relations for their own advantage, and for that of their enterprises. In the new liberalized environment, it became perhaps even more difficult to draw a clear line between personal gain and clever tactics furthering the economic fortunes of one's farm or factory. The currying of favor by one's subordinates, paired with the tendency of leaders to abuse privilege and position, allowed quite complex relations of advantage and abuse to develop. The party then had to clarify these boundaries, in order to rescue its reputation within society.

> It is difficult to imagine how a clear-headed leader can tolerate the fact that others arrange his personal business without supervision, solely with selfless courtesy, and not realize that his affairs not only are taken care of better and more quickly, but also more cheaply than the affairs of other earthly mortals. He who does not realize how much favors have proliferated for him, or who *tolerates* it, has stepped onto the road that leads him to *expect* that his affairs will be arranged faster and cheaper next time around. It is difficult to draw the dividing line between naivete and dishonesty, but in any case this is purely an ethical question, because in the case of a manager, naivete and dishonesty both qualify someone as unsuitable. (*Fejér Megyei Hírlap*, February 26, 1983)

The message for the party was clear: leaders must demonstrate higher virtues. Yet a more universal message was also conveyed. In

Hungarian society one could not expect anyone to do anything selflessly, without concern for his or her own welfare and advantage. At most, one could be expected to bend the rules for one's kin and friends. "It bothers public opinion that they walk a naive manager through the back door and behind him floods the rabble of kin and relatives" (*Fejér Megyei Hírlap*, February 26, 1983).[19] Regulations were seen to be manipulated not only for the advantage of management, but also to privilege their kin.

The use of cooperative farm monies to mount elaborate feasts and parties for visiting dignitaries—the hunting lodge bacchanalia reminiscent of prewar gentry festivities being the preferred form—was also a very common occurrence. Yet despite their regularity, these rituals were also increasingly frowned upon, and criticized in the press.

> Unfortunately, we have plenty of reason to be suspicious of those friendships that are hatched and strengthened around the pub table. This is a shaky past for anyone, and a questionable alliance. Grand promises, deals, and plans, conceived in a stupor, have many times become the focus of police investigation and judicial verdict. In such cases, the only thing that remains from repeated promises and deals is the ordeal of the bier and bearing witness. It is better to stay far away from all this. (*Fejér Megyei Hírlap*, February 26, 1983)[20]

Party officials were concerned with more than just the dubious circumstances in which party alliances and allegiances were forged. The cost and waste entailed in these activities also came under scrutiny. In the increasingly difficult economic environment, such extravagance and luxury could no longer be tolerated. In January 1983 new regulations were passed, reducing expense accounts for entertainment (*Fejér Megyei Hírlap*, January 26, 1983). Such rites continued nonetheless, cementing relations among county officials and local elites, further strengthening ties that translated into social and economic advantage. The fortunes of county officials and local leaders were bound together in these rituals of solidarity, as well as other practices, resulting in strong webs of support and influence

---

[19] The term I have translated as "back door" (*kis kapu*), which actually means "small gate," is used to designate those possibilities within regulations and institutional structures which made it easier to bend the law or go places where others could not.

[20] The ordeal of the bier was a means of determining guilt often used in Hungarian communities. People believed that the corpse of a murdered person would start to bleed if the murderer approached it. Hence villagers were paraded in front of the corpse to figure out who was guilty.

which crisscrossed the county. Hence the interference of county officials in decisions over hiring and promoting personnel was more than just another attempt to wield power over local institutions. It became a way of ensuring that their own personal networks of *protekció* and advantage were maintained and advanced. The collusion of the politically and economically powerful made it nearly impossible to fight corruption or incompetence in office, as the officials responsible for prosecuting the illegalities were themselves often directly involved. In Sárosd a member of cooperative management was sent into early retirement, after his well-connected older brother retired as party secretary. His ouster entailed a long process of very delicate negotiation and machination. However, his subordinate was also kicked out of the party at the same time for not having blown the whistle on the problematic behavior of his senior manager. Of course, to have identified the lawlessness of his superior would have entailed implicating those county officials in dirty dealings who eventually deprived him of his party membership.

How, then, do villagers evaluate these patterns of privilege and patronage? The general attitude is dismay, but not surprise. It is human nature, as Kenedi observed, to take and to offer bribes. All share the assumption that one should further one's own fortunes. Although greater respect is shown someone whose prosperity is built on hard work, the rational course of action when presented with the opportunity to appropriate resources (from institutions, *not* individuals) is to do so. As an editorial in the county newspaper explained:

> No one wants to be dishonest, . . . [it's just that] one doesn't want to come off badly, doesn't want to "lose money" for his integrity. . . . The anxiety that one is not feathering one's nest quickly enough turns some people's heads, the fear that one would get the short end of the stick if one does not take advantage of the possibilities that fall in his lap, . . . (*Fejér Megyei Hírlap*, February 26, 1983)

Villagers became angry, however, when officials in privileged positions exploited their advantages excessively. The manager of the animal husbandry section was well known for his abuse of privilege. As the younger brother of the long-time party secretary, he acted with impunity until he was finally sent into early retirement. He abused his subordinates, spent endless hours "consulting" with his colleagues at the neighboring distillery, and frequently appropriated tools and supplies from the cooperative farm for his private business activities. Moreover, he frequently drew others—

willingly or otherwise—into his schemes, implicating them in illegalities they would later regret. At one point, caught in the act of stealing bricks to build a new pig barn, he was confronted by the head of the pig section, who resented the lies and machinations employed to cover up his underhanded dealings.[21] "I would much prefer it," he was told, "if you just said outright what you were doing. Everyone knows you are stealing. Why pretend otherwise? You are insulting our intelligence."

As Kemény has observed, the ubiquity of shady dealings created a national conspiracy of silence. "Everything is supposed to be kept secret because everyone is in the know" (1982, 355–356). Nearly all Hungarians were involved to a greater or lesser degree in illegalities—the baroque character of these transactions often defying my imagination—yet to sustain appearances of legality required that no one openly discuss their existence. However, the clumsy way in which this manager pursued his dirty dealings offended the sensibilities of the community. At one point, this manager felt threatened in his position at the cooperative farm, having been criticized by county party officials for the poor performance of his unit. The following morning, he marched into the offices of the cooperative farm, claiming that one of his workers had tried to steal two bags of feed from the farm. He demanded that punitive measures be instituted against the wrongdoer. Nearly everyone at the farm found this ploy transparent. He claimed to have found the sacks of feed in the middle of a field, but rather than leave them there so dogs could trace the culprits' footprints or scent, the manager had hauled the feed in his car into the cooperative offices. Other suspicious details also were mentioned by villagers, who were all convinced that the manager had staged the whole affair. But the most telling statement was voiced by one villager who particularly disliked the manager. "If anyone at the farm wanted to steal the feed, they wouldn't have left it in the middle of the field. They would have taken it home." No matter how playful or amusing were the stories people told about their own machinations, when it came to results, everyone was dead serious.

The cynicism villagers expressed toward elites and their abuses was reinforced by constant refrains in local newspapers about the dangers of corruption, undermining confidence in the party and re-

---

[21] The parallels to the corruption scandal in Móricz Zsigmond's famous novel, *Relatives* (*Rokonok* 1932), would almost have been comical, if his actions had not been so pathetically corrupt and damaging to the farm and the larger community.

tarding economic growth. An article chronicling the illegalities of a neighboring council president could easily have been written about scores of officials across the county, with the significant difference that this official had been removed from his position by the party.

> He violated the democracy of council work, the purity of public life; he followed illegal practices that contravened the protection of social property. Responsibility rests with him, because he failed to supervise the keeping of systematic and continuous accounts of the community distillery. He tolerated labor relations that were established there illegally. This contributed to the fact that from 1972 a continuous deficit of large sums developed in the accounting of the distillery. Abusing his office as council president, in more than one instance he granted family members unlawful material advantages. (*Fejér Megyei Hírlap*, February 24, 1983)

This case may have been extreme, warranting intervention by the authorities. But even if local officials in Sárosd committed lesser crimes, villagers regularly assumed that their actions were much more egregious. Since it was presumed by nearly all villagers that those in privileged positions were guilty of corruption, it was difficult for an outsider to evaluate the actual character of corruption and abuse. A long-time observer of rural politics cautioned that I should always halve stories of corruption, explaining that village imagination always exaggerated the degree of criminality and greed local officials were capable of.[22]

Villagers did not dispute the general principle of personal aggrandizement, even though the rapid increase in wealth already visible among boutique owners or other early entrepreneurs in the 1980s was a regular topic of conversation. The easy money that entrepreneurs had did not come from hard work, villagers claimed, but from the facile manipulations of markets. In the final analysis, however, their quarrel was not with boutique owners; it was with the rhetoric that framed and justified actions taken by local and national leaders. It was the rhetoric of the socialist state that villagers found truly offensive.

---

[22] Rumors constantly floated about party secretaries running over innocent pedestrians while intoxicated, and then forcing their chauffeurs to go to jail in their stead. These stories are familiar from all over the former Soviet bloc, suggesting that the recounters of these tales daily experienced the elite's abuse of power and privilege (see Sampson 1984).

The goals of socialism—the right to a job, social and retirement benefits, national health care, decent living standards—were widely supported among villagers.[23] As one villager put it, "I wish this society had come five hundred years ago. We wouldn't have had to work so hard." One villager considered the Communist party itself a welcome change, if only because for the first time simple folk had been encouraged to join a political party, a far cry from the politics of wealthy elites in the interwar period. But the discrepancies between the goals and the realities of socialism and the Communist party were frequently the subject of conversation. While villagers consistently emphasized how much their standard of living had improved since the war, these changes did not lessen their dissatisfaction with the disproportionate allocation of power and wealth in Hungarian society. Similarities between capitalism and socialism were often drawn, as the villagers saw themselves to have been equally disenfranchised in both periods. This was the stated reason for their particular dislike of socialist rhetoric.

Many times, particularly during the yearly rituals surrounding the cooperative farm's annual meeting, people remarked on the absurdity of suggesting that the working class ruled Hungarian society.[24] "If there were real communism, then that would be good. But if it is only good for thirty percent of the people, then that is not communism." As one villager put it, "At least when the manor belonged to Eszterházy, we knew it was his. There was no contradiction between his owning it and our working for him. But now, all they can talk about is how the cooperative is ours, that we are the owners, those in charge. They still run our lives, all in the interest of the working class. It's a sham."[25] Widespread

[23] These attitudes were shared nationwide. "The high welfare expectations of the population can be seen in the results of an international survey on attitudes to inequality and welfare. Public support for welfare was highest in Hungary in a survey which also covered Italy, West Germany, Britain and the Netherlands in Europe, together with Australia and the USA. Some 78 per cent of Hungarians strongly agreed that the state should provide everyone with a guaranteed basic income, compared with 67 per cent in Italy, 51 per cent in West Germany and 20 per cent in the USA" (Swain 1992, 15).

[24] The local party secretary made the distinction quite clear for me when we were debating the legitimacy of the Solidarity Trade Union in Poland after Jaruzelski's military crackdown. "Communism doesn't mean that workers would have power; it means following the principles of Marxism-Leninism."

[25] Burawoy has suggested that a special form of critical consciousness develops among socialist workers, in contrast to the attitudes of capitalist workers. Like the cooperative farmworkers of Sárosd, steelworkers in Miskolc regularly criticized the inanities of socialist planning and frequently commented on the party's hypocrisy. And

disillusionment with political issues was best expressed by villagers when they declared their desire to become members of the Stomach Party. Their concern was for the well-being of their families, and little else mattered. "My father told us as children that we should never get mixed up in politics, because our fate is to work our whole life long." This illustrated not only their disgust with the Communist party, but also their absolutely utilitarian individualism. Most villagers simply did not believe that socialism in its most radical form could ever be realized. "Here we have socialism, and they say that the time will come when everyone will take equally from the basket of plenty. In my opinion that time will never come."

Conclusion

What then were the everyday effects of planning? In a very real sense, planning contributed to the rampant economism of 1980s Hungary. The pervasive concern with economic strategizing may be attributed in part to the ongoing project of economic strategizing at the workplace, in party organs, and in the community. The manner in which planning was achieved consistently separated the actual practices of enterprises and community organizations from their presentation in public forums, and in official contexts. Plans chock full of falsified numbers undermined the integrity of public activities, contributing in one more domain to the split between the public and the private, the artificial and the real. The split between public and private was paralleled by a split between the false scientism of centralized economic planning—which dictated principles of action from on high—and the real knowledge of local practice and hands-on experience. This opposition between artificial science and genuine know-how reinforced local criticisms of technocratic power and privilege. The public/private split also conveyed quite succinctly the political attitudes of villagers, who had long been excluded from politics and so sought control in their homes.

Though the rational calculation of planning itself was not questioned, the practices of working for collective ends was suspect.

---

like their colleagues in agriculture, steelworkers could express their view of socialist justice quite succinctly. "Furnacemen are fond of the joke about the contribution to socialism of three men. 'The first receives five thousand forints a month. He builds socialism. The second receives fifteen thousand forints a month. He directs the building of socialism. The third receives fifty thousand forints a month. For him, socialism is built'" (1992, 128).

Thus modern techniques of approximating reality (such as the numeric calculus employed to construct rational plans) and the ethical worldview that accompanies this innovation (individualism, utilitarianism, rational choice) had come to make sense in the minds and hearts of Hungarians. The economism of planning contributed to the legitimacy strategy of Kádárism, with its appeal to common sense, pocketbooks, and stomachs, but it also seriously eroded the socialist project of the party, however sincere it may have been. All the concessions to markets, prices, economic indicators, and personal incentive added up in the calculus of late socialism to a fascination with capitalism soon to be embraced quite personably by elites throughout the region.

The centrality of economics, and the irrelevance of politics, was understood by planners and villagers alike. The party proclaimed the significance of scientific socialism in ending the reign of bourgeois mystification, in erasing the distinction between appearance and reality. Planning was portrayed as science. The practice of planning, however, took a different form. Through the elaborate process of bargaining, of negotiation, and of falsification it was precisely the separation of reality and its appearance that was central. Ministries, county offices, and enterprises debated and disagreed on the significance of specific numbers, quantifying what was known not to exist. The elaborate machinations entailed in constructing and fulfilling the plan provided bureaucrats and managerial personnel with countless opportunities to manipulate the process for their own personal aggrandizement, discrediting planning, and collective projects generally, in the eyes of villagers. The means by which social goals were determined as a national and local project in fact strengthened notions of divergent interest, and reinforced the validity of pursuing individual advantage over social benefit. The merit of creating a new society, of constructing the "socialist man," was undermined by the conflict between establishing distant, abstract social goals and pursuing immediate, ever-present opportunities for personal profit that characterized local politics.

Meanwhile, villagers pondered the meaning of creating abstract plans, the realization of which merely empowered and enriched those personally engaged in constructing the very goals proclaimed to promote the social good. The entire project was ridiculed by villagers as contrary to human nature, which they believe favored personal utility and personal gain over any social objective. The manner in which planning was realized only confirmed this belief. Perceptions of the cooperative farm, and other socialist enterprises, as institutions only nominally organized to satisfy collective needs led to pervasive alienation with employment in the state sector. Villagers sought to realize their own needs at home, repudiating the collective sphere for the intensely private domain of domestic production.

SIX

The Space to Work

The fields of Harmony Farm encompassed Sárosd. The sea of wheat and blazing expanse of sunflowers encircling the village contrasted sharply with the tiny plots of corn and clutter of sheds wedged behind and between houses. The stark opposition between the wide openness of cooperative lands and the confinement of family gardens was related to another contrast, one of more subtle, but no less cogent dimensions. For within the village, differences could be perceived among houses, among gardens, and among the array of auxiliary service buildings erected to shelter pigs, grains, and tools. These contrasts reflect a process of increasing differentiation within the village community, a process arising from the varying character and degree of agricultural work based at villagers' homes. The contrast between industrial agriculture at the cooperative farm and crafted cultivation at home was thus complicated by divergent patterns of productive activity within the agricultural second economy of Sárosd.

The inscription in space of ideas and action, the materiality of social forms, will provide the narrative structure of this chapter (Bourdieu 1977). By contrasting the construction of space at the cooperative farm, and among five families in the village, I intend to demonstrate how villagers worked in the community, and how apparently similar patterns of work masked different meanings and different purposes. The most obvious distinction was between public and private production, between collectivized and second-economy agriculture. This reproduced the simple geography of socialism: a stark contrast between public and private spheres.[1] But features of

---

[1] Swain criticizes Hankiss's depiction of the stark opposition of public and private domains in late socialism, but acknowledges that his model of the "second society" reflected the commonsense categories of most Hungarians. "His models of first and second society work in skeletal, oversimplistic dichotomies . . . [But] in their daily lives

the landscape might be deceiving. One might also suggest that the heightened significance of opposing domains reveals a landscape profoundly shaped by modern practices and ideologies that are associated with capitalism and commodification: the necessary division of politics and economics, a calculated vision of time and effort, a break between the goals of the collectivity and the pursuits of individuals. Thus, the spatial metaphor of public farm and private yard speaks to far more than the division of land or the construction of homes. The development of oppositions between the public and private spheres, between wages and profits, between the utility of calculation and the value of relaxation constitute significant transformations in time, in purpose, in morality, in meaning.

The map offered below—illustrated with stories of work, leisure, money, and possessions—should be read on two levels. The first is the quite visible pathway, the directions offered by villagers to guide a stranger through a public and private regime of agricultural production. This is the world of a disillusioned cooperative workforce, alienated from collective production and management, seeking their rewards at home. However, as will become clear, the actual practices and beliefs of villagers do not reproduce the public/private divide in any simple way. People are living these ideologies of space and time, public and private, economic and political, in divergent ways. Indeed, local conceptions of public and private domains overlook a variety of pathways that crisscross the community.

Thus, I alert readers to another map, one that suggests a different design. An alternative village topography reveals a number of places that are inhabited by some villagers and shunned by others. Thus, one might recognize that young people share much more in common with the conceptions of work and reward that are articulated in a modernized wage economy associated with the public sphere, than with their elders who conceive of their work in terms of family production and moral community. One could figure out that men and women of different ages can be distinguished in their work and their attitudes toward work. Roads connecting social groups also lead beyond the well-demarcated boundaries of village and class. The political agenda of rejecting public socialist institutions is found not only among villagers, but among urban intellectu-

people were experiencing the schizophrenia of the institutionalized dualism of the first and second economy. The idea of 'second society' struck a chord with everyday social experience" (Swain 1992, 13).

als as well, who have a very different conception of labor and moral purpose than cooperative farmworkers. Thus, the simple sketch of family and farm I offer in the following pages has been drawn with two images in mind. The first depicts the experience of the "institutionalized dualism" of socialism. The second image captures the likeness of social relations constituted in work, relations that differ significantly between young and old, men and women, entrepreneur and wage laborer. Their features suggest a configuration of community different from that which is portrayed in official representation or articulated in protests by villagers themselves.[2]

Cooperative Production: Harmony Farm

With the appropriation of manorial and peasant properties in the final phase of collectivization, the cooperative farm came to dominate the landscape surrounding Sárosd. All of the lands as far as one could see belonged to Harmony Farm, or to comparable institutions such as neighboring cooperatives and the state farm. Crisscrossing the fields at intervals were access roads and rows of trees that served as windbreaks. The monotony of the landscape was broken only by three clusters of buildings (two major, and one minor): the productive center at Jakabszállás, the dairy section of the cooperative, and the former Krén farm. Within the village proper were found the cooperative's granary and storage center, formerly Eszterházy's Puszta, and the administrative offices of the farm. The cooperative farm covered 2,581 ha, of which 2,238 ha were plow land, 172 ha were grassland, 91 ha were forest, and 74 ha were taken up with roads, farm buildings, and other, miscellaneous uses.

The villagers of Sárosd have long been familiar with the social implications of a contrast between large fields and tiny plots. Eszterházy's control of political and economic affairs in the village prior to the land reform was clearly embodied in the absolute pre-

---

[2] It is instructive to compare the experience of dualism, and its use as a tool of social analysis, in a quite different situation, that of colonial development politics in Kenya. Cooper discusses the "dualism of colonial industrial sociology" in which a whole series of classic oppositions come to dominate analysis: the countryside as opposed to the city, traditional versus modern communities, informal versus formal sectors, undeveloped versus developed economies. "What is curious about the discourses of the late colonial and early independence era are their silences. The actual lives of African workers in Mombasa in the 1950s remained unexamined, just as the labeling of a dimension of the urban economy of Nairobi in the 1970s as 'informal' obviated the need to probe what the relations of production actually were" (1992, 239–240).

dominance of his extensive holdings, and those of the Weisz estate, over peasant properties. His castle dwarfed all other homes in the village; his family's private chapel and crypt further heightened a sense of social distance between his family and those of the village and estates, as it signified legitimate isolation from the most solemn collective activity, religious worship.

With collectivization, the short-lived checkerboard of family parcels introduced with the land reform was abolished. The appropriation of all private properties outside village boundaries, and their incorporation into the newly consolidated cooperative farm, restored the grand fields of yore. This clearly represented the first step toward transforming familial-based agricultural work into the hierarchical structures of production that characterized collective agriculture. The appropriation of the means of production by the state constituted the massive and swift proletarianization of the agricultural workforce. As of 1981, the cooperative farm had 319 members: 115 retired members and 204 active workers. Of these, 43 were women, 161 were men; 75 members were under the age of thirty.

Nowhere was the appropriation of manorial properties, and the redefinition of their use, as conspicuous as out at Jakabszállás. A number of buildings—the steward's house, the overseers' apartments, the manorial workers' rooms, the castle—were transformed from residential spaces to rooms for productive activity. The rooms that had once housed manorial workers' families became pigsties; the steward's house and overseers' apartments were used for storing feed for the hog branch of the farm and had been turned into shops where blacksmiths and carpenters worked and the construction crew was based.[3] Yet perhaps the strongest contrast was visible at the castle on the crest of the hill overlooking Jakabszállás. Formerly an elegant mansion, the castle now housed the repair shop and mechanics' crew of the cooperative farm. Once the scene of elegant repasts, the former dining room was dominated by an industrial-sized power lathe; one set of rooms was turned into a huge garage, where the crew worked in the wintertime repairing summer machinery, and formerly intimate spaces had been transformed into the machine shop's supply store. In the back, where many a summer's eve had been spent lounging in the spacious garden, an enormous polyurethane tent structure was erected, designed

---

[3] These rooms only finally became available to the cooperative farm after 1968, when the last family of former manorial workers was able to acquire a house in the village and leave the manor for good.

for summer repairs on tractors, combines, and trucks. The temporary "summer tent," which the farm planned to replace with a building for year-round use, signaled the ongoing process of transforming productive relations, for the mechanization of agricultural production was far from complete. The ascendancy of the repair crew among the various productive brigades, following the increasing importance of mechanization, was expressed in plans to build them a completely separate lunchroom, and to improve the dressing rooms and showers available for their use.

The shift from labor-intensive to capital-intensive agricultural production was evident in contrasts between old and new hog barns. Old barns inherited from the manor were squat structures, adjoined on one side to an open pen where the hogs were fed. Dirt floors were spread with straw, which though warm, had to be cleaned frequently. These barns required much more labor than the new barns, which had elaborate systems for the mechanized removal of manure constructed into their concrete floors. Feeding in the new barns was also facilitated by running a little car on rails down the middle of the aisle, from which feed could be shoveled into the pens. These features reduced the number of workers required to staff the barn, but unfortunately they also had adverse effects on the pigs. All members of the hog brigade agreed that the older barns were much better for the pigs. Huddled together on the straw, pigs kept warm; in structures with concrete floors, the incidence of pneumonia was much higher. Pigs were easily injured on the grates covering the manure troughs, and in the too narrow aisles through which they passed when transported. But these were necessary changes, as the absolute number of workers in agriculture was not expected (or encouraged) to increase, much to the dismay of the managers of the hog brigade. As one villager commented, the pigs would just have to be reeducated, as all citizens of the socialist state had been, into new forms of production and consumption.

Another sign of these processes of mechanization and capital-intensive agricultural production was the gradual elimination of access roads and windbreaks between large fields. As of the mid-1970s, new attitudes toward the proper economy of scale to which cooperative farms must aspire took shape. In many cases the policy shift, based primarily on analogies with industrial-sized American farms, prompted county and state officials to consolidate contiguous cooperative farms, which brought myriad problems. The policy was misguided from the start, having been based on superficial impressions regarding economies of scale in American farming.

And even if the analogy would have held for individual enterprises, it certainly did not hold for the structure of the economy at large, thereby further exacerbating organizational and infrastructural problems within the agricultural sector (Juhász 1983).

For individual cooperative farms, the problems ranged from the relatively prosaic concerns with increasing dependence on petroleum products, since the workforce had to be transported long distances to work (at a time when petroleum costs were skyrocketing), to serious social clashes, when communities were forced to work and make decisions together, complicating traditional intervillage animosities. Although Sárosd escaped this fate, the management had nonetheless decided to alter the boundaries and sizes of fields by removing a number of access roads and windbreaks, ostensibly to accommodate larger machinery. That the cooperative farm owned only three machines that could conceivably require larger fields for their more efficient use was secondary to the perception that economies of scale at the cooperative farm needed to be altered. Until the reorganization of field boundaries, the physical layout of fields at the cooperative farm had maintained the field structure of earlier manorial farms. Hence the elimination of old roads and windbreaks was just as much a statement about the modernization of the cooperative farm over prewar large-scale production, as it was a statement about expected changes in mechanization and the social relations of production in the future.

The organizational structure of the cooperative farm in its early years was quite simple. There were three centers of production—Jakabszállás, the Puszta in the village, and a site at the end of Main Street. These sites grew out of the cooperative farms that had once been distinct entities. For this reason, they represented very separate social communities: the manorial workers of Jakabszállás, the day laborer and harvester families of the poor end of the village, and the peasants of Main Street. Members were assigned to one of three sites, performing a variety of tasks in crops and animal husbandry under the supervision of the brigade leader in charge of the site. At that time management consisted of the cooperative president, agronomist, party secretary, bookkeeper, granary boss, dairy farm boss, and the three brigade leaders.

The changing composition of management and the workforce was also visible in spatial configurations. When allocating space for different branches of production in the 1960s, cooperative leaders designated Jakabszállás as the site for hogs, sheep, and horses. Barns and other buildings from the former estate were available to house these herds. The former Krén farm, consisting of little more than a small house and larger barn, was selected as the site for bulls. Initially allowed to graze, bulls were eventually stabled year-round, primarily to save labor. To accommodate the growing dairy section of the farm, new barns were built in the 1960s just beyond the forest cooperative outside the village, near to the granary and storage center formerly associated with the Puszta, and close to the grazing pasture at the peasant end of the village. (The only other pasture within the village, adjacent to the "day laborer" neighborhood, was used solely to graze privately owned pigs and a few geese.) Since villagers rented pasturelands for their privately owned cow stock, the site of pasturelands was important. Villagers living far from the cow pasture complained about this disadvantage. The selection of sites for cooperative farm centers was undoubtedly influenced in part by the concerns of the heavily peasant management of the 1960s, who were able to designate that cow pastures be at their end of the village.

Increasing mechanization of agriculture and the rise of managerial power within the cooperative drastically altered the character of work for the villagers of Sárosd. A small managerial elite complemented by a workforce adept in a variety of tasks was replaced by a much more specialized workforce and much more developed managerial community. In 1981, the farm included the following work brigades: the tractor driver brigade, two brigades working in the machine shops, two brigades in animal husbandry, a brigade of all-purpose workers in crops, and a construction brigade. Management was divided into two groups: the operative directorate and elected directorate (the so-called "social leadership"). The operative management team was divided into several different groups. The head agronomist managed the production unit, which included crops, animal husbandry, plant protection (e.g., fertilizers, insecticides), the machine shop, and construction. Accounting was administered by the bookkeeper. Another unit was devoted to personnel. The cooperative farm also employed an internal inspector and a legal counsel. The vice president of the farm was charged with miscellaneous tasks beyond production and accounting, such as protection of property, labor safety, fire protection, and civil defense. He also maintained relations with other enterprises in the village, and political organs, such as the town council.[4] (See figure 4.2 for organization of the cooperative farm in the 1980s.)

---

[4] This position was held by the former party secretary, who had been removed from his position to comply with recent laws on party positions. The law required that a

The elected directorate included a number of different committees of varying sizes. The full membership itself gathered at the annual meeting, considered by management and the party as the ultimate forum for enterprise democracy. (Farm members outside management did not share this opinion.) Four committees were elected by members at the annual meeting: the Control Commission, the Women's Committee, the Arbitration Committee, and the Selected Committee. Committee members of the Control Commission and the Selected Committee were chosen by secret ballot, whereas only the president of the Women's Committee and Arbitration Committee were elected by secret ballot, the other committee members being selected by a show of hands at the annual meeting. The Control Commission had five members, whose task was to supervise cooperative affairs, an independent body designed to check management's excesses. The Women's Committee saw to it that the regulations issued by the party and the Agricultural and Food Ministries concerning women's protection at the workplace were followed. The president of this committee took her position seriously, often advocating for women's issues at meetings of the Selected Committee. The Arbitration Committee dealt with disciplinary problems, adjudicating grievances between workers and management under advisement by the farm's legal counsel. The Selected Committee, composed of fifteen members, was also elected by secret ballot at the annual meeting. This committee was also referred to as the "Social Leadership" (Társadalmi Vezetöség). They assisted management in making policy decisions for the farm, meeting monthly to discuss issues and questions raised by management and sometimes by workers as well. The Selected Committee was also broken down into three additional committees: the Social Committee, the Campaign Committee, and the Household Plot Committee. The Social Committee was concerned with social and cultural issues. They arranged cultural programs for members, such as excursions, short trips, or evenings of operetta in Székesfehérvár. They also dealt with social problems, for example assist-

---

certain level of the workforce be party members before the enterprise could justify paying a separate salary for its own party secretary. Harmony Farm did not meet these requirements. Although his job description sounded formidable, I most often saw him sweeping the hallway of the cooperative farm offices. It was well known that the current president of the cooperative farm had been handpicked by the party secretary, who still wielded enormous power over decision-making, despite his general ignorance of things agricultural (and even though the president had an advanced degree in agricultural science). In other words, the position of vice president was probably created to pay the former party secretary a salary in the last few years before his retirement.

ing members who had fallen ill and who had no family support, or supplementing particularly low pensions of members with occasional bonus payments. The Competition Committee was tied to the Socialist Brigade Movement, which organized campaigns in the workplace to spur production and ensure plan fulfillment. The Household Plot Committee considered questions concerning wages for shepherds and cowherds for private stock, and also issues arising from the distribution of land within cooperative acreage for private purposes, for instance, the corn fields, alfalfa and potato plots that each member was allotted.[5] They also figured out who among the membership would be working their corn lands privately, and who would be choosing to have corn harvested for them. These calculations determined where privately cultivated corn lands would be situated within the lands jointly owned by the cooperative.[6] In addition to these committees, cooperative farms were encouraged to establish delegate assemblies to enhance enterprise democracy. Every brigade in each branch of production selected a representative, who was to convey membership attitudes and opinions to management, and to relay decisions to the membership which had been made on their behalf by management. This innovation was intended to give greater voice to membership in co-

---

[5] The amount of land distributed to members depended upon the size of the private gardens surrounding their houses; one *hold* (kh) was allocated each member, minus additional properties. The head of household had the size of his or her backyard garden deducted from acreage designated as household plot lands (*háztáji*) within cooperative properties: e.g., a house garden of 800 square fathoms would be deducted from the 1,600 square fathoms of *háztáji* provided by the cooperative farm. Any additional family member was allotted a full 1,600 square fathoms (or 1 kh). Thus some families with a number of cooperative farm members were able to farm several acres under the aegis of cooperative farm properties.

[6] Many more members chose in the 1980s to have their lands worked for them by the farm, receiving 25 quintals of corn as their allotment, even though if they did all the plowing, hoeing, cultivating and harvesting themselves, their yields would be closer to 40 metric centners/kh. In earlier years, farm members had worked corn plots with their families, preferring to spend more time and effort on their own to ensure that they got higher yields. Even though in those days people worked hard to get the most out of their land, they could not necessarily offset inequalities that were promoted by the party elite. Sly tricks were played on members when lands were distributed for private cultivation. Everyone picked a number designating a plot from a hat, giving the impression of impartiality in plot distribution. However, the slips of paper designating plots were folded differently, once over for plots of land which were known to be particularly rich, twice over for poorer soils. That way those whom the cooperative leadership wanted to reward were told to feel carefully which slip of paper they were selecting. Others claimed that when household plots were measured out by a man's walking out the distance, the length of one's stride could give one larger plots than others.

operative affairs, but no one took it seriously. Finally, in years past, the cooperative farm had also had a committee to deal with youth affairs, but since the numbers of young people had grown substantially among the membership (reaching over a third of the active membership), administration of their specific concerns were transferred to the Communist Youth League (KISZ).

Harmony Farm was very late in introducing any auxiliary enterprises to its production profile. (See table 6.1 on auxiliary enterprises at cooperative farms.) Auxiliary enterprises were seen by management to be necessary only in those regions where the quality of the soil or other local conditions hampered good agricultural production. Since the quality of the land was excellent, management resisted establishing an auxiliary enterprise, thinking it would be shameful to concede to the need for additional income. However, negotiations were under way during my stay in the village to finance a concern in Budapest making venetian blinds. This would not have provided any local jobs, but it was hoped it would provide a steady flow of cash into the farm's coffers. Another nonagricultural project was also eventually initiated, to provide employment for women who had been employed in the garden brigade. The cooperative farm established a sewing shop, where women were employed to do piecework, sewing bed linens on contract with a large textile factory.[7]

A description of the site and evolution of administrative offices also provides insights into the changing character of managerial power at the cooperative farm. Initially, the administrative center of Harmony Farm was in the large house formerly inhabited by Eszterházy's estate steward. Appropriated, as were all other buildings of Eszterházy's estate, the steward's house came to shelter the cooperative farm's administrative center, the village library, the culture center, and the movie house. Certainly the continuity in space of managerial control from manor to cooperative farm seemed appropriate to many villagers. Akin to the state's appropriation of private lands for the cooperative farm, the cul-

---

[7] The demise of the garden brigade significantly reduced the number of women working in productive branches of the cooperative farm, thus following the general trend of "defeminization" of Hungarian collective agriculture. "Large enterprises producing on a reduced base of labor power preferred to offer men jobs demanding higher skills, for whom the variety of jobs and the increase in wages made employment opportunities that were accessible locally or nearby increasingly attractive once more. Women, however, were gradually squeezed out of large-scale agriculture" (Répássy 1987, 125). This process contrasts sharply with the feminization of agriculture common in Romania (Moskoff 1978).

ture center represented an appropriation of local initiative and cultural concerns by the state. In the early 1950s, all local associations—peasant circles, reading groups, sports clubs—were disbanded and outlawed, to be replaced by the institution of the culture center.[8] Thus, sharing the premises with the culture center also seemed appropriate, joining culture and production under state control.

By the mid-1970s, however, cooperative management decided to move its offices to a house formerly occupied by a private shoemaker, arguing that more space was needed to accommodate the administrative needs of the farm. Not so incidentally, this change in residence corresponded in time to the shift in increasing control of the farm by management staff. The move signified both a repudiation of relatively democratic structures and the rise of a technocratic elite supported by the party and the state. For though institutionally independent entities, the contiguity of village institutions of production, education, and leisure nonetheless suggested their integrity, and so general importance, to all villagers. Once the administrative offices of the cooperative farm became isolated, the solely productive, indeed specialized, character of their work was emphasized. Furthermore, what began as a simple peasant house with three rooms grew enormously, with ten office spaces, a meeting room, and a gracious parking lot, to accommodate the cooperative president's personally chauffeured car and the other cars maintained for management personnel by the farm.

The rapidity of the growth in administrative personnel was frequently the object of discussion among cooperative members. Nonoffice workers—those who considered themselves directly involved in production—found the number of administrative workers abnormally large. The term used to refer to, and denigrate, administrative personnel meant "those of the office" (*irodista*). The expression, focusing on the space where work was performed, and not on the character of the work itself, conveyed the cooperative membership's dismissal of such activities as not even being work

---

[8] Manned by a director trained in "socialist cultivation" (*szocialista művelődés*), the culture center was organized to raise the consciousness of villagers through edifying cultural and political programs. Programs sponsored by the culture center, however, were poorly attended. Most villagers explained that they were too busy to participate, but that they also found the cultural fare—poetry readings, folk music programs, county theater groups—quite uninteresting. With adults busy working at home, the club had essentially become a social haven for teenagers. The one exception to poor attendance at the club was when the tobacco consortium held a meeting on new cultivation methods, which was well attended.

TABLE 6.1. Ancillary enterprises within agricultural cooperatives: number of units engaged in nonagricultural activity

	1968		1971		1972		1973		1974		1975		1976	
Food supply industry	916	(20%)	1,681	(25%)	2,273	(31%)	2,790	(37%)	3,116	(40%)	2,481	(37%)	2,279	(37%)
Milk processing	68	(7%)	472	(28%)	595	(26%)	434	(16%)	343	(11%)	255	(10%)	176	(8%)
Meat processing	71	(8%)	158	(9%)	179	(8%)	173	(6%)	185	(6%)	187	(7%)	190	(8%)
Chicken, egg processing	—		14	(1%)	24	(1%)	20	(1%)	21	(1%)	14	(1%)	22	(1%)
Wine production	5	(1%)	437	(26%)	486	(21%)	513	(18%)	475	(15%)	385	(16%)	363	(16%)
Alcohol fermenting	198	(22%)	235	(14%)	283	(13%)	267	(10%)	273	(9%)	257	(10%)	249	(11%)
Pickling	—		105	(6%)	113	(5%)	97	(4%)	80	(3%)	70	(3%)	55	(2%)
Baking	21	(2%)	48	(3%)	58	(3%)	57	(2%)	64	(2%)	60	(2%)	59	(3%)
Wood processing	346	(8%)	716	(11%)	914	(13%)	1,028	(13%)	977	(12%)	931	(14%)	933	(15%)
Extraction, supply of bldg. material	920	(20%)	1,131	(17%)	1,158	(16%)	1,093	(14%)	949	(12%)	821	(11%)	747	(12%)

	1968	1971	1972	1973	1974	1975	1976
Supply of other indust. products	628 (14%)	847 (13%)	809 (11%)	791 (10%)	707 (9%)	707 (11%)	712 (12%)
Indust. services	1,745 (38%)	2,309 (35%)	2,164 (30%)	2,137 (27%)	2,131 (27%)	1,748 (26%)	1,459 (24%)
Total industrial units	4,555 (100%)	6,684 (100%)	7,316 (100%)	7,839 (100%)	7,880 (100%)	6,688 (100%)	6,130 (100%)
Farms engaged in transporting	1,853	2,328	—	—	—	—	—
No. of shops	1,818	1,509	1,497	1,507	1,388	1,365	1,362
No. of restaurants	409	710	752	779	823	843	1,291

Sources: Swain 1985: 198

Notes: These figures refer to all cooperatives, rather than "producer cooperatives." Very few ancillary enterprises are attached to looser cooperatives. The figures in parentheses refer to percentages of all industrial units except for the subgroups of the food supply industry, which are percentages of the food supply industry as a whole.

at all. Opinions often voiced about the superfluousness of so much labor expressed more than just the usual contempt for nonphysical labor. A large proportion of those working in the offices as secretaries and bookkeepers were close relatives of upper management, thus reinforcing the members' views that jobs had been created to accommodate the needs of kin, rather than the requirements of farm management and administration. Overall, cooperative members found the elaborate structure of administration and management ill-suited to the tasks of agricultural production. This feeling was reinforced not only by their memories of early organizational patterns of the farm itself (not to mention the manorial estate of prewar years), but also by the highly successful productive profile of the second economy, which was a family affair. Like planning and other socialist innovations, the managerial reorganization of the cooperative farm seemed inappropriate to the tasks set by agricultural production, which had long been pursued by simple organizational units like the family. One villager complained about the transformation of the farm into a large enterprise (*üzem*), a transformation he found rather unnecessary. "The president of the cooperative is the biggest peasant in the village. If someone is a director, that sounds better, but the president of the cooperative, well, that's just a peasant. . . . This is no longer a pigsty; now it's a 'bakterium' or fattening farm. They make a big deal out of everything." The personal ambitions of management, rather than any requirements of production, were seen to drive organizational change.

The political and economic authority of the cooperative farm over the village was expressed in its literal encompassment in space. The state, having appropriated the lands and imposed collective production, manifested its presence in the affairs of the villagers by its dominion over village territory. Within the village, however, pockets of autonomy could be discerned, pockets circumscribed by the boundaries of family households. It is to the description of these arenas of nonstate activity that I now turn.

Family Production

To illustrate the character of work in the home, I have selected five families in Sárosd to describe in detail. I have made no rigorous attempt at representing types of families, nor have I tried to construct a picture of average families in the community. I knew all these families by having met one of their members at the cooperative farm, whom I then sought out at their homes, where I observed them at work with their families. Several of these families became important havens for me in the village: offering comfort, solace, support, and provocative inspiration as I negotiated the complex social relations of farm and family in Sárosd. All the families portrayed below had some association with the cooperative farm, although there were many villagers who were employed outside cooperative agriculture—in local stores, the state farm, the forest cooperative, and the numerous industries of Székesfehérvár, Dunaújváros, and even Budapest. I do think, however, that the various patterns of domestic activity I describe—particularly in terms of agricultural work—were characteristic of families throughout the village.

The Pál Family

The Pál family lived comfortably in a house built at the turn of the century.[9] Stretching lengthwise alongside the interior courtyard, it lay perpendicular to the street, which was faced only by the tiny parlor window peeking through a wrought-iron fence. All other doors and windows opened onto the interior courtyard; this pattern was repeated among all the houses built during this period, whose long porches bathed lazily in the rays of the morning sun. The house initially consisted of only three rooms: the parlor, a storage room or pantry, and a kitchen, which sprawled between them. Since the 1960s the Pál family had enlarged the house, adding another room next to the parlor, pushing the kitchen and pantry back farther from the street, installing a bathroom, and building additional storage facilities out next to the pigsty. The parlor and additional room served as bedrooms for the extended family, with the middle room also housing the television for family viewing. The front gate opened onto a lengthy courtyard, which was designed for processing goods and accommodating the traffic in people and animals moving back and forth from pastures and fields. Off to the side was a wine cellar, used for cold storage, and a well, boarded up since a villagewide system of street pumps was installed in the mid-1970s. The courtyard was busy with animals and storage facilities. The pigsty, usually housing three to four sows and their many piglets, was found behind the house, followed by an open stall for a cow and two calves. A rack two stories high for drying tobacco abutted the lean-to in which the cows

---

[9] I have changed the names of all family members, out of deference to their wish for privacy.

were kept. Finally, a large haystack completed the courtyard's western side, ending at the garden gate. On the opposite side of the yard one found the family privy and a chicken coop. In the late fall months, one compartment in the chicken coop was stuffed with dried tobacco until the tobacco consortium finally signaled to the growers that their time had come to weigh in and sell their crop.

Six adults lived in the Pál family household. Uncle József,[10] once manager in charge of the dairy section of the cooperative farm, had since retired. Before collectivization his family had owned nearly 6 ha (10 kh), so although his house was located on the peasant street of the village, his family had never been wealthy. In his youth, before the war, Uncle József had been a servant for other, wealthier peasant families. Aunt Anna, his wife, also came from a relatively poor family, from an ethnic German community to the south. During the summers she worked the land, harvesting wheat and other crops, and in the fall she would perform domestic service, sometimes as far away as Budapest. She moved to Sárosd during the war, working as a servant for a local Jewish family. Most of her family, along with an estimated 240,000 ethnic Germans, were exiled after the war.[11] They all lived in West Germany, finding a decent life there as autoworkers. Aunt Anna never sought a job in the socialist wage economy, prefering to work at home. Aunt Anna and Uncle József exemplified the hard work ethic that characterized poorer peasant families, perched on the edge of subsistence farming and working hard to stay there. The wide-ranging activities their family engaged in within the second economy, and their attitudes about dignity in labor and the shame of leisure, demonstrated this ethic every day.

Aunt Anna and Uncle József had three children: two sons and one daughter. Their daughter, Éva, worked in the kitchen at the local old folks home, was married, and had two young daughters.

---

[10] I will use the terms "Aunt" and "Uncle" when referring to the elderly members of this and other families, rather than "Mr." and "Mrs.," as this usage reflects the practice within the village. The choice carries further significance, however, as a woman was rarely referred to or addressed by her husband's name, and if his name was mentioned, it was commonly followed by her maiden name, for example, Mrs. Barna, Szabó Juliska.

[11] "In the course of the protracted operation, which ended in 1950, only very inaccurate data of the numbers of those who departed are available. Different works record the numbers of the deportees as between 177,000 and 240,000 persons.... Of the total of 239,000 persons, 170,000 ended up in the current territory of West Germany, 54,000 in East Germany and 15,000 in Austria" (Pető and Szakács 1985, 25).

Her husband, Oláh József, worked as a tractor driver at the cooperative farm. Anna and József's son Ferenc was the section leader of the hog farm at the cooperative. He acceded to the position of party secretary when the former party secretary was hired as vice president of the cooperative farm.[12] Alongside his many official duties, he was also enrolled in a correspondence course leading to a college degree in animal husbandry, traveling every several months to Kaposvár to take exams. While I was living in the village, he married, moving to live with his wife in Székesfehérvár. However, he spent at least six days a week at his parents' home, eating meals during work hours, and doing his share in all the private work the family pursued. The Páls' second son, József, helped around the house and was employed as the cowherd for the privately owned cows that grazed in the pasture at the end of Main Street.[13]

The Páls' garden was devoted primarily to growing feed corn and potatoes, but approximately one-sixth was taken up with garden vegetables. Aunt Anna grew onions, tomatoes, carrots, parsley, celery, turnips, lettuce, squash, peppers, radishes, horseradish, poppy seeds, sunflowers, melons, and between the rows of corn, several kinds of beans. Bushes of raspberries and red currants flanked the garden. Several fruit trees dotted the garden; in years of good yield, Uncle József concocted some moonshine plum brandy to supplement his wine stock. The family also owned land at the end of the street, the last house plot before the forest and pasture at the perimeter of village residence. They grew tobacco there, though the plot's ostensible purpose was as a site to build Ferenc a home once married.[14] The house to the east of the family dwelling was

---

[12] Ferenc took over the position without salary, when regulations changed regarding enterprise-level salaried party secretaries. When I asked another cooperative member why this had occurred, he answered wryly: "If there is no herd, then you don't need a shepherd."

[13] Main Street was also often called Peasant Row. Its official name, however, was a prime example of dogmatic party thinking. It was named March 21st Street, in honor of the 1919 Soviet Republic. A more appropriate name would have been March 15th Street, since on three separate occasions it had been an important date for the village. In 1848 villagers joined actively in battles for independence, remembered nationally on March 15; in 1919, villagers were said to have established the first cooperative farm on March 15; and in 1945, the village was liberated from the Germans on March 15. However, since this date was so strongly associated with nationalist traditions in Hungary, the party insisted on commemorating the completely irrelevant date of March 21.

[14] It was very common for a family to own an additional plot of land within the village, which they then cultivated to augment their own garden and household plot at the cooperative farm. Since there were serious constraints on buying properties for

owned by Uncle József's cousin, left to him by his parents. Since his cousin lived in Dunaújváros, the family had taken to cultivating this generous garden as well, which gave them additional land to grow corn for their livestock. The Pál family was heavily involved in second-economy activities: many pigs, cows, tobacco, and corn. This was possible not only because several grown children lived at home and could help out, but also because they lived in a part of town that was afforded generous plots with access to the canal at the end of their garden for easy watering.

The absolute centrality of the kitchen in the living and working of this home indicated the significance of production for the domestic unit. Simply, the reproduction of the family was demonstrated in the processing of goods in and for the kitchen. Other rooms were purely supplementary, and existed to satisfy a very restricted category of acts: sleeping in the bedroom, and ritual events in the parlor, such as funerals, engagement parties, or private interviews with a foreign anthropologist. All other activities—from eating, writing letters, making sausage, shucking corn, and weaving baskets to receiving neighbors and kin—occurred in the kitchen. The centrality of the kitchen evinced, furthermore, the prominent role of Aunt Anna in managing work carried out around the house. Aunt Anna and Uncle József may have shared many tasks, but it was the final responsibility of Aunt Anna to initiate, pace, and carry through all projects. Her power emanated from the kitchen and directed all work to replenish the material basis of that power: a stocked larder.

None of their garden produce was ever sold, all being consumed by the family fresh from the garden, or canned for the winter months. Accordingly, the Pál family's eating habits demonstrated a definitely seasonal character. Vegetables and fruits were eaten in season, and some, like tomatoes and cucumbers, were processed for canning. From December through May, pork was consumed: in sausages, smoked ribs, and chops. Chicken was next on the menu, eaten through the summer until the fall, when deliciously juicy ducks, force-fed until their livers reached nearly a kilo in weight, became the staple on the Sunday table. Needless to say, the only foodstuffs bought at the local store were spices and bread. No one baked bread at home anymore in the village.

---

private agricultural use, these plots were usually registered with the local town council as the future site of a home for the family's growing children. In many cases houses would eventually be built on the plot, but families could cultivate this land for many years without having to pay punitive taxes or other duties.

Although agricultural activities based at the Pál home were focused in the kitchen, giving the impression of a house devoted primarily to subsistence farming, the family was very active in cash-cropping and in raising livestock for sale. When too many eggs accumulated, Aunt Anna would sell them to the privately owned grocers near the town hall, or to families preparing for a wedding. Like most families, the Páls rarely butchered more than two pigs a year. Their contract with the cooperative farm ensured a favorable market price for their hogs, if they could deliver at least twenty a year; they usually supplied between thirty and fifty. The Pál family also had been one of the most successful and steady producers of tobacco, despite young Ferenc's critical attitude toward smoking. Aunt Anna's mother traveled from West Germany every fall for several months to visit, but also to help out the family in stringing up tobacco for drying. She frequently complained about the hard work they put her through, but she returned every year despite her complaints. The extensive landholdings they controlled, combined with plentiful labor resources of the family, made it possible for the Páls to save hundreds of thousands of forints. Indeed, Ferenc was able to pay for his apartment in Székesfehérvár in cash.

Although the Pál family had a television set propped up in the corner of the bedroom, there was no room or space in the house solely devoted to leisure; every room had multiple purposes. The parlor was not such a space either, as it was used as a bedroom, as Ferenc's study, and as a site for formal occasions. There once was a bench out in front of the house, where older people would gather to talk in the evening. The bench disappeared when Uncle József had a new wrought-iron fence built by the blacksmith; television had since become the substitute for conversation on winter's evenings.

The Róna Family

The Róna family shared many attributes with the Pál family, most important of which was a strong work ethic. This was demonstrated in a very active domestic economy, including a thriving tobacco-growing concern. Aunt Kati and Uncle István's house was found at the other end of the village, in an area once dominated by Eszterházy's manorial estate, accounting for their smaller garden and courtyard. The house, like that of the Pál family, was originally a long house that faced the inner courtyard, though it had since been augmented with a new kitchen and bathroom, and the former parlor had been widened and broken into two separate rooms. Much of the rebuilding was to accommodate the family of their younger twin daughter, nicknamed "Cutie." During my stay in the village, Cutie and her husband were in the final stages of building their own home close to the train station. Prior to its completion, the Róna family had used the plot to grow additional feed. Their tobacco fields were rented from the town council, located just down the street alongside the cemetery. Soon after I left Sárosd, Aunt Kati decided to install a telephone, the only private phone in the village. (Telephones could be found only at the post office, town hall, and cooperative farm office.) She wished to be able to keep in better touch with her other daughter, who lived in a suburb of Budapest. Aunt Kati also had a beautiful flower garden, flanking the driveway and fronting the house.

Uncle István came from a large family of former manorial workers and was employed as a tractor driver at the cooperative farm. Aunt Kati came from a peasant family of comfortable means, who had moved from Dunaföldvár to make a new start by buying land just outside Sárosd in 1947. They had sold everything—house, tools, horses, cows, land, and 5 ha (9 kh) of land—to buy nearly 10 ha (17 kh) on the border between Sárosd and Seregélyes. She never neglected to comment upon the irony of buying land on the eve of collectivization. Her mother's protracted illness and early death took an additional toll on the family's ability to survive in the onerous Stalinist years. Kati married István in 1956, clear in her mind that marriages founded on joining property were obsolete. Throughout her marriage, she provided the stimulus to work hard and to accumulate for the family and future. Aunt Kati was employed at the cooperative farm during my stay in the village, although she would have preferred to have kept her job at the light steel factory in Székesfehérvár. She found the work environment in the factory far more civil—agreeing with party leaders that industrial workers had a more advanced consciousness than agricultural workers. Managers at the steel factory were much more respectful to their staff, she believed, and pay was more commensurate with effort. However, she had found her husband spending far more time than she would have liked at the pub across the way during her absence from the village, and decided to find employment where she could keep an eye on his drinking.[15] She

---

[15] The classic, and crucial, case of the wife's control in the household is the pacing of her husband's drinking. Juhász Pál has suggested that one can determine the social background of wives in any village community by observing their behavior toward their husbands' drinking. If they pace and regulate it by allowing their husbands to drink

earned a minimal salary as gatekeeper at the pig section of the farm, but spent her time productively, knitting and embroidering. Her efforts in private work made up for her relative idleness at the cooperative farm. The daughter who lived outside Budapest was employed comfortably in the offices of the national railroad. Her other twin daughter was also employed by the railroad, but in the train station at Sárosd.[16] Both of her sons-in-law were from villages far to the south of Sárosd. One was a construction worker, who made a decent income building houses in the suburbs of the capital city, while the other worked at the steel works in Székesfehérvár, constructing his own home on his days off. The daughters each had two young sons.

Although their garden was relatively small, the Róna family grew all of their produce in the backyard. Like so many of their neighbors, they grew onions, tomatoes, carrots, parsley, celery, turnips, lettuce, squash, peppers, radishes, horseradish, poppy seeds, and melons. Their garden being too small for corn, the family always chose to hoe and harvest their household plot allocation of corn lands themselves, which assured them much higher yields than the twenty-five quintals guaranteed by the cooperative farm. They also supplemented these stores with corn grown in the house plot where Cutie was building a home near the railroad station. The tobacco fields they cultivated demanded more labor than the local resources of the Róna family could provide, so Aunt Kati enlisted the support of her siblings and their children in hoeing and harvesting the crop.

Animals were kept behind the house, in barns and sties kept fastidiously clean by Uncle István. Pigsties were found immediately behind the house, just beyond the small kitchen that had been built onto the rear end of the house. Despite limited space, the Róna family raised fifty pigs a year. Across the yard, coops

---

between tasks around the house, then one can assume that they come from a peasant background, where women organized daily work schedules. If they do not interfere with their husbands' drinking, allowing them to visit the pub and while away hours with their friends, then one can be certain that they were of day laborer or manorial worker background, since in those families men and women did not monitor each other's work, but were drawn apart in seeking day labor or manorial work outside the home.

[16] The pressures of working at a salaried job, building the house, tending the garden, and raising two small boys eventually took its toll on Cutie. She complained of constant fatigue, was home on sick leave for months on end, and had been prescribed tranquilizers for her difficulties. Her complaints about problems with "nerves" was a common one among women in the village who bore so much of the burden of keeping a family afloat.

housing chickens, ducks, and geese were found. These were built up against the rear wall of the garage, where the family car was kept and tobacco was dried during the fall months. Rising high above the chicken coops on the far side was a haystack and additional drying racks for tobacco. A corncrib flanked the haystack, tucked into the near corner of the vegetable garden in the back of the yard.

More than most anyone else in the village, Aunt Kati fashioned herself as a peasant woman. She took great pride in her peasant background, and had quite disdainful views of the work habits of former manorial workers and day laborers. Unlike so many others in the cooperative farm, she frequently spoke up, voicing her dislike for her boss, the infamous manager of the animal husbandry section. She was constantly investigating possibilities to augment their income, and did everything in her power to achieve her goals. She learned to drive a car, which was relatively rare among the women of the village, in order to shuttle tobacco back and forth between her house and tobacco fields. She worked her family hard, but they benefited from her diligence.

The Galamb Family

The third family lived in the section of town referred to as the new streets (*új sor*), house plots sectioned out during the minimal land reform of 1924. These houses shared the same architectural style of the Pál and Róna family homes—long houses perpendicular to the street, facing the courtyard. But the garden plot was small, much closer in size to that of the Róna family than of the Pál family on Main Street. Many of the families who lived in this part of town had moved there from the manor during the 1930s, choosing the joy of insecure day labor over the pain of secure manorial contracts. This was true of Galamb József's father, who happily turned his back on the world of the manor in the early 1930s to move into the village. When I moved up the street from the Galamb family, having left a room on Main Street behind, Uncle József commented: "Welcome to the 'proli' part of town." Aunt Juli added, "[Your former landlord] up on Main Street still looks down on us 'cause we didn't have much land before the war."[17] The

---

[17] Uncle József found this sort of class prejudice outdated and silly. Discussing another former wealthy peasant, he said: "They still look down on people, although they don't live any better than we do now. He can't eat any more than I can. If he does, he'll just vomit."

Galamb family house still had only three main rooms, but had enclosed part of the front porch and added a new bathroom with bathtub and toilet. The main room (former parlor) was where Uncle József and Aunt Juli slept, as did their daughter Terike. The middle room, once a kitchen, served as the bedroom for their older daughter, Juliska, who lived in Székesfehérvár but who often came home on the weekends to visit with her parents. The dining table was found here, as was the stove that warmed the house on winter's eves. A room in the back, Terike's bedroom while she was married, had been abandoned, as she was by her husband. The summer kitchen out in the courtyard was used year-round by Aunt Juli, leaving the middle room open for conversation and dining.[18] The television, primary source of entertainment for the family alongside the radio, was set in the far corner of the parents' bedroom. Unlike many of his fellow villagers, Uncle József liked to read, although his reading was restricted primarily to the winter months, when tasks in the garden and at the cooperative farm were less demanding. He was particularly fond of stories about American Indians, and he often recalled Winnetou's adventures along the Yellow River for me.

Although Uncle József's parents had been very poor before the war, they had succeeded in acquiring several acres of land over the years. Uncle József had been rounded up by the Russians at the end of the war—despite having been too young to serve in the army—and was sent to Siberia. He worked in mines and forests for three quite debilitating years.[19] Even though he was gone during the land reform, he had been awarded 5 kh in absentia. Aunt Juli came from a family of poor peasants as well, but had received land in the land reform, so could bring several acres to the household when they married in 1949. Altogether this gave them 7.5 ha (13 kh) to farm, plots scattered all over the perimeter of the

---

[18] Over the last twenty years villagers have taken to building what are called "summer kitchens" (*nyári konyha*). During the warmer months, all cooking and other untidy projects are done there. At the same time villagers have begun to decorate interior kitchens, making them much more similar to the living rooms or leisure spaces of newer houses (see discussion below).

[19] His near contemporary, Pál József, had the good fortune of being captured by Americans while he was fighting on the Western front, and spending his year as a prisoner of war working at the camp pharmacy. This gave him access to precious materials, which he could then sell to augment his own rations. Like so many others who returned from prisoner-of-war camps in the West, Uncle József returned fatter, healthier, and with pockets full of chocolate bars. Galamb József was lucky to come home at all, having watched thousands of his compatriots die in the mines and forests of Siberia.

village proper. They struggled during the 1950s, moving in and out of the cooperative farm as economic policies shifted and their fortunes in farming ebbed and waned. Eventually Uncle József became brigade leader of one of the regional centers of the cooperative farm in its early years. Because of disagreements with management during the shift to technocracy, he quit his leadership position and became a pigherd at Jakabszállás. By the time I arrived in the village, he had been promoted to brigade leader of the pig unit. Aunt Juli, unlike most women in the village, never worked outside the home, raising her two daughters and tending to livestock. Their daughter Juliska worked at the Videoton in Székesfehérvár, making televisions and other electronic equipment. Her husband, Tóth Károly, was a carpenter, employed at a large construction firm in Székesfehérvár. He moonlighted on the weekend, hoping to save enough money eventually to build a house. They lived in a two-room apartment, financed with a low-interest loan from her company. Their two sons, Karcsi and Róland, spent the summers in Sárosd with their grandparents. Their second daughter, Terike, worked at the deli counter of the large department store in Székesfehérvár. The family was never wanting in salami or good bread, both rare commodities in the village.

In the front of the yard was a boarded-up well, no longer needed, since the family had hooked up their plumbing to the street system several years before. Behind the house was a garage for Terike's car, which was flanked by the pigsties where the family kept a modest number of pigs—usually two sows and a few piglets. Next to the pigsty was a small storage area, where Uncle József kept his feed mixer and various tools for gardening. Three sheep, kept as much for their grandsons' amusement as for the wool they provided, were housed beneath the hayloft. Across the yard behind the summer kitchen were coops for hens and rabbits. Although the family would eat rabbit meat on occasion, most of the animals were sold to the consumer cooperative in town.

In the garden all the requisites for the kitchen were to be found, as well as corn and fruit trees and bushes. Occupying a sizable portion of the garden were Uncle József's cherished grapes—rows upon rows of both wild and domesticated—upon which he doted, their cultivation being his favorite pastime. As an added bonus, he made a small wine supply each year. Across town the family owned an additional plot of land, intended as a future site for a house for Terike, but used in the meantime to grow additional corn and beans. The Galamb family produced enough to sustain themselves and their daughters' families and children. However, they produced far less in the second economy than many of their neighbors, in part because the efforts expended by Uncle József in his jobs at the cooperative farm prevented him from building up a substantial private economy, and in part because of the absence of grown children working in the village who could supplement the labor of their parents.

The Kovács Family

The Kovács family had lived comfortably in their new home for several years, having been one of the first to purchase a lot and build in the new neighborhood called the Vineyard, once the site of Eszterházy's small vineyard at the Puszta. Built in the new "cube" style (as it is derisively called throughout the country), the house was completely square; the windows of its front rooms faced the street unabashedly. A barely submerged basement, housing the garage, thrust the main floor up half a story, which was reached by an elegant stairway leading straight up to the front door. The square structure occupied the full breadth of the lot; the fence was broken only to provide entrance to the garage and the stairway.

Ferenc came from a poor family. His father had been a harvester at the manorial farm, but after the war his father worked in road construction. None of his relatives had ever been members of the cooperative farm. He started out working at the state farm in the crop brigade, but then joined the cooperative farm, because the pay was higher. He worked for Galamb József, who suggested after six months in the crop brigade that he try working as an assistant to the blacksmith at Jakabszállás, and then was apprenticed to the blacksmith in the village center.[20] Although he could have earned more money working as a blacksmith in industry, he chose to stay at the farm, finding city life and the pace of factory work alienating. He had begun to consider opening up a private

[20] The blacksmith in Sárosd was quite famous, not only as a good blacksmith, but also for his hobby: putting horseshoes on eggs. This strange hobby was a test of skill and adroitness, inspired by someone telling him in his youth, "You may be a good blacksmith, but you could never shoe a dove or an egg." He proceeded to do so, sending decorated eggs to English prime ministers, Horthy, and Brezhnev. He became so famous for his hobby that he began to compete with a colleague from the Soviet Union, head of a large automobile factory, who had also taken up the craft. For years they competed, putting more and more tiny shoes on an egg. However, his Soviet colleague conceded defeat after visiting him, realizing that the blacksmith in Sárosd did not shoe hard-boiled eggs, as did the Soviet master, but the empty shell. Unfortunately, the blacksmith was never awarded the coveted title of Master of Folk Art, an oversight he attributed to having never joined the party.

shop, to do decorative work such as making fences and performing other tasks for the community. His wife, Ibolya, was trained as a pharmacist, and worked at the pharmacy just down the street from their house. Both Ferenc and Ibolya were committed party members.

The lot was relatively shallow, allowing only for a small vegetable garden and a few fruit trees. A small, makeshift pigsty cluttered one corner of the backyard; it was clearly not in use. The pigsty was the only remaining trace of the long years spent working and saving toward building the house. Ibolya had promised herself that once the house was built and furnished, she would no longer keep pigs. Her salary and Ferenc's wages as a blacksmith at the cooperative farm were adequate for their needs.

The interior of the house contrasted sharply with the homes described above. A long hall stretched from the front to the back entrance, cutting the house in two halves. On the left side was a bedroom, occupied by Ibolya and Ferenc's young boys (aged five and six). Proceeding down the hall one found a bathroom and WC, elegantly decorated with painted tiles Ibolya bought at a specialty shop in Budapest. The washing machine and centrifuge were also kept here. The farthest door opened onto a combined pantry and supply closet, the only room in the house not centrally heated. The kitchen, thrust into the back righthand corner, was well provisioned with appliances, including a small refrigerator and a new gas stove.

The front room, though it doubled as Ferenc and Ibolya's bedroom, was first and foremost the living room. A photo-mural depicting an autumnal scene in some unknown mountain range covered one entire wall, in front of which was situated a couch and delicate coffee table. Another wall was obscured by a combined wardrobe, chest of drawers, and decorative shelves. Underneath the huge picture window facing the street was another couch, the second of two beds in the room, and in the corner was the latest model color television set.

The house of the Kovács family contrasted with that of all three previous families in several crucial ways. The most striking difference was the shift in the relative value of rooms. The kitchen had effectively been demoted or devalued; bedrooms for family members were separated out from one another; and a room primarily devoted to leisure and relaxation had been included in the house's design. No sheds or outbuildings took up space in the yard, with the exception of the decrepit pigsty. The shift in functions for rooms, and for the house as a whole, was conspicuous. The demotion of the kitchen and the absence of outbuildings most clearly indicated the absence of any agricultural or craft activities based at the home. The reliance upon wage labor for providing the family's sustenance was not only evidenced in the absence of spaces intended for domestic productive activities; it was also clear in the prominence of space devoted to leisure activities made possible by the bounding of the working day characteristic of wage-labor contracts. Finally, the demotion of the kitchen to just another room, sharing with all other rooms a specific and limited function, conveyed most clearly the process whereby work had lost its absolute quality, and had become increasingly commensurate with other categories of action or objects, such as leisure and money. Like commodities, these rooms were icons of singularity, defined by virtue of their exclusive function and their negative relation to all other rooms.

The transfer of productive activities to places outside the home also had repercussions for the relations between spouses in this house. No longer structured on sharing productive work in the home, the relations between young couples like Ibolya and Ferenc centered on sharing their free time together. The amounts of free time enjoyed by Ferenc and Ibolya differed, however. Although Ferenc helped Ibolya with the children on weekends, she was still the primary caretaker, occasionally assisted by her mother. With the exception of stoking the basement furnace in the winter, which was Ferenc's job, all other tasks within the home fell on Ibolya's shoulders: cleaning, cooking, washing, ironing, planting, hoeing, and so on. The disproportionate division of tasks within the home reflected the demotion of many young wives within the family's hierarchy and domestic division of labor. Unlike Aunt Anna, Ibolya had no immediate control over Ferenc's time or energies. She bore the primary burden of providing the family with all that it needed day to day, in addition to her responsibility of earning an income outside the home to sustain the level of consumption and leisure established by the family. The frequent absence of women from social occasions such as sports matches or cultural events organized by the cooperative farm or culture house was related in part to their more intensive "second shift" at home.

The identification of members of the family with their own rooms also suggested that family members were separated from one another in their respective tasks. The children's primary task was school attendance, which lasted at least through the eighth grade. Their participation in domestic activities was far more limited than was the case prior to the war, when children were taken out of school after four grades to assist their parents at home or seek wage labor at the manor. Also, the practice of raising children had become oriented to promoting their leisure, rather than engaging them in heavy workloads at home before their late teens.

Very different patterns of consumption from those of the Pál, Róna, or Galamb families were evident in Ibolya and Ferenc's house, patterns that range from investments in machines that eased Ibolya's domestic chores to elaborate decoration. Without a large garden and animal stock, most of the food consumed by the Kovács family was bought at the local grocery store. Ibolya grew a few basics in her garden—such as tomatoes, peppers, lettuce, carrots, onions, parsley, and raspberries. But with the exception of sausages, which neighbors and kin would offer after winter pig butchering, all the meat consumed by the family was bought at the bargain butcher shop next to the police station. Thus, the eating habits of the Kovács family were far less constrained by seasonal differences than other families. The sole constraint was cost; since they did not grow most of their own food, the grocery bill represented the most significant domestic outlay in the Kovács family.

The character of decoration within the house itself was provocative. Devoting a separate space to a combined bathroom and WC, and embellishing it with elaborate decoration, signified the new role of hygiene and comfort in the village. The prominent display of artifacts acquired during leisure activities indicated the scale of money and time invested in them by the Kovács family, as well as the symbolic importance of these activities to Ibolya and Ferenc's own sense of identity. These patterns of consumption reinforced the overall tendency of the Kovács family to be defined in terms of its participation in markets: markets in commodities, such as food and furnishings, and markets in labor, by means of which they gained sustenance and identity.

Ibolya and Ferenc lived in their house comfortably, yet they lived in their space quite differently from many other families in the village. In contrast to neighbors and kin, Ibolya and Ferenc defined the character of their home life in opposition to their productive activities in the workplace.

The Gerevics Family

The house built by the Gerevics family was similar in overall design to that of the Kovács. Square in construction, it sat just off ground level; the garage stood off to the side of the lot, separate from the house. Built near the forest bounding one corner of the village, the lot selected by Jancsi and Juli for their new home was smaller than the lots on Main Street, but certainly larger than the lots in the Vineyard. They had originally planned to expand the house where Jancsi's parents live, but the elders objected to being relegated to rooms in the back of the house close to the chicken coop. As a solution, Jancsi and Juli simply started over, building a new house several streets away. Much remained to be done on the house, including the exterior plastering, but recent monies and energies had been spent on building a cow barn and pigsties. The cow barn was built to accommodate six cows, though they had yet to buy any. Jancsi did own a donkey, which was the butt of jokes from his neighbors. This little animal pulled a small cart, which was quite convenient for late-night excursions into cooperative lands to pick up a few extra corncobs and other goodies. Adjacent to the cow barn was the pigsty; four sows and their recurring batches of piglets noisily asserted their presence here. Wedged between the sty and haystack was the chicken coop, populated in separate compartments by chickens, ducks, and geese. On the far side of the barn rose an enormous stack of hay, two stories high, hand-mown by Jancsi. Across the yard, behind the garage, there was a small shed where Jancsi kept his tools, bags of feed, and the electric grinder he used to concoct his own special feed mix for the livestock. A small tobacco canopy serviced the drying crop. The Gerevics's garden was similar in composition to that of most other villagers. One corner was devoted to kitchen vegetables and fruit bushes; the rest was filled with corn, and a couple of plum trees. The tobacco grown and harvested by Jancsi and Juli was planted on a plot just beyond the edge of the village, a short distance from their house.

Jancsi had worked at the cooperative farm since graduating from elementary school, not counting a brief stint working at a factory in Székesfehérvár and his absence during military service. Uninterested in studying for a trade, he was drawn to working with horses and a certain independence he enjoyed while working at the farm. With mechanization, the substantial horse herd had been eliminated. Clever with his hands, and one who treasured a flexible schedule, Jancsi was hired at the hog section of the cooperative farm as an all-around repairman. He was on hand to repair machinery and to solve other problems that developed for workers in the barns and feed stores. The rest of his time he spent constructing gadgets for work at home, and resting up his energies for late-night projects. Juli worked as a quality control inspector at the glove factory in the middle of the village. Though not very remunerative, it guaranteed her a pension, and was flexible enough to allow her to shop at the grocery store across the street during working hours and get home in time to take care of the children at the end of the day.[21] Both Juli and Jancsi came from poor backgrounds; Jancsi's parents were manorial workers at Csillag Manor. With the exception of one year of domestic service in Budapest, his mother had never left the village. Jancsi's father, having retired young after having contracted tuberculosis from the dairy herd, spent his latter years sitting quietly in his kitchen, occasionally listening to the radio. His idleness contrasted sharply with his son's busy nature, always on the lookout for a deal, a scam, a new project. Juli came from a large family, most of whom had moved to Budapest and other towns to seek employment. Jancsi and Juli had three children. When she finished eighth grade, Klári chose to find a job immediately, in a factory in Székesfehérvár. Their second daughter, Réka, acquired training as a seamstress. Because she had long been plagued by difficulties walking (despite a series of painful and ultimately unsuccessful operations), the family thought it best she find a career that she could pursue at home, if necessary. Their son János was still in school.

The interior layout of Jancsi and Juli's house differed in some respects from Ferenc and Ibolya's. The center of the house, which could be entered by the front door facing the courtyard, was devoted to a living room, where the family relaxed, greeted guests, and watched television together. Three bedrooms opened from the center living room, as did the large bathroom and WC. The kitchen occupied the fourth corner of the living room; the back door, opening out onto the cow barn and sheds, also provided entry to the kitchen. In fact, nearly everyone entered the house through the back door, seeing the kitchen as the primary entryway into the household. A small kitchen pantry opened out to the right of the back door. Though the family ate in the kitchen, and occasionally lingered over a meal and discussion, the living room had become the new center of social activity and relaxation.

---

[21] Pine notes that in Poland, women consider state-sector employment as little more than a source of benefits. "[W]omen view state work totally instrumentally. They feel no moral obligation to the state, and no compunction to work well or to be productive. This frees them to take advantage of facilities available, and to use the workplace as the location for a variety of tasks necessary for daily home life, and deals aimed at helping the domestic economy" (1993, 236).

Many similarities held between the Kovács and the Gerevics family homes. The relative equality among rooms was evident, each room having very specific functions; none commanded the general catchall function of the Pál family's kitchen. Indeed, no room doubled functions. Jancsi and Juli had their own bedroom, rare even among city dwellers. Though the television was a prominent object in the living room, Juli and Jancsi did not find much time to watch shows, preoccupied like their neighbors with work around the house.[22] The demands of livestock and cash-cropping precluded active involvement in sports, travel, or other leisure activities.[23] In this respect, their lifestyle was much closer to that of Aunt Anna and Uncle József than to that of Ibolya and Ferenc.

It was significant that the center of agricultural activities based in the Gerevics home had moved out of the kitchen and into the work shed in the yard: Jancsi and Juli left the house to work in the second economy. Like all the other rooms within the house, the site for commodity production was separate, manifesting its special function against other activities performed in its proximity. Moreover, the use of sheds in the yard for processing, and the everyday tasks associated with the second economy, was directly related to the demotion of the kitchen, and concomitantly, the demotion of female labor in the home.

Money for Time

The picture I have painted of homes cluttered with activity differs markedly from an image of vast, uninhabited collective fields. This portrayal reproduces the moral sentiments of villagers in Sárosd. Yet it is a deceptive picture. For just as crucial to understanding village life in the 1980s is a depiction of divergent intentions, understandings, and consequences. The pervasive distinction between public and private domains in socialism was especially visible to villagers in the discrepancies between public fields and private gardens. How people worked in the public domain—as members of the cooperative farm—differed markedly from their patterns of diligence on the home front. The socialist sector was universally criticized as poorly paid, badly organized, and extremely alienating. Villagers fled cooperative production, spend-

---

[22] The particular character of agricultural work, and the constant activity villagers engage in, is mocked in a local joke. "What do cooperative farm members do [make] when it rains? Children, because otherwise there's no time."

[23] The primary reason cited by villagers for not attending Sunday church services was the heavy demand of their private work on their time.

ing their most precious hours working at home. Yet significant differences can be discerned in second-economy activity. It is to these oppositions—of public and private, of money and time, of purpose and profit—and to their complex confusions, that I now turn.

The boom of Hungary's second economy during the late 1970s and early 1980s attracted much attention. In the Western press, journalists were quick to proclaim the rise of capitalist tendencies. Among socialist planners and economists throughout the Soviet bloc and in China, debates were mounted on the advantages, and disadvantages, of Hungarian policies, which were characterized by an increasing reliance upon so-called market mechanisms and growth in nonsocialist sectors of the economy. Despite the fact that much attention had been devoted to defining the "second economy" (Gábor and Galasi 1978; Grossman 1977; Kemény 1982; Róna-Tas 1990; Sampson 1985–86), the use of the term was often vague. Originally it was coined to refer simply to activities outside the state sector, understood both in economic and legal terms. For this reason the second economy had often been used as an umbrella term for all sorts of illegal or black-market practices. Yet many of the practices defined in the second economy had been legal for decades. The difficulty of precision arose, therefore, from a confusion surrounding what the concept covered: an illegal domain, or economic activities characterized explicitly by private ownership or pursued outside the direct control of state bureaucratic agencies. In short, the second economy was often treated as a residual category of socialist production and ownership, as if the primary economy itself were a stable and unchanging body of laws and practices. This was far from the case, either in the theory of socialism propagated by the state and Communist party, or in the even more complex domain of historical practice, as earlier discussion of policy shifts has demonstrated.

To add to the confusion surrounding the concepts of a primary and second economy, two other terms were used to cover those activities that were always patently illegal. These were the "third" and "fourth" economies: the former refers to bribery and corruption, the latter to theft (Hankiss 1979; Kemény 1982). In practice, of course, it often was very difficult to distinguish between theft and corruption. Indeed, in villagers' eyes, it was sometimes difficult to distinguish between theft and planning. Following extensive changes in the early 1980s, notably in legislation permitting a wide variety of private economic pursuits, private enterprises grew in size and complexity; the implicit distinction between employment in the private versus the state sector informing the concept of a second economy became blurred. No longer were people just working with their close kin in the backyard, but they were being hired by sophisticated enterprises. Computer consulting firms, maintenance brigades within state-owned factories, law and architectural firms, boutiques, produce marketing agencies, and small factories were all examples of private businesses then sanctioned by state agencies. The growing number of terms used to refer to this wide range of activities highlighted the absence of analytical clarity in this area, hindering our ability to untangle the complexities of economic activity across and between all sectors of the national economy. I use the term "agricultural second economy," then, as a descriptive shorthand to refer to agricultural labor, and supplementary activities such as rental cartage and technological services, not controlled by state farms and cooperatives.

According to the 1981 survey of small-scale producers compiled by the Hungarian Central Statistical Office, 42% of Hungary's population engaged in some sort of small-scale agricultural production, contributing thereby 33% of the nation's gross output in agriculture (Central Statistical Office 1982b, 7–8). Yet only 10% of the land cultivated in Hungary had been devoted to small-scale production (Baló and Lipovecz 1988, 700). Szelényi's figure on acreage as calculated from a 1982 survey by the Central Statistical Office is somewhat higher.

> Almost 90 percent of the rural and 30 percent of the urban population grew agricultural products in 1982. In this country of 10 million inhabitants, there are some 1.5 million minifarms which are large enough to qualify as "enterprises" in government agricultural surveys. These 1.5 million small farmers cultivate slightly more than 12 percent of the arable land and produce 34 percent of the gross farm product. (Szelényi 1988, 31)

The capacity to produce large amounts of agricultural foodstuffs on small areas was in itself not unique, although in Hungary it represented a significant increase in intensity of production over that characterizing agriculture in the 1930s. This was true even though the level of mechanization in domestic production had essentially remained unchanged. Indeed, the level of intensification in private production was at a much higher level in Hungary than in Poland, despite the fact that in Hungary, unlike in Poland, there had been no market in land for decades (Juhász and Magyar 1984). Intensification in the Hungarian second economy signaled a shift toward commodity production. "[I]n the last two decades, . . . commodity production has more than doubled, while . . . total production has

TABLE 6.2. Distribution of Hungarian households pursuing agricultural small-scale production (%)

Type of household	1972	1981
Working class	30.4	31.2
Cooperative peasantry	22.4	11.2
Households with two incomes	11.8	8.9
Nonmanual occupation	9.2	18.1
Petty commodity producer, small retailer	5.2	3.7
Inactive	21.0	26.9

Source: Répássy 1987: 127

increased by only 50%. This tendency is coupled with the increase in specialization on family farms" (Elek 1991, 86).

Nearly three-fourths (73%) of those engaged in such small-scale production were employed elsewhere full-time, and so pursued domestic agriculture outside the primary economy (Central Statistical Office 1982b, 15). Only 20% of the country's workforce was employed in agriculture, so many of those actively participating in the agricultural second economy were employed in industry or in other, nonagricultural sectors of the economy (Central Statistical Office 1982a, 8). Figures published in the county newspaper from a 1980 national survey gave the following breakdown: "70% of industrial workers, 90% of physical workers in agriculture, 25% of white collar workers and 40% of retirees work in the supplementary economy" (*Fejér Megyei Hírlap*, January 25, 1983). (Table 6.2 shows the distribution of households in 1972 and 1981 pursuing small-scale production according to primary occupation.)

Figures on the monetary value of produce and family consumption patterns demonstrate the significant role second-economy agriculture played in supplementing incomes and improving living standards for many families.

> After deducting expenses, an average rural family earns 1765 Forint (Ft) a month (this figure, also includes the more than 10 percent of rural families that do not grow food). That is a handsome sum, if one considers that the average wage in industry in 1981 was 4332 Ft a month. One word of caution: a significant proportion of this income is the estimated value of the agricultural goods consumed by the families themselves. In 1981, small producers still consumed 38 percent of their own products. . . . Although they were responsible for 34 percent of the gross farm product, they produced only 25 percent of all agricultural commodities. (Szelényi 1988, 31)

TABLE 6.3. Time-budget of Hungarians in 1977 and 1986: numbers of minutes on an average spring day

Activity	1977	1986
Primary occupation	217	191
Conferences	1	2
Secondary nonmanual work	0	3
Secondary manual agricultural work	2	9
Household farming	60	74
Occasional agricultural work	1	2
Totals	281	281

Source: Róna-Tas 1990: 10

Swain's figures for the economy as a whole demonstrate the substantial contribution second-economy income became in comparison to primary wage earnings. "[N]et income from work as a proportion of the population's total income fell steadily as opportunities in the 'second economy' developed, from 80 per cent in 1960 to 70.2 per cent in 1976 and 64.9 per cent in 1986" (1992, 168–169).

Finally, time-budget studies also show that commitment to private work increased between the 1970s and 1980s.

> Time budget studies conducted in 1977 and 1986 show that income earning activity outside employment increased from 63 minutes per day to 88 minutes per day on the average. . . . As, at the same time [the] regular work-week was shortened, the proportional increase is even larger. While in 1977 22 percent of all time spent earning income took place outside state employment, this figure by 1986 was 31 percent. (Róna-Tas 1990, 9)

These figures clearly captured the significant increase in nonagricultural activities initiated with the reforms of the 1980s. However, as table 6.3 shows, the increase in the time commitment to household farming was also substantial. It is important to note the commensurate drop in time devoted to primary occupation, which reflected not only a reduction in the workweek but also presumably a greater commitment to more lucrative income-earning activities outside the socialist sector.[24]

---

[24] The distinction between time spent at work on one's primary occupation and on personal business has long been blurred, the most frequent examples being women doing their grocery shopping during office hours, or people running bureaucratic errands during work time. But as Róna-Tas has argued, this problem was exacerbated in the 1980s when "the extra labor input in the private sector came at the expense of

The state and party often heralded the close relation between large-scale production and home-based production in agriculture.

> The primary basis of small-scale production is individual or family work. The value created by such work occurs by using means of production owned in part by persons, but in larger part owned as social property or state or cooperative property. In agriculture there is a uniform price system or price level in effect for means of production and produce. This creates the objective bases for the integration of petty commodity farms, for the fair division of labor between large enterprises and small producers. It produces possibilities for developing an interest system that is capable of mobilizing the capacities of both large enterprises and small. (*Fejér Megyei Hírlap*, December 22, 1983)

However, despite all the innovations in reforms, even in the 1980s some qualms about private economic activity still arose. After listing the growing strengths of the private sector, the author of the newspaper article above felt compelled to quell fears that the freedoms enjoyed in the private economy would endanger the socialist sector.

> In agricultural small production—despite the strengthening of its entrepreneurial character—the danger of violating basic socialist principles and especially the principle of distribution according to work is negligible. The restriction on landed property and regulation of the use of land gives a guarantee of that, as well as, and above all, the system regulating small production which creates balanced market relations and so works among those conditions. The integrative activity of large enterprises also serves as a guarantee. That is the way of production and organizing work that directly connects small producers into the socialist planned economic system and through its particular automatism regulates distribution. All of this does not inhibit, but instead encourages the interest and security of small producers according to the principle of integration. (*Fejér Megyei Hírlap*, December 22, 1983)

The claim often made by the state in the 1980s that private production was fully integrated into the socialist sector may have

---

cutting effort in the socialist sector. . . . A study conducted by the Central People's Oversight Commission in 1983 revealed that half of the 8,400 surveyed clients attending to their own private business at various service outlets were doing so during the time they were supposed to be at work . . . One half of those absent from their jobs did not even bother to ask for permission" (Róna-Tas, in press).

been somewhat exaggerated, and tended to minimize the tremendous contribution in time and energy made by villagers to the national welfare. Some analysts have preferred to emphasize the state's strong dependence upon small-scale producers. Clearly, the state relied upon the agricultural second economy to complement and augment production in the socialist sector, and to bolster its agricultural exports.[25] However, recognition of private effort should not lead us to neglect the growth of supports that planners and party officials were willing to extend to the second economy after 1976. Stores selling grains and feed, often operated by the local cooperative or state farm, had become more and more common. Sárosd had two stores, one on Main Street and another on the other side of the village, near the Galamb family home. Having easy access to feed grains certainly enabled villagers to achieve the level of animal husbandry that characterized the second economy and contrasted sharply with the agricultural policies of other socialist states, most notably Poland (Juhász and Magyar 1984; Nagengast 1991). Cooperative farms and retail agencies developed a variety of marketing contracts, offering more favorable prices than local private markets, and at times increasing the sales price if a set number of animals, such as pigs or rabbits, were to be purchased over a specific period of time. These measures clearly facilitated county and state planning, as they encouraged a certain level of production. It could also be argued that by encouraging agricultural production in the second economy—much of the produce being consumed at home, as Szelényi's figures show—the state was free to export a much higher proportion of its agricultural products than its neighbors in Eastern Europe.[26]

---

[25] The increasing significance of animal husbandry in the second economy, and the willingness of state and cooperative farms to rely on private producers to bear much of the cost and effort of raising animals (pigs, cattle, poultry), essentially vindicated those in the early debate on collectivization who had advocated that animal husbandry remain in private hands (Donáth 1977, 168). Policy makers had come to recognize the advantages of small-scale production for raising animals, in terms of labor costs, stabling, and even health concerns.

[26] Consumption patterns were not only affected in village communities. Industrialization drew many rural dwellers into the city during the 1950s and 1960s, but most families did not sever relations with kin back home. Thus second-economy production fed many city dwellers, who returned after visits to their home villages (sometimes explicitly to work the land and harvest crops), weighed down with heavy baskets of fruit, meats, and vegetables. If an urban family did not have country cousins, and had not cultivated a garden at their summer cottage, then they would often develop elaborate networks to ensure that they could buy food directly from people in the countryside, to circumvent problems with high prices or scarcity which would regularly affect markets in the city.

In the early 1980s, the value and advantages of private entrepreneurial activities were often touted by state agencies and party officials. Despite the growth of regulations encouraging initiatives in private production, many Hungarians were hesitant to take advantage of these new opportunities. In Sárosd, villagers were quite skeptical of the innovations introduced with the 1980s reforms. Having witnessed changes in party policies time and time again, they had no confidence that these innovations would not also be revoked in a few years' time. In the face of such widespread skepticism, the state and party worked hard to encourage private entrepreneurs, demonstrating that both scale and organizational innovation were not barriers to growth and success. The second economy in agriculture had long been a prime example of limited resources being used to produce quite substantial rewards. Yet the state was clearly interested in encouraging a variety of new crops and animals, to augment export.

> Bees, bunnies, snails, weeds, geese, turkeys, and many other trifles: without these we would surely be poorer. And if for a moment we consider the heart of the matter, then it quickly becomes evident that we would not be just a little poorer. Because these days it is very easy to prove the old saying that lots of small things add up in the end. . . . The real problem is that we tend to think in grand schemes, we would like to redeem ourselves, and save the world, all at once. Yet patience creates a lot of convertible currency—if we could rewrite the proverb of yore according to current requirements. (*Fejér Megyei Hírlap,* January 14, 1983)

Resistance to developing new organizational forms was also criticized.

> Man's nature is such now that if he doesn't like something, or he cannot accomplish something, then he explains it and justifies it by reasons outside himself. Simply put, he offers excuses. These days, this soon becomes the pretext for the argument that not everyone has the ability to undertake business. And so say believers of this view: "not everybody can organize a *gmk* (economic work community), specialized group, a company for civil law, a small enterprise." . . . This is a false, self-assuring logic, because it can divert attention from establishing small businesses that society would have need for. And it is even more mistaken, because it reduces the concept of entrepreneurship to small businesses. One must embark on business everywhere; it is possible everywhere. Indeed, first and foremost one must do so where one single idea or initiative can immediately

bring billions, where the use and development of the greatest productive capital goods may and must be improved, namely in large industry. (*Fejér Megyei Hírlap*, April 8, 1983)

The party had reversed its role. After years of centralized economic and political control, it had turned from stifling individual initiative to encouraging it. Yet it was an uphill battle, since the party had to combat attitudes—fostered by its own policies and practices—of cynicism and disinterest. Villagers were quite unconvinced. Why wouldn't this be just another flip in the many flops of socialism?

To assist entrepreneurial activity, technical innovations that had been developed for large-scale production were extended to petty commodity producers. Articles in the county newspaper advertised new publications explaining standards regarding contractual arrangements with consumer agencies buying vegetables and other produce (*Fejér Megyei Hírlap*, April 29, 1983), and promoted participation in a lending library program developed by the Fejér County Federation of Hungarian Small Animal Breeders to receive publications through the mail (*Fejér Megyei Hírlap*, November 28, 1982). A representative of the tobacco consortium visited Sárosd to discuss new varieties and strains of tobacco, and to show a film on more advanced harvesting techniques. As another example, it became increasingly common during this period for state and cooperative farms to have cattle raised on contract with private groups—families or individual entrepreneurs. Accordingly, prefab barns, initially designed for large farms, were altered for use in private production.

> There is room for 10–20 animals in the stable, depending on their use. [The stables] have open side walls, are furnished with a feeding trough and automatic fountain, and can be assembled from prefabricated concrete parts. . . . The price and cost of building the small stables are repaid in two years. There has been a great need for this design. The majority of old farm buildings had become obsolete and gone to ruin. They are putting these [stables] at the disposal of small farms at the best time, because more and more large-scale farms are fattening up their steer and bull-calf progeny in the stables of small farms. (*Fejér Megyei Hírlap*, March 20, 1983)

In the early 1980s, strong lobbying was waged in some quarters to provide villagers with adequate machinery for small-scale production, but its manufacture in the socialist sector was never initiated.

Villagers were left to depend on the ingenuity of local mechanics and machinists to build the tools they needed.

Shifts in the character of private agricultural production, combined with the push to make private production more entrepreneurial, altered the way in which many Hungarians viewed second-economy agriculture. The overall character of production was seen to have changed, away from the limited character of subsistence production of the peasantry to petty commodity production.

> The small production that has become general today cannot be compared in terms of its economic goals, operating form, and production procedures to the small peasant farms of earlier Hungary. Indeed it cannot be compared to traditional *háztáji* farming either, which was a characteristic component of Hungarian agriculture at the end of the 1950s and in the years following agricultural collectivization. ... Small production for subsistence is in decline as an economic goal and operating form. More and more it relinquishes its position to production oriented to petty commodity production and the market, that is, to production explicitly centered around income. While in subsistence production of old virtually only the peasantry took part, now every stratum of society, from the government minister to the retiree from the cooperative farm—nearly 5 million people—participates in small-scale petty commodity production and garden subsistence production. (*Fejér Megyei Hírlap*, December 22, 1983)

No longer were debates mounted among sociologists and economists on the demise of the traditional peasantry or on the particular strategy of worker-peasants in the socialist transformation of agriculture. Villagers engaged in second-economy agriculture were now all seen as budding capitalists. Hence, the heavy involvement of Hungary's rural population in the agricultural second economy was often assumed to be motivated by a simple wish to increase one's cash income. The president of the cooperative farm in Sárosd echoed these sentiments, when discussing how to encourage more villagers to keep milk cows. His attitude was that the only way to really succeed was to abandon propagandistic slogans and use money as an incentive. "One must make business. In this respect the two systems [socialism and capitalism] do not differ. . . . One cannot persuade peasants with losses, only with good business. We have already learned this here." The poor record of agricultural production in Romania and the Soviet Union showed that they had not learned the lesson of business.

Yet not all felt comfortable with a vision of a thriving second economy as indicative of a new industrious, money-conscious peasantry.[27] An editorial in the county newspaper suggested that an unrelenting work tempo could be debilitating. Listing the many activities that filled the busy hours of the weekend—hoeing in the garden, cleaning the house, adding a wall of bricks to the new house, writing a new computer program, repairing cars in the garage, working another shift in the private brigade at the factory—the author feared that Hungarians were not resting up for the ardors of the coming week, sacrificing their health to heathen gods. "The seventh day is your own, you do with it what you will. But on the day of leisure and free time, set aside a few minutes to ponder how long one may sacrifice the seventh day on the altar of the god of money without one becoming the victim?" (*Fejér Megyei Hírlap*, April 24, 1983). The author of the newspaper article may have been more concerned about Hungarians' commitment to work in the primary economy than about emotional or physical health. Nonetheless, he assumed that the pressure to work, to create, to be active was motivated by a wish for more money. Comparable sentiments were voiced by the nun who lived in Sárosd at the parish, assisting the priest in local religious education and playing the organ during church services. She worried that all people cared about now was earning money. She tried to convince villagers that they should rest and not always be working so hard. Her concern was less with emotional health than with spiritual well-being. "What do they need a second story on their house for? They shouldn't be building a house here; they should be building a house in heaven."

Pigs became the icon in the national imagination of a money-driven second economy. The raising of pigs, perhaps the most common and visible product of the second economy, was thought by city dwellers to bring villagers high profits, exactly because it was so common. Virtually every village household kept pigs, their numbers exceeding those in collective production. The distribution of pigs between socialist and private stables according to figures from 1983 gave the advantage to private producers: "[A]t the end of the old year 4,670,000 fell to [small producers], while large

---

[27] The claims that villagers were motivated by base monetary reward resonates with a long tradition of urban prejudice against the peasantry. Urban dwellers have often accused peasants of being overly thrifty, even stingy, with their money, indeed, even suggesting that peasants fleeced city dwellers, getting rich at their expense. This is the flip side of village prejudices against the lazy and indulgent city slicker, who makes money but never works for it (see discussion below).

farms kept 4,368,000 hogs. With regard to piglets, last year 54% were in small farms, whereas this year it was 56%. In other words, the proportion is shifting in favor of small farms" (*Fejér Megyei Hírlap*, March 22, 1983). Yet despite the ubiquity of pigs in private courtyards, one would have been hard-pressed to justify their presence as a money-making venture. Villagers generally estimated that the profit on fattening a pig—which required nine to ten months of morning and evening feeding chores—came to approximately 500 forints, which probably equaled four days of wages by the male head of household (assuming an hourly wage of 17 forints and an eight-hour shift). Asked why they kept pigs when the payoff was apparently so low, everyone remarked that it was nice to receive the total sum of money, that is, for twenty to thirty pigs, all at once, as if this enhanced the value of the money received. Others explained that they needed the money for their grandchildren. But this explanation was inadequate. As one villager explained, "Well, of course we work for our children. But it is also in our blood. That is what we are like." Clearly, more than simple monetary concerns were involved in second-economy activities. A complex set of motivations fueled the particular character of the second economy. Some of these motivations were shared by all villagers, others distinguished men and women, still others, older and younger villagers.

In the village, honor and integrity were bred by hard work. Great pride was taken in the diligence villagers demonstrated in their everyday affairs. I was told time and time again that the villagers loved to work, despite all the pain and drudgery they endured. Indeed, they made a virtue of necessity. "We loved to work, but we always had to." "We worked under duress, but we loved it anyway." Villagers would laugh and nod in agreement with the proverb, "The best thing about peasant work is relaxing" (*A legjobb a parasztdologban a pihenés*). Yet complaints voiced in quiet winter's evenings about the stress of hard physical labor were forgotten as spring approached, and villagers itched to get back into the fields to sow another year's crops.

Villagers spoke proudly of their work, echoing the party's tarnished claims on the sanctity of physical labor. Interestingly, the emphasis in the villagers' concept of work on the production of material objects was similar to the Marxist-Leninist concept of productive labor propagated by the Hungarian state and Communist party, and so was often perceived by villagers to mean the same thing. The congruence between the villagers' understanding of productive labor and the term as popularly constructed in the national press and party documents, however, differed markedly from Marx's view. Productive labor was generally understood by villagers and ideologues to mean physical labor, the direct production of material objects, and was not defined in terms of the extraction of surplus or with specific regard to production for exchange, as is true for Marx. This simplistic definition of productive labor explained the villagers' disdain for administrative personnel, who, it was said, spent long afternoons polishing their nails or drinking at the pub. Similarly, villagers frequently and angrily denounced policies that continued to reward bureaucrats, policemen, and soldiers over the truly productive labor force, themselves, and their fellow physical laborers. Recent policy shifts that effectively enshrined and further empowered management over workers were problematic in their eyes. So too were policies promoting entrepreneurship, which was not seen as a productive act, but as a way of making money. As was so often said in the village, "we work so we have no time to make money." The opposition between moneymaking activities and the production of goods was clear in their minds. The increasing role of markets and of moneymaking ventures in Hungary was seen by villagers to contradict a central goal of socialism: the abolition of market practices and ideologies.[28]

Despite a general respect for hard work and a pride taken in industry, these qualities were not expected of villagers when they worked in collective production. People paid close attention to how hard one worked at home, encouraging diligence and industry. However, this did not translate into a hardworking cooperative

[28] Antimarket sentiments, or criticisms of entrepreneurial wealth heard frequently in the village during the 1980s, can be attributed to a combination of two factors: (1) local ideologies of the honor of physical labor, and (2) the long campaign in certain sectors of the party against the foibles of unbridled consumption (see Róna-Tas, n.d., for a thorough discussion of these debates). The view often expressed that income inequalities were increasing during this time, however, cannot be substantiated. As Róna-Tas has demonstrated, second-economy income reduced inequalities rather than increased them (1990, 142–144). Swain corroborates this finding. "The ratio of highest to lowest decile for per capita income moved from 5.8 in 1962, to 4.6 in 1967, to 5.0 in 1972" (1992, 191). Éltetö and Vita conducted a study in the early 1980s that extended the finding of studies conducted in the 1970s that substantial increases in income equalities could not be identified as arising from second-economy sources (1987). One possible explanation for the widespread discomfort with moneymaking ventures in the village—beyond the impact of state ideology and local cultures of moral work—is the correlation Swain identifies between educational levels and poverty (1992, 202). Since many villagers did not pursue advanced education, they were at a disadvantage in the altered entrepreneurial environment of the 1980s, when education and skill played a larger and larger role.

workforce. If one worked as hard in the public sphere as at home, one was called a fool, and lost the respect of fellow villagers. Workers at the farm would tell others who were perceived to be working too hard to slow down. Punctuality, like diligence, was not expected of workers in collective production. One day I was late for a job I had promised to do for the garden brigade of the farm. Jumping off my bike, I started to run to join my fellow workers. The women laughed, and shouted to me across the field, "No one is ever late to the cooperative farm!" When I asked villagers why they worked harder at home than at the farm, they would answer simply: "Because it's mine." In other words, integrity in work was only truly achieved when one owned the products of one's labor.

There was a strong resistance in the village among both young and old to sanctioning inactivity. Diligence was highly valued, especially in private time. Public diligence was always suspect, but its neglect in private time was considered scandalous. The near universal response to my queries about free or leisure time (*szabad idö*) was that it did not exist. The frequent denial of leisure time by villagers could certainly be substantiated by figures often cited for the amount of time spent by village dwellers nationwide on "productive" activities outside their salaried workplace.[29] (These figures vary from 2.5 to 4 hours daily.)

Despite the busyness observable in households throughout the community, the denial of leisure time by villagers represented more than simply a statement about their domestic obligations. It also expressed their ambivalence about qualifying their own actions as inactive or "unproductive." This was clear in the marked contradiction between how villagers defined their own activities and those of other communities, most notably city dwellers. Villagers represented themselves as the untiring, and unsung, laborers of the nation, whose "productive" labor may have been lauded by the regime but finally went unrewarded. Few in the city knew, villagers complained, how much work was involved in providing them with food and drink. Urban dwellers, symbols of inactivity, were purported by the villagers to enjoy leisure regularly, such as by spending long hours watching television, and to be totally ungrateful for the sacrifices villagers bore for their welfare.

---

[29] The near impossibility of conducting interviews at any time except in the depth of winter also reflected the villagers' reluctance to take out time from gardening and other projects to spend several hours talking.

Yet villagers also spent long hours watching television, especially in the winter and slow agricultural seasons. Their reluctance to qualify their own behavior in the same terms as that of city folk suggests that the existence or possibility of leisure time in the village was very problematic. Two incidents made this very clear. In the first case, I was derided by one of my colleagues in the women's garden brigade, who also happened to be a neighbor, for subscribing to more newspapers than anyone else on the entire street. When I asked if she subscribed to any newspapers, she replied that she had neither time nor money for such things, the implication being that she tended solely to her domestic duties. Another woman then jumped in, explaining my behavior by saying that I had no television. "Nonproductive" activities—like reading or watching television—were classified as forms of education, or as cultivation, as one villager explained to me. No one would ever have chosen to label them as leisure.

In the second incident, Pál Anna recounted her first vacation to me. Known for her industriousness and firm hand at home, many were surprised to learn that she had agreed to leave home for two weeks to take advantage of the inexpensive vacations offered at the new resort hotel built collectively by the cooperative farms of Fejér County at a southern hot water mecca. She confessed to me upon her return that she had been afraid of the trip, assuming that she would be impatient and bored by the second day. On the contrary, she found it very enjoyable, and wished that she had had the opportunity to relax and be waited upon when she had been younger. The unexpected pleasure she enjoyed in leisure did not finally alter her own patterns of work at home; she resumed her strenuous schedule, and continued to impose it upon her family. Nonetheless, she had come to find that the fear of inactivity she held was unjustified. Her experiences at the cooperative's resort, where the prevailing theme was that cooperative workers deserved the rest they enjoyed there, persuaded her not to be ashamed of leisure.[30]

---

[30] The most common explanation offered in the village for not taking vacations was that regular feeding and tending to livestock prevented one from leaving home. Such practical constraints, coupled with the shame of leisure, were strong barriers to taking vacations. However, the mere cost of vacations may have also been a serious impediment for some families, even though many industrial enterprises and cooperative farms offered generous subsidies for resort vacations. Bokor's comparison of the incomes and living standards of "deprived and non-deprived" populations in Hungarian society suggests that poverty could be a factor. "Not only did the 'deprived' get less income from the first economy, they had to be prepared to work for less in the second, spending more time working, to earn lower incomes. Bokor also found that

These two anecdotes demonstrate the general ambivalence surrounding leisure or inactivity. Only one other response to the questionnaire brought as much unanimity of response as the denial of leisure time. Everyone agreed, and without hesitation, to the truth of the proverb "Work/object is not shameful" (*A dolog nem szégyen*). The difficulty of resolving villagers' concerns about nonproductive activity arose from the very moral basis of their work. For the fear of inactivity was a fear of social death, a fear of the loss of esteem within the community. The fear was much greater among the elderly.

The absence of a category of leisure time among the aging population of Sárosd may be explained by their absolute commitment to "possessing activity." To possess action was the goal of action; the value of activity derived from its constant realization by acting, working incessantly to substantiate, materially and socially, these very principles. The cycle of action was closed; nothing was commensurate with "possessing activity." The crisis faced by aging villagers highlighted how irreplaceable "possessing activity" was. The loss of the ability to work, to contribute in even the most restricted ways to the household, was considered a tragedy among the elderly, and perceived to a be a loss of social maturity or adulthood. To work day in and day out in physical labor was considered human nature. It could not be replaced. To lose one's ability to work was shameful, and could even bring thoughts of suicide. In anecdotes, the idle elder was classically portrayed as a child. One elderly woman mentioned to me that her children had forbidden her to do any more work in the garden as of the coming season. She seemed resigned to this, but another elderly woman, hearing her story, expressed amazement and anger. "If I have to crawl on my hands and knees, I will still go. No one will prohibit me from working in my own garden!" Younger people might have enjoyed the pleasures of leisure with fewer trepidations, but they shared with their elders the social value of constant activity, through which one's honor in the community was created and established.

Yet the compulsion among younger people to work long hours beyond their salaried working day arose from concerns and attitudes very different from those held by their elders. The very opposition of salaried versus private work itself was significant: partici-

---

while the non-deprived spent 32 per cent of their free time working and 30 per cent on holiday, the deprived spent 55 per cent of their free time working and only 2 per cent on holiday" (Swain 1992, 222).

pation in the agricultural second economy by younger couples was intimately related to their own participation in the "primary" wage-labor market, or state sector. This was true in several ways.

In strong contrast to aging villagers, young people (young men in particular) were very conscious of the monetary value of their labor. One often heard young men debating how to spend their time most profitably on weekends. Discussions revolved around how much money could be earned for any particular project, the base sum usually being the hourly wage at their respective jobs in the primary economy. The desire to work for cash often clashed with the expectations of family and close friends, who often asked young men "to help" (*segíteni*) them in all manner of tasks, most frequently house construction. Since the only reward for a full day's work was a hearty meal and ample consumption of wine or beer, it was seen to be a less advantageous, but often necessary, choice. Since many couples were either actively building, or preparing to do so, young men could not forgo helping out, for fear that they would be left in the lurch when their time came. Hourly wages also figured prominently in decisions taken about where to seek work in the primary wage sector, or whether to change jobs to increase one's base rate of pay. The major exception to this was the revival of interest among the young in working at the cooperative, where closeness to home, and access to grains and other agricultural products, became important factors. Thus, distinctions among kinds of labor were being made with reference to their explicit monetary value, a type of contrast not employed by elderly villagers.

Money, moreover, was not the only quality or characterization used to distinguish kinds of work. In the primary economy, all jobs were classified according to the role of skilled knowledge in their execution. But this did not exhaust the number of ways in which work could be defined. The number of terms used to describe different kinds of work performed in village communities illustrates the complexity of relations established around work. New expressions for the kinds of labor demanded by projects gauged the degree of skill, the level of brute effort, social relations in work, temporal considerations, forms of remuneration, and whether and how these jobs were performed in the socialist or private sector.

General terms for working (*dolgozni, melozni*) begin a long list of various expressions for work in the primary and private sectors. Terms such as "to travel to work" (*eljár dolgozni, bejár dolgozni, munkába jár*) all conveyed a change of place or movement as central to the wage relation. The word *munka*—a general term for work—itself connoted movement, since by the twentieth century it designated work done outside the home, unlike *dolog*, which referred in the most general terms to one's activity in every possible context. General terms independent of the site of work focused on working long and hard; strenuous effort was required (to push: *hajtani*; to toil: *gürizni*; to "slave": *robotolni*, related to the term *robot* for feudal service). To work as a worker, without any skill in general terms, was conveyed in two expressions (*keccsölni, bulcsázni*). Terms such as working overtime (*túlórázni*) emphasized temporal components. There were also terms used to refer to borderline activities, such as turning out private projects during one's work time using factory machines and materials (*fusizni*); this term was also used to refer to short-term projects one did without official permits. The term referring to black-market activities (*feketézni*) has much more of a commercial connotation, and with the exception of "black transport" (*fekete fuvar* or *svarc*)—commonly used in the village to describe activities using cooperative farm equipment and horses—the term for black-market activities was rarely used for agricultural activities.

In the late 1950s, a term was coined to define work done privately (*maszek*, an abbreviation for the term used to describe the private sector in the 1950s—*magánszektor*).[31] To work privately (*maszekolni*) was also differentiated according to the kind of work done under these circumstances. One term referred to doing private work as a semiskilled worker, assisting the skilled workers building a private home (*culáger*). The term *kontárkodik* referred to someone doing skilled work for which he or she had not officially been trained, for example a carpenter doing electrical work. This was common in Hungary, though patently against the law. The term "to help" (*segíteni*) was usually used when one had enticed friends, relatives, or other acquaintances to work on a project involving no skill or payment, but merely a contribution of labor to be returned at a later date. This term harkened back to labor exchanges in the 1930s and 1940s, when time was exchanged for time, with no reference to skill or monetary compensation. Finally, one still heard terms inherited from the prewar period, such as "day labor" (*napszám*) or "large band projects" (*kaláka*). While day labor or hourly wages were understood to entail monetary payment, sharecropping (*részesmüvelés* or *akkordmunka*) was done only in ex-

---

[31] Unlike so many other awkward neologisms, this term was not coined by party bureaucrats. It came from a cabaret skit, in which it appeared alongside abbreviations for a number of new state-owned enterprises and factories (Róna-Tas, personal communication).

change for produce. Significantly, such work was more apt to be defined as women's work.

The generalized attitudes of younger men, such as toward the monetary value of labor or its relative skill, were clearly related to changes in the organization of work, and how work was accorded value in the 1970s and 1980s. These changes also found voice in anxieties expressed by middle-aged villagers, people who had lived through the transition to a more industrial tempo, about the oppression of time and of the clock. One frequently heard complaints about being enslaved to the clock (Thompson 1967). Being at work on time and catching the bus on time were experiences that were seen as contrasting sharply with the temporal constraints of their elders. The intensity of modern life was juxtaposed to an idealized vision of the carefree peasant sauntering out to the fields to put in a day's labor, in his own time and at his own pace. Young people sometimes balked at the suggestion that they did not work as hard as their elders, a common theme among older people. Elderly villagers were heard to complain that young people left the village to work in the factory because one did not have to work there. And on top of that, younger people were seen to tire easily. "We never got tired when we were young. We worked eleven or twelve hours, and sang to and from work." Young people bristled at these charges. "But what they did in eleven or twelve hours, we have to do in eight. And there wasn't the nervousness, the tension about having to be somewhere on time, I might be late, I'd better hurry." People felt strongly that the nervousness arising from constant preoccupation with time or the clock was very unhealthy. Although the drudgery suffered by older villagers was acknowledged, it was finally dismissed, since villagers, young and old, still shared the value of physical labor above and against all else.[32]

Implicit in these discussions of new kinds of time and different sorts of effort was the observation that work in the public and private realms had differentiated in ways unknown before. In the interwar period, peasants who lived off their land devoted their

---

[32] The strong value of physical labor, and the accompanying distrust or depreciation of "intellectual" activities, was very common in the village, but stronger among the elderly. A classic anecdote told by the writer Móricz Zsigmond demonstrates the absolute distinction between physical and mental activity. Asked by a peasant what he did for a living, Móricz invited the man to watch him as he wrote one evening. After four hours, the peasant finally lost his patience and demanded to know when Móricz would begin his work. I had comparable experiences when, time and again, elderly villagers would ask me when my vacation would be over.

attention entirely to the running of the farm, a project firmly situated within the household economy. Poorer families left the household to find employment, but their paltry holdings and limited household economy did not command much effort or produce many goods. Now adults were engaged in two fully developed economies: the wage-labor state sector and the private second economy. Two separate worlds, both demanding endless attention and effort, were often too much to bear. Women, in particular, commented upon the difficulties of having to work all day at a job away from home, and then return to another full day's work around the house.

The sensitivity of villagers to new "kinds" of time was telling. The relevant category of time for young people was the "shift" (*müszak*) within the working week, in contrast with the weekend. With the introduction of the eight-hour workday in all enterprises in 1982, the workweek was generally only five days long.[33] Cooperative farms were a significant exception to this change; shifts were still ten hours long on the average. The cooperative sector generally lagged behind the industrial and state farm sector in changes regarding work time, because the principle of self-management allowed cooperatives to determine their own work times. Since in the mid-1970s members were still not being paid overtime, they usually chose to increase their primary work time to increase their salaries. Thus, when shifts were changed from twelve to ten hours between 1968 and 1973 for the industrial, construction, and managerial segments of the working population, cooperative members were still working much longer hours at their workplace, in addition to the average of an additional 2.6 hours daily in home-based tasks (Donáth 1976, 24–25). Despite the differences in seasonal demands on labor time, policies stipulating changes in the workweek at cooperative farms by 1985 were eventually implemented. Some questioned the wisdom of introducing a shorter workweek in agriculture, in which the work demands fluctuated radically over the seasons. However, one villager looked forward to the change. He complained about the cooperative farm's practice of denying workers days off during the summer months when seasonal work was quite demanding, and then insisting upon people taking their time off in the winter months when work was slack. "You can't do this anymore. This is an enterprise (*üzem*), not a farm (*gazdaság*). It is written up as an enterprise. . . . You can't do

---

[33] The school week had been reduced from six days to five at about the same time, adjusting the school child's time schedule to that of the adult world.

things like in the old days, forcing people to work constantly when you want them to." In short, the villager demanded consistency. If one justified radical changes in organizational structure and differential payment schedules and perks for management on the basis of a factory model of production, then one had to be consistent in treating workers as if they too were employed by a factory rather than a farm.

The notion of a shift explicitly bound time and wage, and emphasized bursts of activity, connoting alternate periods of intense activity and of rest. While younger people clearly attributed far greater autonomy and control to agricultural workers than was the case prior to the war, their perception of a day undifferentiated or unsegmented did capture a notion of time in demise. The collapse of time and money characteristic of wage-labor contracts—the bounding of a project or task in time for a unit of money—was foreign to the everyday expressions of work and time employed by the elderly.[34]

Elderly villagers did not speak of tasks or work in terms of shifts. The incessant rivalry in diligence in which they engaged was seamless; each project flowed into the next. The endless cycle of the seasons was the predominant conceptualization of time. Incidents were dated and remembered with double or collapsed reference to the agricultural demands of a particular season and the corresponding saint's day(s); even the organization of agricultural tasks was framed within the religious calendar. For example, an event in early summer would be remembered with reference to the wheat harvest, which is presumed to begin sometime around Péter-Pál (June 29). When reconstructing a story, people would often query each other about what work they had been doing at the time: planting beans, shucking corn, trimming fruit trees. Each event would conjure up the season, the moment, the event. Work and time were thus intimately related, and experienced as moving, as processual, as cyclical.

Within this stream of actions and objects ever in transformation, periods of time or specific events were demarcated in terms of the Catholic saints' calendar. In May, late frosts were said to occur with uncanny regularity around the three successive saints' days of Pongrác, Szervác, and Bonifác (May 12–14). After the tenth of August (Saint Lörinc), when melons tend to become overripe, villagers

[34] Even the prewar migrant labor contracts were defined explicitly in terms of their temporal dimensions and monetary constraints, "monthly workers" (*hónaposok*) and "intermittently paid workers" (*summások*).

cautioned one another not to eat melon, citing the proverb "Lőrinc has pissed into the melon" (*Belepisílt Lőrinc a dinnyébe*). Before the war the great guild feast of shepherds, attended by shepherds from all over the country, was held in Székesfehérvár on the day of the patron saint (Mihály, September 29), clearly demarcating their shift from summer pasturing to winter stabling.

Saints' days lived on in the continuing practice of celebrating name days, which were observed even if a child had not been baptized.[35] New popular names, not specified in the religious calendar, were now assigned by state agencies to days in the year, but the large majority of names still derived from the saints' calendar. For example, the celebration of one's name day was still distinguished according to the specific St. John, St. Mary, or St. Francis one had been named after within the saints' calendar. The only other exceptions to ignoring the saints' calendar in public life were the obligatory editorial on the upcoming wheat harvest around Péter-Pál, and the saint's day appropriated to the secular calendar, St. Steven's Day (August 20), which was renamed Constitution Day after World War II.

Young women straddled two worlds of time, their work and their time bearing similarities to the structures of both their parents and their husbands. The never-ending cycle of domestic work endured by women, unsegmented and undifferentiated by days, characterized descriptions of the prewar household. Certain ethnographic accounts (Fehér 1975; Kára 1973; Némethy, E. 1938) have suggested that women's labor in the household was rigidly differentiated according to days in the week: washing on Monday, baking on Saturday, and so on. Elderly women scoffed at this idea, saying that they did things when they could, often rising at 2:00 A.M. to get the bread baked before other chores would vie for their attention. The discrepancy between ethnographic accounts and the experiences of women in Sárosd may be traced to the poverty of the community before the war, resulting in a higher

---

[35] In line with its policy of undermining the role of the Church in village life, the state actively encouraged parents to forgo the ritual of baptism and to celebrate their child's birth by participating in a ceremony called a "name-giving" (*névadás*). Very similar in structure to civil wedding ceremonies, name-giving ceremonies were held at the cultural house and administered by employees of the town hall. If, for reasons of party membership or a simple desire to live within the "new rules" of social life, parents chose to participate in a name-giving, this did not preclude their taking their newborn to another village or town far away to have the child baptized as well, but in secret. This was also the case with marriages: political or personal problems with undergoing a religious marriage ceremony in Sárosd were circumvented by secret weddings in other villages or towns.

proportion of both men and women seeking wage labor outside the home than would have been characteristic of villages inhabited by wealthier peasants. This would account for the strains on time women recount, as their work at home would have to supplement that of absent husbands, and also be organized around their own absences from the domestic sphere.

Despite the overwhelming participation of women in the labor market (in 1980, 70.8% of women between the ages of fifteen and fifty-four were wage earners in the primary economy), the significance of nonsalaried tasks to a woman's workday had not waned (Bálint 1983, 141). The primary identification of women was with their domestic activities. The burdens that women bore in the household have been referred to as the "second shift" (Goven 1993; Kulcsár 1985; Moskoff 1978, 1982; Volgyes 1985; Volgyes and Volgyes 1977; for comparative material within Eastern Europe, see Wolchik and Meyer 1985). The dominance of nonwage work in the identity of women meant that it was much rarer for women than for men to speak explicitly of the monetary value of their labor, or of the skill they brought to a particular job. If they did comment on these issues, it was only in reference to their salaried workplace, never to their domestic or private activities. In projects like canning or garden work, the equation of time with money was completely absent. The imperative was to avoid spending money, regardless of how much time would be required to fulfill this commandment. When asked, women would laugh at the suggestion that they would be better served by buying food than by preparing it themselves (with the exception of the new breed of villager represented by families like the Kovácses, who were still a small minority).

The weight of domestic responsibilities did not always rest lightly on women's shoulders. Though resigned to their inferior position in the family and in the workplace, they did not find it just. More than once at the cooperative farm, the garden brigade was brought in to complete a project men had started, but not finished. They were quick to point out that they were being paid less than their male colleagues, to put right a task that men had been incapable of doing properly. "Will you write in your book about how hard women work? Here only women work, men do not. Men do not know how to work. Everything that is accomplished is done by women." These comments were considered universally true, not only characterizing the organization of work at the farm. Men were seen to spend their time talking and drinking. "Men know how to do only one thing: make children."

The particularly onerous burden of village women's responsibilities in second-economy activities was represented in discrepancies in time commitments to household labor between cooperative farm women and those in other fields and urban communities.

> Women working as manual workers in agriculture—employed or as members of cooperative farms—spend 5.2 hours of the total 11.5 hour work time in their average daily schedule on income activities. The larger half of their work time is devoted to their own farm, domestic chores, taking care of family members. . . . Women working as members of cooperatives spend twice as much time on domestic chores as women in industry or other employment (3.1 hours). (Kovácsné Orolin 1976, 216)[36]

Attitudes among young women in the village toward the time they spend in domestic activities echoed these same concerns. Weekends or holidays were not seen as times to rest, but as time to catch up with chores around the house. Furthermore, the pressures on women to devote all their nonsalaried time to domestic chores were great. This was clear, for example, in the particularly strong sanction against women using the library. It was understood that a woman could always find work to do in the home, and should not be permitted the leisure of book reading. The few women who did frequent the library packaged the books in newspaper to carry them home in disguise. Otherwise, their neighbors would have made veiled, and not so veiled, comments on their irresponsible behavior. Although men often spent all of their weekend, "non-shift" time working around the house or moonlighting in the village, it was considered reasonable for them to take time off on occasion to attend a soccer game, holiday events organized by the cooperative farm, or even to meet friends at the pub for a drink or two.

---

[36] "Cooking is the most time-consuming domestic chore, amounting to nearly one-third of total housework time. In the case of industrial workers or employees it is one hour, for cooperative members two hours. The lack of workplace canteens on one hand, and generally the absence observed in cooperatives and village stores of so-called oven-ready food items, explain the difference. So too cleaning demands more than one hour a day—not a full hour in the case of industrial worker women and employees—but here is included keeping the front yard of the domicile clean, and heating and carrying water. There are no characteristic differences in the case of urban and rural women in the time demands of washing and ironing. The use of washing machines is extraordinarily widespread. . . . Shopping is the only domestic chore where village wives save time in comparison to urban wives. All sorts of things grow in the garden round the house, which the urban woman is forced to buy" (Kovácsné Orolin 1976, 217).

While women defined themselves in terms of the work they performed around the house, this did not necessarily translate into their identification with the increasingly market-oriented work of the second economy. It was very difficult to carry on the scale of agricultural production I have described for the Pál family without the full-time labor investments of at least one spouse. Cows were especially demanding in this respect; cash-cropping solely in grains or tobacco, with limited animal husbandry, was more manageable alongside two jobs. In cases where full-time home work was demanded, it was usually the woman who left her job and not her husband. This decision was based on two factors: (1) a poorer wage scale was generally offered to women, and (2) as the primary child-care giver, women found it convenient to be at home full-time with young children. Yet as second-economy activities became more and more oriented to cash-cropping or providing skilled labor, women were excluded as primary actors.

> In the case of farms producing for traditional subsistence, the participation of women dominates. . . . With the growth in the intensity of work done in small farms and the increase in producing for market, an increasingly stable division of labor develops between spouses, and decisions regarding managing production are transferred gradually to the men's "hands." In small, specialized farms producing for the market the role of men is determinative with regard to the volume of work performed and even more so concerning the forming of strategy to be followed. (Répássy 1987, 128)

This was evident in the village in the shift of second-economy activities out of the kitchen and into the yard. The location of productive activities outside the house proper—the domestic sphere so closely associated with women—reflected the primary association now of commodity production with the male domain, and men's effective control over the process.

What can we conclude, then, about the character of work in the village? What sorts of differences can be discerned in people's attitudes toward work and their participation in different kinds of work, differences that influence participation in both the state sector and second economy? Notable differences existed between the young and the old in their attitudes toward work, and hence their motivations for participating in the second economy also differed. This could be discerned in the kinds of products chosen by the young and old to cultivate or raise, and the relative degree of intensification. The older generation tended to stick with traditional staples of the kitchen, essentially expanding production beyond subsistence. Instead of raising three pigs, they would raise ten; they would plant three times as much lettuce or onions, taking the surplus to market. Young people, on the other hand, calculated the character of their production explicitly in terms of time commitments, capital investments, and return on labor. Traditional staples like pigs were abandoned because they required so much work for so little return. They turned to raising musk, at least as long as the price stayed high, or to planting tobacco, since, as Róna Kati said, "You don't need to feed it." Theirs was a capitalist logic of utility and economic rationality, principles that were entirely absent in the structuring of domestic affairs by the elderly. Younger women, though participating in decisions to abandon pigs or take up tobacco, were much less apt than their male fellows to define their time in terms of money. Their attitudes toward time, money, and labor had less of a calculated quality, testifying to their stronger identification with domestic tasks than with their wage-labor positions.

The success of the Hungarian second economy rested, then, on the unique combination of two increasingly divergent ideologies of labor and identity. The incessant activity of the elderly, paired with the sharpened logic of utility characteristic of the young, resulted in high levels of production and a growing diversity in the range of agricultural pursuits.

Yet clearly not all villagers chose to express their sense of personal and social identity through labor in the second economy. For some it was the repudiation of additional work outside the primary workplace which defined their lives. Ibolya and Ferenc gained autonomy by structuring the time outside work toward the fulfillment of their personal pleasures and aspirations. Another villager comes to mind whose waking hours were preoccupied with managing his affairs so he could get away every weekend to hang-glide. Thus, the private sphere of villagers was paramount as the domain where social identities were forged and honor achieved. This was true even when the particular activities they chose to pursue differed substantially: from raising pigs and growing tobacco, to coaching soccer teams or hang-gliding.

Across all domains of production—in the first or second economy—marked differences characterized the motivations for and meaning of work by members of the older and younger generations. The differences evident between generations, here defined simply as those above and below the age of forty-five, were clearly related to their divergent work histories. Those below the age of forty-five spent their entire adult working lives in jobs outside the home and family, specifically in institutions controlled by the state and party. Those above the age of forty-five had come of working age prior to collectivization, and in some cases prior to the pressures to find jobs outside farming during the push toward industrialization of the 1950s. The younger generation talked very explicitly about their labor as a commodity. It had a monetary expression, which they took for granted. As a commodity, it was comparable to other commodities, that is, one's labor time could stand for or represent other objects. Very simply, labor had come to be defined in terms of its relation to other quantities or forms; it was comparable and convertible to other forms of value. One could work toward a vacation, as some villagers did; each cow anticipated in Juli's stable had already been identified with future purchases she had planned; Tóth Károly could refuse to take on a job in private construction, because he hadn't been offered the monetary value he understood to define his labor ("it isn't worth the effort"). The process of work became identified with its product, or defined in terms of its monetary or temporal value, for example, wage or shift.

Thus, value was substantially altered. Value was now expressed in relation to other qualities; it was now increasingly defined as a relation, a utility, an instrument, a means.[37] This stood in stark contrast to the conception of possessing activity among the elderly in which value was represented in the denial of comparability: one labored to realize one's self, admitting no intervening or commensurate factors. One was one's activity. The younger generation, on the other hand, discussed the monetary value of their time. Time, money, object: value lay in circulation, comparability, exchange, a far cry from the proud, intensely immobile possession embraced by their elders. A grand grid of comparability and exchangeability—of numbers, bottom lines, bits of time, and spurts of labor—structured the social universe of the younger generation.

---

[37] "Although social labor is always a means to an end, this alone does not render it instrumental. As noted, in precapitalist societies, for example, labor is accorded significance by overt social relations and is shaped by tradition. Because commodity-producing labor is not mediated by such relations it is, in a sense, de-signified, 'secularized.' . . . [C]ommodity-determined labor is, as concrete labor, a means for producing a particular product; moreover and more essentially, as abstract labor, it is self-mediating—it is a *social means* of acquiring the products of others. Hence, for the producers, labor is abstracted from its concrete product: it serves them as a pure means, an instrument to acquire products that have no intrinsic relation to the substantive character of the productive activity by means of which they are acquired" (Postone 1993, 180–181).

Thus, it was clear that the concept of possessing activity had been transformed, or, one might say, the nature of work had been transformed. The concept of possessing activity was still salient for the younger generation, as illustrated in the widespread denial of leisure by even the young. But for the younger generation, the object to be possessed was no longer activity, it was the commodity, the relational fetish of the market place. Living work, a process lived by different members of the community in different ways, had subtly altered its contours, its purpose, its meaning.

Despite the commodification of labor and its influence on attitudes toward work and identity among the young, possession or control of self, albeit the self as commodity, remained central. The second economy constituted for villagers a realm of control, an arena of mastery in which to define their selves against the state, to resist as fully as possible the appropriation of labor and of self. The family—the site of second-economy activity in village life—became the true autonomous organization in socialism, challenging the state's ability to eliminate all organizations outside its direct control.[38] The process of constructing a new collective society resulted in the firm embrace of private lives, further substantiating the distinction between public and private as necessarily opposing spheres (see also Hankiss 1984; Rév 1987), and further emphasizing the political significance of autonomy and resistance to control by the state (Róna-Tas 1990). This form of political alienation and resistance was expressed clearly by the Hungarian writer Konrád:

> Anyone who defines himself by the eight hours he spends earning his bread has given a profound inner affirmation of dependence. . . . Autonomy, if it means anything, means that I am not identical with my status. I sidle away from it. I step into it as into a costume. . . . I know many people—machinists, cabinetmakers—who work contentedly in a little shop next to their homes, turning out no more than their collective requires. When I ask them why they work at home, they smile: It's worth more than anything else to work without a boss. . . . Free time doesn't mean idle time; it consists of those hours of which we ourselves are the masters. This is our real life, our most precious possession; our free time most resembles ourselves. (Konrád 1984, 201)

Gerevics János attempted to distance himself from this common sentiment when we discussed his desire to work full-time outside

---

[38] Pine notes the same in Poland: "[I]n the Podhale region where I worked, and possibly in other regions as well, the family provided both in fact and in ideology an alternative and often oppositional social structure to the state" (1993, 230).

the state sector, privately. Even though he hated his boss, the scorned younger brother of the party secretary, János stated simply that this was not his motivation for wanting to leave. He argued fervently that his motivations were totally pecuniary, explaining that a desire to escape control by others was fruitless. "If I were a private farmer, then even the wind could order me about." Few, however, have spent as much time as János preparing strategies for leaving the world of managers and bosses behind for the drudgery of following the wind.

The broad cultural significance of mastery over self clearly also has relevance to political philosophies throughout Central Europe. Prominent in this genre is Bibó István's treatise (1960) on the Third Road toward national development, which eschewed the political traditions of East and West. In more recent years, the rebirth of interest in civil society and the concern for autonomy from the state have been articulated by writers in Hungary and Czechoslovakia, and by the theorists and participants of the Solidarity movement in Poland (Havel 1988; Konrád 1984; Michnik 1985; Pelczynski 1988; Vajda 1988). It is important, therefore, to emphasize the strongly political motivations that inform second-economy activity, political motives that weave through the politics of family life and reinvigorated a politics of privacy and intimacy in Hungarian society. This was a sentiment felt across class and across region, a passion that reached far beyond the walls of private homes in Sárosd or elite apartments in Budapest, beyond the boundaries of Hungarian society to all of Eastern Europe.

In his book *Socialist Entrepreneurs* (1988), Szelényi wished to account for the very small number of entrepreneurs (1–3%) who, in the 1980s, had increasingly diverged in their economic profile from their fellow villagers. His explanation for the shifts in patterns and attitudes of work over the last forty years rested on a theory of interrupted embourgeoisement. The development of capitalist farming discernible among certain segments of the peasantry between the wars was cut short, he claimed, by the command economy between 1948 and 1960. With collectivization, and shifts in specific policies toward agriculture in the 1970s, the opportunity for villagers to return to private entrepreneurial activities arose once more, leading families previously on a capitalist trajectory to resume petty commodity production.

Szelényi's approach to the rise of petty commodity production in the second economy was certainly a provocative one, and was quite congenial with other theories current in late socialism about the necessity of rejuvenating civil society and reinvigorating a thriving private economy. However, its goal was limited: to explain the particular history of the small entrepreneurial elite within the agrarian community of the 1980s. I have attempted to offer a different picture, portraying the general phenomenon of a thriving second economy, albeit one that demonstrates quite different patterns of work and attitudes toward money, time, and labor across the village community. Describing the rapid growth of petty commodity production in the 1970s and 1980s, I have emphasized the combination of two diverging patterns of work: the "rivalry in diligence" among the elderly and "labor as expedient, as utility" principle among the young. I have attempted to illustrate that complex shifts have taken place within the community regarding agricultural practices and conceptual notions over the past forty years as a direct result of the state's attempt to alter the character of economics and politics in Hungary, and the strong resistance mounted by villagers to this onslaught. In the course of this battle over land and labor, home and family, younger villagers have come to share the ideology of capitalist production and engage in capitalist practices that Szelényi identified as cultural capital left over from a prewar capitalist world. By demonstrating the generalization of capitalist forms of thought in the countryside, my portrayal also clarified how families not descended from prewar "incipient capitalists" could jump on the entrepreneurial bandwagon, and flourish in the private sector. All in all, the alliance between generations in the second economy during the 1970s and 1980s was a short-lived phenomenon. In the radically new context of reprivatization, it will be important to keep this in mind.

Yet Szelényi's analysis is limited not only by its focus on a small group within the village community, in itself a legitimate concern. His notion of cultural capital leaves little room for the constitutive role of social process, and so offers little insight into the complex historical dynamics that characterize this period. Szelényi's vision of cultural capital being hoarded in incipient capitalist families while they bide their time waiting for a change in economic policies seriously underestimates the impact of the socialist period itself, a complicated process refiguring and restructuring the conceptual worlds and practical activities of Hungarians themselves. No matter how rigorously rural Hungarians may have resisted state efforts to collectivize and restructure economic practices, they also actively participated in the reorganization of agricultural production, and actively engaged in new forms of wage labor, in new structures of production, in the rise of new markets within and outside the state sector. These organizational innovations, these quite pragmatic activities, were essential to the substantial shift in the attitudes and practices of rural Hungarians. Socialism cannot be construed simply as an interruption in an otherwise straightforward developmental trajectory, but must be analyzed as a significant historical period in the lives of rural Hungarians (and, one might add, in those of their urban compatriots). The great irony of the socialist state's impact is to have generalized attitudes across the entire populace which are usually defined as capitalist, creating a significant shift in attitudes toward work, toward labor, toward money, toward time, and toward commodities unprecedented in the Hungarian countryside. These were not latent concepts or unused practices; these were new ideas and novel practices.

Conclusion

Encompassed spatially and politically by the cooperative farm, families turned to activities at home to realize their personal ambitions. The affairs of the cooperative farm were frequently ridiculed, the alienation of state employment decried. The cultivation of private gardens became the measure of social respectability, villagers reveling in their resistance to control by the state and the party.

But the broad consensus among villagers over the moral superiority of private work concealed notable differences in how they worked and why. Simply reproducing an opposition of public and private worlds—so dominant in 1980s Hungary—would overlook significant variations within the community. Families constructed their houses in various ways, just as they lived their lives in divergent fashion. The motivations and practices of the older and younger generations showed the greatest disparity, but significant differences could also be discerned between men and women, and between those in the younger generation who pursued private entrepreneurship, and those who sought leisure. In short, dynamics of generational change, incipient class divisions and gendered forms of differentiation were all at work.

Villagers actively constructed these differences, but did so in a context dominated by the socialist state. The state, in hopes of reforming socialist economic practice, fostered economistic attitudes and a spirit of private effort, contributing to a general environment of utilitarianism and individualism, particularly among the young. Thus, while proclaiming a historical role in creating a new society devoted to collective goals and socialist ethics, the state and party actively undermined their own claim to legitimacy. All the while, villagers, for whom the concerns of state, of Lenin, and of Marx were of very little interest, celebrated the dignity of labor and the solidarity of autonomy.

SEVEN

Conclusion

So we are left with the curious image of highly individualistic, utilitarian actors born on the road to collectivism. The simple explanation frequently offered for this contradiction between socialist principles and socialist practice is that party officials and planners were disingenuous. They had no intentions of radically altering society, only of restructuring it to their own personal advantage. Certainly this reflects a view commonly held in village life. Another variation on this theme is that since the Stalinist period, the party consistently mounted a retreat from true socialist principles, only to back themselves into a dead end. At times I myself have held this view, seeing every compromise in socialist policy to be a capitulation to capitalist categories of human nature, incentive, and price mechanisms. Neither of these descriptions captures the complex process of social restructuring that we may now define in the past tense as the socialist experiment in Hungary.

I wish to offer a different perspective. Socialist theorists and politicians in Hungary committed themselves, and the nation, to building an alternative to capitalism. However, the manner in which this goal was pursued seriously undermined its being achieved. The methods employed by socialist politicians and planners to alter the character of economics and politics contributed substantially to the growth of many of the attitudes and institutional configurations they had pledged to eliminate. While they may have discarded a profit motive, socialist economists introduced alternative means of calculating intention and constructing reward in the elaborate project of planning. While to all appearances they sought to change class relations and hence empower the proletariat, socialist officials reinvigorated the hierarchies of industrial enterprises. Although the ostensible goal of collectivization was to alter relations of property, production, and authority, cooperative farms were handed over to former peasant elites in the early 1960s, enshrining them as managers over once manorial servants. For many years the complex forms of work units and auxiliary benefits such as *háztáji* plots were the primary sources of income for cooperative farm members. This minimized the use of money in determining wage scales. With time, however, monetary wage forms became the dominant means of evaluating one's social position within an enterprise, industrial and agricultural. Indeed, some would claim that the institution of work units itself initiated the calculation of time and effort central to the monetarization of labor in agriculture.

Yet the convoluted story of socialism in Hungary cannot be told only as a tragic, failed attempt at social planning. The intentions of social actors—in this case, planners and party elites—cannot be accorded such great weight in the telling of social history; far more is at play. The version of history that emphasizes the failure of socialist planning portrays villagers and others not responsible for policy-making as victims of the foibles of Communist party malfeasance. This too is a familiar theme, voiced repeatedly around kitchen tables in village and town, and now easily printed in the press since the transition. In the story told in the previous pages, we have seen time and again how villagers forcefully resisted Communist party policies and the intrusion of state agencies in their lives. Clearly, the social process issuing in the extreme utilitarianism and economism of village social life must be seen as the direct product of the actions of many actors: of the lowly pigherd as well as the powerful party secretary. Politicians and planners consistently attempted to find novel solutions to the problems of economic development, and their innovations were met with equally creative solutions among workers, managers, families, friends. I do not wish to suggest that all actors were equally equipped to inflict suffering or to withstand the party/state's onslaught. The pains of forced deliveries, of collectivization, of managerial ascendancy are well documented above. Yet also well documented are the valiant efforts among villagers to retain their dignity and self-respect.

How have I explained this curious history? I began my analysis with a description of agricultural work and relations among villagers and agrarian workers in the interwar period. Labor occupied a central role in the conceptual and practical world of villagers, structuring relations among groups within the community. The centrality of labor to one's identity and honor within the community was manifested in labor-intensive patterns of agricultural production. Though they were familiar with markets—both in labor and in commodities—villagers were uncomfortable with the market's role in destabilizing moral commitments to place and threatening habits of diligence acquired though controlling one's own. Struggling for possession of one's labor, land, and family was a losing battle for most of the villagers of Sárosd before the tragedy of World War II.

In the early socialist period (1945–1956) a command economy characteristic of the Stalinist period was established. Central to this project was the rapid industrialization of the economy, reducing the role of agriculture in the overall profile of Hungarian economic production. The state worked diligently to sever the ties between property relations and production, nationalizing industry and commerce, and trying, though unsuccessfully, to force agrarian workers to abandon their property to collective control. The workforce underwent a massive process of proletarianization, achieved under conditions in which labor markets were suppressed. The results of these policies were two-fold. As a consequence of the policies attempting to separate out production and property, the identity of workers defined by their actions as *laborers* was strengthened. This reinforced the labor-centered value system of agrarian communities already familiar in pre–World War II society. The second consequence of the massive proletarianization of the workforce was to initiate a process whereby many adults were drawn into a daily regimen of working in large, hierarchical enterprises. Their labor was submitted to control by others, which secured their subordination to formal organizational hierarchies demanding labor discipline. Efforts to instill new attitudes toward work were not very successful, but programs to raise productivity and to reduce patterns of absenteeism and tardiness in the early years of socialism introduced new ideas to the workforce. Workers were now acquainted with innovative concepts of time and notions about commitment to an industrial regimen of diligence which were alien to their own work histories in agriculture prior to 1948.

Thus, despite the absence of active labor markets, workers employed by the state were drawn into industrial settings whose structuring of time, space, effort, and identity had many features similar to industrial production in market-based economies. Cooperative farm members and workers employed at state farms escaped the most intense forms of socialization into modern, industrial work regimens, sharing with those still working their private lands a calendar of effort and intention in rhythm with the agricultural seasons. Yet the struggle to reorganize agriculture collectively did bring new regimes into being, and hence new problems arose. The loss of lands and independence among the cooperative peasantry and state farm workforce, as well as the debilitating experiences of socialist industrial workplaces, produced anger and dissatisfaction. Villagers regularly resisted the socialist state's intrusion in their lives, in small acts of defiance and in the much grander rebellion of 1956.

Following the revolution, the long-term approach to agriculture changed. In the short term, collectivization was finally achieved, forcing virtually everyone engaged in agricultural production to join a cooperative farm, find a job at the state farm, or leave for industry. Early cooperative production was stymied by widespread resistance among the workforce, prompting concessions to preferences for family-based production structured through sharecropping contracts. With the reforms, the long-term consequences of an altered approach to agriculture were initiated. Agricultural enterprises were restructured in crucial ways. Reforms of the economy, designed primarily to alter industrial production, had an influence on cooperative production as well, rehabilitating its value in a socialist context, and making possible the growth of a vibrant second economy in agricultural production. Over time, production was increasingly mechanized, changing the character of everyday work at the farm and influencing the composition of the workforce. New notions of skill and education were introduced with the technocratic revolution in cooperative farm management and organization, changing the character of local politics at the farm. Finally, money was introduced as the primary means of remunerating the workforce, and decision-making over wage policies became an additional arena of disenfranchisement and dissatisfaction between management and the manual workforce.

What have been the consequences of these policies for village life in the last decade of Hungarian socialism? I divided my discussion of everyday life in the 1980s in two, first addressing the politics of planning, then turning to the economics of space. I considered planning in general terms as an exercise in the scientism of politics, and in particular, as an everyday practice in village life. The discussion of planning was prompted by a simple question: why must the plan always be fulfilled? The elaborate machinations associated with fulfilling plans at all levels of society were explained in terms of the peculiar situation that arises when the state represents itself as a science. The scientific project of historical materialism, embodied in the Marxist-Leninist party/state of

Hungary, required a steadfast commitment to calculation and calibration. Yet the implementation of planning—the assertion of abstract, and quite often distant, social goals for the polity—was realized in the very private schemes of managers and party elites whose plans for personal aggrandizement were successfully achieved despite their clash with the purported social goals of the project. The glaring disparity between social goals and personal gain was offensive to villagers not only as a prime example of party hypocrisy; the very project of planning itself was considered unnatural. It seemed contrary to human nature, or so villagers said, that abstract and impersonal social goals should be elevated over immediate personal consideration and private gain. In other words, the notion that value was born in immediate, and quite ephemeral patterns of social interaction had become hegemonic. Equally hegemonic, then, was the notion that personal advantage was important and meaningful in the construction of social relations. These views of transient value and individual purpose stood in stark contrast to the official representation of planning. Planning in this view emphasized the development of long-term goals, which would benefit the interests of large groups over those of individuals. Since the everyday practice of planning was often quite at odds with its official representation, the attitudes among villagers that planning was neither feasible nor desirable were reinforced. The numerical computation associated with planning may have been foreign to the deliberation of value and meaning for villagers, but rational calculation—the cautious weighing of advantage and consequence—was not. Hence I have argued that the rampant economism and utilitarianism of the early 1980s were related in part to the practice of planning, despite the consistent attempt of Hungarians (be they villagers or urban social scientists) to distance the cultures of private and public life.

By all accounts, the chasm between collectivized agriculture and private farming widened in the 1980s. The opposition of public and private production was quite visible in Sárosd: houses and barns were constructed differently, people worked with each other in quite different configurations, villagers remarked upon the differences they felt working at the cooperative and at home. The social configuration of public and private diverged, accounting for the differences in which economic ideologies of utilitarianism and individualism were lived. As the second economy grew more and more commoditized, this world was defined by the increasing role of men's control of family production. The refiguring of patriarchal control within the family bore similarities to the gender segmentation of active markets in labor characteristic of state employment. Reinvigorated patriarchy in the second economy was represented, however, as a return to traditional values, and seen as a legitimate means of structuring economic action within the home. So too, the possibilities for personal achievement and individual gain differed in the first and second economy, although in both arenas such concerns fueled individual interactions and institutional practices. People worked much harder at home, believing that personal ownership made a difference in how and why they worked. Quoting the expression "A farmer's eyes fatten his livestock" (*Gazda szeme hizlalja a jószágot*), villagers explained that they had a material interest in their possessions, and a passion for success in their own affairs. Their fondest desire was to be left alone, to work with their loved ones at their own pace.

Yet the easy contrast between public and private worlds masks as much as it reveals. Seen as opposing and quite different spheres of activity by villagers and analysts alike, the first and second economy shared important qualities of intention, production, and meaning. The increasingly commoditized universe of the first economy influenced the character of the second economy, which also became more and more commoditized. Attitudes and practices bred in socialist production, such as the equation of time, money, and labor, structured second-economy agriculture as well. So too, attitudes bred in an industrial universe about the convenience of a wage and the legitimacy of leisure led to new behaviors among young families who repudiated second-economy work entirely. Socialist planning and economic practice thus contributed in significant ways to the development of individualism and utilitarianism in rural communities, practices most firmly embodied and notions most quickly articulated by the younger generation of villagers. Thus, I have argued that the character of what would often be called capitalist attitudes and behaviors was in many ways the product of socialism in Hungary.

The final irony of the socialist experiment in Hungary was that the great success of Kádárism in developing creative means of addressing problems within the economy contributed to its eventual demise. This was actively facilitated by the entire population, whose consistent efforts at distancing themselves from public affairs not only forced changes in policy, but also contributed to the particularly vibrant character of late Hungarian socialism. Do the heightened utilitarianism and vigorous private economy of village life make the prospects for rural communities rosy in the transition? I think not.

The Transition

What does the story told in these pages portend for villagers in the years ahead? My brief comments are based on short visits to Sárosd and Budapest several times since 1989, as well as on sources that allow me to observe urban politics from afar. I wish to discuss potential changes from the view of those living in Sárosd, addressing the manner in which the transition has been perceived and understood by villagers. These comments will give a very different picture from that which has been offered by those promoting market economies in Eastern Europe, or those who look to entrepreneurs to transform Hungarian society. There is a class bias in much of the literature on the transition which is subtle and unremarked. It is important to provide alternative voices to the optimism assumed about the transition, voices of folks who do not welcome these changes with glee and who will probably not share in the wealth generated in new businesses or the advantages provided by the new politics.

The transition from socialism is defined by two major changes: a shift to a multiparty system, and the reintroduction of a market economy. The reintroduction of a market economy is widely hailed as an important and necessary innovation to rectify the problems of socialist planning and economics. Among formerly socialist states, Hungary is understood to be particularly well situated for this shift, since (1) the patterns of reforms had introduced some market pricing of labor and commodities, and (2) Hungary had a vibrant second or private economy prior to 1989. Yet one might ask whether the reforms have prepared Hungarians for a full-fledged market economy. Does the existence of a thriving second economy mean that Hungarians will easily slip into the bustle of the market? First one needs to ask whether the second economy was indeed a uniform phenomenon, or whether there were differences between the agricultural second economy and the private enterprises in other sectors which arose in the 1980s. In fact, although the agricultural second economy was the forerunner of much private economic activity in Hungary, it also had a special role in the economy at large, and so warrants special consideration. I believe that observers have misunderstood the character of the agricultural second economy in significant ways, seeing its participants to have embraced far more developed market behaviors and attitudes than has actually been the case. Accordingly, the expectations many have that rural Hungarians will welcome a swift transition, and accommodate easily to its rigors, are also misplaced.

I have argued throughout this book that labor was commodified during socialism. People, most notably younger men, came to understand by the 1980s that their labor time had a monetary expression. Paired with this new understanding was the quite visible comfort people expressed with individualism and utilitarianism. Villagers, however, were not schooled in the clever techniques a successful entrepreneur requires to negotiate in a full market economy. They did not regularly take risks in their private work. They were frequently discouraged by state agencies, or by their local proxies, from embarking on innovative projects. Or if local officials were supportive of their ideas, numerous other barriers to production and marketing existed. Those who were successful, the subject of Szelényi's analysis, were the exception. Not having lived in a vibrant market economy has made the transition difficult for many rural dwellers. They have struggled to make sense of this new world, to accommodate themselves once again to the grand restructuring of society and economy taking place. In short, the expectation following 1989 that Hungarian villagers would quickly transform themselves into astute entrepreneurs was unfounded, for this would have required a transformation in belief and action that is scarcely possible considering the recent history of the Hungarian socialist economy.[1] Let me illustrate these general points more thoroughly.

Since the reforms of 1968, Hungarians have become used to changes in prices for goods bought in stores, and changes in the price of labor at the workplace. But the degree to which villagers had to accommodate to price increases was limited. The price of goods in stores may have varied somewhat, but basic staples like bread, milk, and other primary foods had remained relatively fixed for decades. People expected that seasonal foods would vary

---

[1] Indeed, I would argue that many of the difficulties that Hungarian villagers are experiencing, and the fears that they express, are not unique. Most people living in capitalism are not entrepreneurs; surely most Hungarians will not become skillful capitalists overnight. The trepidations that rural Hungarians have voiced about the workings of the market stem in large part from their experiences in socialism. But I would hazard to guess that many people in strongly market-oriented economies share many of these concerns, finding the ins and outs of economic growth to be baffling. After all, a whole field of specialists thrive on explaining the arcane character of capitalist markets to us lesser mortals (and a slew of wealthy practitioners make money by figuring out its strange logic before everyone else does). In other words, I suspect that many of the problems I will discuss below—dissatisfaction with impersonal markets, fears of unemployment, dismay at the prospects of increasing poverty—are not limited to Hungarian villagers, but may haunt many people in capitalist economies across the globe.

in price, such as fruits and vegetables, and that rare commodities, such as stereo components or foreign cosmetics, could fetch a high price. In village communities like Sárosd, however, the relative fluctuation in seasonal food prices had very little effect, since villagers produced all but the most basic of foodstuffs within the household. Money earned in the state wage economy or in private, income-producing activities could be set aside for more substantial purchases, such as additional feed for livestock, clothes for the children, or bricks for home construction.

People also learned that the value of their labor in the wage-labor market could vary, a factor they could manipulate to their advantage. In the 1970s and 1980s, workers would regularly change jobs, to increase their salary or improve their benefits. In the 1980s, widespread innovations in organizing work within industrial settings—"enterprise business work partnerships"—clearly constituted a concession on the part of authorities to pressures among workers for higher wages and greater control over the process of work itself (Stark 1986; see also Stark 1989). While some workers pressed for wage concessions, others opted for better benefits. I noted the example of young men choosing to join the cooperative farm in full knowledge that their wages would decrease, but that the loss would be offset by greater freedoms in working in the village on weekends, and by gaining substantial supports through local agencies for their private farming.

Changes in price, either in commodities or in labor, could be offset in village communities by relying on the second economy in agriculture. Families could sell their privately produced foodstuffs to earn extra income, and villagers (primarily men) could sell their labor to neighbors and kin on weeknights and on the weekends to supplement their state-sector incomes. As the preceding analysis points out, the particularly active quality of the agricultural second economy in the 1980s was the result of two very different attitudes toward work: the rivalry of diligence of the elderly, and the increasingly commoditized work habits of the younger generation. This propitious conjunction of diverging worlds will no longer hold in the transition. An important pillar of production in the second economy will soon be gone: elderly villagers who have borne great sacrifices to produce basic foodstuffs with little concern for profit and time investments. Young villagers will no longer be able to depend on their elders to provide a steady flow of pigs and a bedrock of grains, such as corn and alfalfa.

Discrepancies in the character of the agricultural second economy were not only manifest in the divide between the elderly and the young. Divergent patterns have also been observed among younger villagers in their commitment to hard work, sacrifice, and entrepreneurial ingenuity. A few villagers have come to rely on their incomes from their primary wage-labor jobs, and engage in private production of foodstuffs only to supplement their kitchen budgets. They have not expressed an interest in entrepreneurship or the new business climate. But there are those among young villagers who have chosen to pursue an active entrepreneurial path—people who are taken with the possibilities of higher incomes and greater freedom. In this new environment, they will have to be more attentive to market fluctuations, willing to take risks, and creative in developing niches for their products. But theirs is not the path chosen by most villagers.

The relative freedoms in selling goods and labor which villagers enjoyed in the 1970s and 1980s are often pointed to now in the transition as an important training ground for capitalism, for entrepreneurship, and for individual prosperity. But what exactly were the lessons villagers learned in the agricultural second economy? The large majority of villagers marketed their produce through the cooperative farm or the cooperative marketing agency. The state encouraged villagers to avail themselves of these contracts, as it facilitated planning and channeled private production through state hands, often to lucrative advantage in foreign markets. For villagers, contracting with the cooperative farm or marketing agency was easy and quite reliable. Agents of the state would literally drive up to villagers' front gates to pick up their pigs for transport. This was neither an environment characterized by fluctuating prices nor one in which villagers actively took risks in developing a product and offering it to an unknown clientele with hopes of success. Villagers contracted with the cooperative farm for delivery long before the pigs were ready for sale, perhaps even before the piglets were born. In the 1980s one could certainly choose to sell animals or produce at private markets in towns or even in village communities, but in Sárosd (and in many communities across the nation), the local cooperative farm and marketing agency were preferred. People were far more interested in a guaranteed price than they were in the marginal differences in prices one could achieve selling at private markets held in various sites across the country.

Communities across the nation do differ in patterns of commercialization and commodification. There are certain areas that have a very different profile from that of Sárosd, such as those in the southeast where communities of Bulgarian truck farmers have actively participated in markets in produce since the turn of the century. In the 1980s, villagers from this region regularly sold their produce in Budapest, never hesitating to embark on a five-hour trek to reach market stalls of the capital city early in the morning. Villagers in Sárosd considered the trip to Budapest too much trouble, even though the city was an easy one-hour's train ride away. Great differences could also be noted within Fejér County in village commercial culture. In the village adjacent to Sárosd, villagers regularly sold melons to passing motorists, a habit they had no doubt acquired from the days of a thriving peasant economy prior to the socialist period. Past experiences seriously influence patterns of economic calculation and activity; rationality, after all, is a local product.

In fact, the tendency to take risks in the economy, or to avoid them, to create markets or to ignore them, was influenced not only by historical patterns of the prewar period, but also by the patterns of socialist reforms themselves. Villagers had come to expect the sometimes capricious changes bureaucrats and politicians would make in reform plans, outlawing what had been legal only days before. These changes could carry quite heavy penalties. Or, as we have seen in Sárosd, local party leaders would prevent the implementation of reforms encouraged in Budapest. Thus, people often held back from taking risks when opportunities presented themselves. In 1983, one villager had been toying with the idea of opening a pub in his basement, since new regulations permitted private firms greater leeway. He was short of money, and the prospect of drunks in his new house gave him pause. So he had thought about building a small shack alongside his new house. He needed to get further training in the restaurant business, to supplement his degree in commercial sales. But as we discussed the various concerns he had about the project, his father-in-law, recently elected to the town council, piped in. "Maybe five years from now we can really do this. But then, no one knows what the situation will be like then. Private business is really in full swing now, but how does one know whether they will tighten things up again, so that the state wouldn't have any use for it any more, or whether they will rescind those regulations?"

Willingness to take a risk in a more active market economy has not characterized villagers in Sárosd, or, I would argue, many other villagers across the country.[2] During socialism people lived

---

[2] Laki argues that a number of factors have strengthened the conservative approach of entrepreneurs throughout the entire second economy in Hungary (1993).

within an institutional context that mitigated the negative consequences of market risks or economic downturns. The benefits and often measly salaries in the state sector were secure, come what may in reforms extending the private economy. Contracts with marketing agencies and the cooperative farm may not have returned as great a profit as one might have hoped, but they were secure. And, with the exception of the few local entrepreneurs committed to increasing their profit margins, most villagers were content with earning a stable income from their private activities. Some may have tried raising tobacco to earn a prettier penny than they earned on pigs, but these were often short-termed experiments, enabled by the steady flow of cash from the ever ready rabbit supply and prosaic pig population. After all, it is worth remembering that few calculated the cost of their labor in private activities, obscuring the mechanisms of determining a "fair" price for their time and effort.

What did people do with the income earned in the private economy? With few exceptions, people consumed it. They used it to build houses for themselves and their children, to buy cars and television sets, to install plumbing and central heating. In the early years of the agricultural second economy, villagers devoted much money toward improving their basic standard of living. Indeed, in the early years of the agricultural second economy, during the 1970s, villagers were able to close the gap in living standards between country and city to a remarkable degree, lessening inequalities that had divided urban and rural inhabitants since the 1950s (Manchin and Szelényi 1987; Éltető and Vita 1987; Róna-Tas 1990).

But with time, even as these discrepancies were eliminated, villagers continued to invest their money in comfort and consumption, as other possible uses for their savings were severely constrained. Very few options were available to extend production in the private economy. Severe restrictions were placed on the purchase of land. Tools and other goods suited to the scale of private agriculture were virtually unavailable. The second economy in agriculture was quite labor-intensive, a pattern that stemmed as much from the absence of adequate tools as from a work ethic that promoted the use of brute strength over the expenditure of money on tools. As a result, monies were not devoted to investment in production or sales. There were occasional examples of entrepreneurs building elaborate hothouses to grow flowers or vegetables out of season, but their numbers were small and their influence on the patterns of consumption and saving was minimal. But even if one had accumulated a serious nest egg during the socialist period, these savings were generally insufficient to set up a business in the new economy.

The particular character of the agricultural second economy in the 1970s and 1980s was a fleeting phenomenon. Attitudes born in socialism, however, are less apt to be discarded. The pervasive "naturalization" of the economy in socialism—the flat, straightforward character of labor, of material goods, of production—continues in important ways to influence villagers' attitudes toward the construction of value in society. The moral character of labor and of goods produced through that labor issued as much from their necessary everydayness—from their central role in social experience—as from their utility. This is perhaps most evident in people's concern with selling agricultural goods. The collapse of the agricultural market within Comecon and with the former Soviet Union has hit villagers hard. Villagers complained to me that they could not fathom what they would do now that their wheat was still not sold (that is, committed to a buyer) as late as March, although this was long before the July harvest. Such expectations may seem unusual to those in long-capitalist economies where many—farmers and businessmen alike—know that prices can change substantially between planting and harvest. This possible discrepancy is unfamiliar to Hungarian villagers, who assume that brokers have a permanent interest in buying produce. For villagers, it is also inconceivable that the price of wheat—set by who knows who and who knows where—could fluctuate so radically that it would undercut the sale of wheat entirely, or that overproduction could mean bad prices. After all, wheat is the staff of life, the most treasured of all foods. More wheat in Hungary always meant greater prosperity. Before the war, more wheat meant more bread. After the war, more wheat meant that cooperative farm members would fulfill the plan, and possibly earn export monies abroad. Now more wheat can mean low prices or no buyers entirely. Formerly congenial members of Comecon (Hungarians, Poles, and Slovaks) outbid one another in the Middle Eastern markets, beating each other out of sorely needed markets for livestock and meat. Meanwhile, the Common Market keeps its doors tightly shut. Food has always been needed, villagers believe, and when people are starving, even more so. They wonder how it could go unsold. A clash of absolute value and market caprice, in which the market dominates, is simply incomprehensible. The workings of the market do not appear to make any sense to villagers, whose visions of farming have always been drawn from the infinite bounty of goodness and health the earth provides.

The notion that value inheres in goods independently of market forces is a belief found not only among village dwellers. Parliamentary debates in 1990 raged over the incompetence of government officials who sold state properties under their "true" value. When an economist, now sitting in Parliament, attempted to explain that the value of a property is determined by the price one can get for it, he was summarily dismissed. Yet it would not be entirely accurate to portray Hungarian villagers as unfamiliar with the volatility of market transactions. They knew full well, far better perhaps than their capitalist brethren, that goods could be quite scarce, and require much effort to obtain. But the relation between scarcity and price was not as simple as the usual laws of supply and demand would dictate; nor was one's ability to acquire scarce goods totally dependent upon the size of one's pocketbook. As Kenedi observed, pulling strings was a crucial technique in planned economies. The price of a scarce item might include bribing a store clerk, or it may not, if the clerk was one's cousin. Certain privileges were available only to high-ranking members of the party; wealth alone could not facilitate access. Hence elaborate strategies were required to negotiate the strongly personalized character of economic transactions in socialism. Who received a bribe, and when, could vary tremendously in the course of a transaction, dependent upon a variety of factors often quite specific to the individual, or even to the moment the transaction was taking place. The lesson drawn by most Hungarians, then, was that the value of a commodity may shift and change, but in direct relation to one's individual qualities and personal influence. Theirs was not a world dominated by impersonal markets and uninterested parties.

People's impatience with the market and pricing mechanisms these days stems in part from the state's role in managing the economy for so long in socialism. During the socialist period, market transactions were highly personalized. In an analogous fashion, national politics under the Communist party were eminently personified. The anonymity of market forces is quite at odds with the concrete, and quite visible, figure of the state: planning, projecting, protecting the economy. The state had an embodied character ("Uncle State"), clearly reified in everyday discussions and in daily newspapers. Planning was described as a deliberative, thoughtful process of structuring the economy. This representation, though often at odds with the process of planning (as we have seen), nonetheless has significantly influenced how villagers see the state's job in controlling economic factors. This representation was further abetted by the centralizing tendencies of the Marxist-

Leninist party, which claimed full responsibility for the economy. Villagers are perplexed about the state's reluctance now to intervene in economic affairs. Why doesn't the post-Communist state do something to halt spiraling costs, they wonder, or simply step in to stop inflation, or eradicate the chaos of unemployment? It is very difficult for people to reconcile themselves to a state committed to liberal market policies.

Villagers' discomfort with the workings of the market are related in part to their ambivalent attitudes toward money. The strong ethic present in the 1930s which placed a moral value on hard physical labor above all other activities was strengthened during socialism. Work was defined in terms of producing goods, a simple equation only muddied by the intervention of monetary considerations. This is not to say that in certain contexts money did not have its proper place. Selling one's goods on the market for a tidy sum, or finagling a higher piece rate from one's boss was appropriate and certainly respected as clever and shrewd. Such transactions made sense. But the more abstract transaction of making money from money, the predominant view of those engaged in commercial activities or anything not seen to be immediately productive, made less sense. Although labor may have been commodified, the capitalist project of moneymaking on its own terms was foreign to villagers. This attitude was also strengthened in socialism, when the shenanigans of capitalist financiers were parodied and bourgeois exploiters were regularly excoriated. Thus, it is with skepticism and dismay that villagers watch the rise of new banks, stock markets, and monied ventures, sure that those behind these activities are up to no good. Indeed, the success of such ventures reinforces the view strongly held in socialism that those engaged in honorable activities of production lose out to those in powerful and privileged positions. One more time, those who work with their hands will be excluded from wealth and prosperity. The growing gap between those who work and those who become wealthy, a gap that underscores the opposition between honorable and disreputable activities, is conveyed in the oft heard expressions "Those who work don't have time to make money" (*Aki dolgozik, annak nincs ideje pénzt keresni*) and "You can't become wealthy with honorable labor" (*Tisztességes munkával nem lehet meggazdagodni*). The activities of hard work and the machinations of moneymakers inhabit two separate worlds, and cannot be bridged.

Of course, many people are not working, and with limited financial means, cannot turn their leisure into profit. A number of significant industries in nearby Székesfehérvár, such as Videoton, have reduced their workforce substantially; some have closed down entirely. Harmony Farm is still operating, but is constantly on the edge of bankruptcy. Management appears to have lost confidence in its own abilities to manage, in the possibilities for agriculture, and in the meaning of its work; sensing its ambivalence, the membership has become even more critical of management. Members continue to expect decent wages and the same benefits package, even though there are few resources available to sustain these. The fact that nearly every neighboring cooperative has been dismantled, or has seriously decreased the size of its workforce, has not tempered their demands.

The threat to cooperative farming nationally has also been a severe shock for villagers in Sárosd. People had long acknowledged the advantages of cooperative production, even if their appreciation was heightened by the easy fit it made with private farming. But the happy accommodation between first- and second-economy farming during the 1980s would not alone have been sufficient to win villagers' support for collectivized agriculture. The end of drudgery in work—achieved by massive mechanization—was extremely important to many villagers, especially those who had spent their youth in the painful rhythm of work from sunrise to sunset. When asked in the early 1980s whether they would have preferred to return to full-time private farming, villagers responded unanimously in the negative, since they understood farming to be far too debilitating an activity. One might imagine a world in which small-sized machinery could have lessened the burden of private farming, but it was not available to villagers at the time, nor has it become widely available since.

Yet the availability of small tractors would not have convinced many of the advantages of full-time private farming. Villagers also approved of the economies of scale represented in the division of labor between cooperative grain production and private livestock raising. With the specter of decollectivization on the horizon, villagers scoffed at the idea of cultivating grains in plots of two to ten acres, the bare minimum for a subsistence farm in Sárosd prior to collectivization. The complex institutional form of cooperative farming was considered the modern way to pursue agricultural production and commerce, not least because the advantages of large-scale production were touted continuously in newspaper, radio, and television broadcasts during the socialist period. People agreed that modern economies of scale were appropriate. Support for modernized farming did not prevent cooperative members from criticizing the way in which farming policies had actually been implemented at the cooperative farm, or the unwarranted privileges management enjoyed, or the way management had padded employment registers with friends and family. Nonetheless, many aspects of collectivized or large-scale agriculture were seen as reasonable. So when the Smallholders' Party, which advocated decollectivization, came to campaign in Sárosd in 1990, they were laughed out of town.[3] In the 1990 elections, the village voted overwhelmingly for the Free Democrats (SzDSz),[4] a party whose liberal policies were not so attractive to villagers, but which voiced concerns about working Hungarians not articulated by Smallholders, Christian Democrats, or the nationalist Democratic Form (MDF).

The subsequent policy of decollectivization, promoted by the Smallholders and implemented by the coalition government (1990–1994), has been met with anger and disbelief in Sárosd. Many Hungarians believe that the one successful branch of the formerly socialist economy—agricultural production—has been dealt a severe blow. The advantages of cooperative production have been ignored, and ancient claims to property ownership glorified. Even if many villagers agree with the ultimate value of private property, the actions taken by the Smallholders seem all out of proportion to the current status and future of Hungarian agriculture. Sárosd's history as a poor community of manorial workers and harvesters differs from those villages where a strong peasant culture before collectivization is frequently recalled in memories of independence: independence experienced not only in farming, but also in local politics. In such communities, the Smallholders held more sway, although often among the elderly, no longer able to bear the burdens of private farming. Yet the reluctance of Sárosd's farmers to embrace private farming was in fact the majority view across the nation; in a poll taken in 1993, "over ninety percent of co-op members said that they would not opt for a breakup of the cooperatives into small-sized individual holdings" (Agócs and Agócs 1994, 33).

---

[3] The one villager who found their proposals attractive had been a wealthy landowner prior to the war, his family owning nearly 1,000 kh. He attended party meetings of the Smallholders in neighboring Seregélyes, a community that was far more congenial to their platform for having been a strongly peasant community prior to the war.

[4] Rhymes recited during the Easter Monday rituals, in which young girls are sprinkled to ensure their fertility, reflected SzDSz popularity in Sárosd. "This little cologne/Doesn't hurt anyone/Long live the Party/of Free Democrats!" (*Ez a kis kölni/Senkinek se árt/Eljen a Szabad/Demokrata Párt*). I strongly suspect that in previous years, these stanzas were recited to laud the Communist party, whose name was easily replaced by SzDSz.

The manorial history of Sárosd also plays a strong role in the fears people have about their own futures. Many are worried that they will lose all control over their lands to outsiders. If the current state of the claims process is finalized, then nearly 80% of the lands in Sárosd will have been claimed by those living in Budapest and other cities.[5] For most villagers, this raises the likelihood of becoming manorial workers once more, toiling on the soil at the behest and whims of others. The distress this prospect occasions is only made worse by the knowledge that those who have laid claims to the land are mostly city dwellers, who have no understanding of or commitment to agricultural production.[6]

What can villagers do, if the cooperative farm folds, and few jobs exist in the city? The specter of unemployment, and the actual problem of being unemployed, is particularly difficult in a society where avoiding work (*munkakerülés*) has long been offensive to villagers and for years was considered a crime by the state. Few rural residents have confidence that economic changes will constitute an improvement for those outside privileged financial circles or those who have turned political connections into economic fortunes. Some have suggested that villagers faced with unemployment should embark on their own careers. Yet few villagers have substantial savings to invest in new businesses, much less the confidence or courage to do so in an unstable and uncertain environment. In the early 1990s, the conditions for acquiring loans to start up a business were extremely difficult, and quite unreasonable. Collateral for most loans consisted of 150% of the actual value of the loan, which made it impossible for most to actually acquire any start-up capital. Friends of mine whose daughter had become a seamstress thought of buying a sewing machine to improve her business. This would have freed the daughter and mother from working at the local glove factory, where together they earned a monthly salary of 5,000 forints, a sum that was below the minimum wage set for only one wage earner. When the bank suggested they put up their foreign car as collateral, they couldn't believe their ears. They would not have needed to ask for a loan if they had been wealthy enough to buy a car. My friends' poor opinion of

---

[5] This figure was cited to me by a member of the town assembly.

[6] At certain points in the negotiations, proposals had been suggested that only those now working the land should be allowed to make claims to landownership. This suggestion was vetoed, though it would clearly have reflected the sentiments of most members of the cooperative workforce across the nation. Indeed, it would have legalized the slogan promoting land reform in 1945, "Land to those who work it!" (*Azé a föld, aki megműveli!*).

bankers and commercial interests was strengthened: they were completely out of touch with village life.

The difficulties of acquiring training and finding a job have affected young people the most severely. While advanced schooling may be respected in the abstract, few parents in Sárosd have encouraged their children to continue on to higher education. During the socialist period, teachers and intellectuals were often poorly paid, particularly in contrast to trades like carpentry or plumbing. Villagers knew trades paid better, particularly in the private economy. This knowledge was reinforced by the prevalent attitude that held that manual labor was more important than intellectual pursuits. While the gendering of occupations made it preferable for girls to work in offices and schools, they were not encouraged to aspire to intellectual pursuits either, especially as it might interfere with their responsibilities to husbands, children, and family. While in socialism it was common for a boy to leave grammar school with hopes of becoming a painter, carpenter, welder, or barber, now young boys are stymied, unsure where their fate will take them. Young girls may turn their predilection for white-collar employment to their advantage if they have the wherewithal to learn a few foreign languages, skills necessary for the secretarial jobs opening up in international trade and business. Otherwise they too can ponder where they might turn for job opportunities.

The collapse of a secure economic future for villagers since the transition has contributed to a widespread sense of anomie and cynicism. As recounted in these pages, many villagers supported the ideals of socialism—economic justice and democracy—but did not believe that these would ever be implemented. They strongly supported the social welfare benefits of the system, benefits that had been sorely lacking in the prewar world. Thus, the few tangible advances that they could identify with socialism are now eroding around them rapidly. Indeed, their fate as the poor and exploited appears to be worsening.

The second pillar of the transition has been the introduction of multiparty politics. But villagers do not expect much change here, either. Rural inhabitants are keenly aware of the ability of local elites to forestall and circumvent political and economic change. Long witnesses to the power of ensconced privilege, villagers are dismally aware of the intransigence of local authorities, even in the face of directives from central organs of the party and government. The disproportionate control of local potentates (the "small king syndrome") is widespread, and there is no indication that alternative parties will breed a new race of public officials to replace the old. After all, we know that "even saints line their own pockets first." Thus, rural inhabitants have no assurances that the current changes occurring in Budapest will constitute a serious change in local affairs, much less in economic policies and opportunities, which they consider of far greater importance than politics per se.

Nor are political attitudes quick to change. In the village, people would express their disinterest in all things political by claiming to be members of the Stomach Party. Although primarily a statement about the importance of one's material well-being, which includes the ability to provide for one's family, such a comment also conveys the common attitude that people wish to be left in peace, undisturbed by state agencies or party officials. Forty years of Marxist-Leninist party politics, in which all forms of local initiative and popular participation have been either prohibited or crassly manipulated, have clearly left their mark. Moreover, political practice prior to the Second World War could hardly be characterized as democratic. Despite a tradition before the war of active party politics, open parliamentary bickering, and biting editorials, back-room compromise frequently resolved public disputes, leaving most of the population unheeded and unrepresented. Hence, the general attitude among industrial and agricultural workers that politics is the exclusive domain of the upper classes stems from a long history of elite politics, be it among aristocrats, Communist party bosses, or even dissident intellectuals and leading figures from alternative parties.

Villagers have watched television broadcasts of parliamentary proceedings with much interest, but this has also reinforced their sense of exclusion.[7] In the first year of parliamentary politics in particular, the fascination of parliamentarians with issues like the national symbol or with righting socialist wrongs seemed far less important to villagers than the pressing economic problems they faced every day. Many villagers do follow political events closely, but with the sense that they have once again been left out of decision-making. Unfortunately, many of the political parties have not paid sufficient attention to reaching out to their constituencies. Parties have held many local meetings in communities

---

[7] The parliamentary deliberations have been broadcast regularly in Hungary since the transition. One can see the broadcasts televised in grocery stores and family homes, but the most loyal parliament watchers are the elderly and retired, who have time to follow the broadcasts day after day. Younger villagers and city dwellers find this a luxury they cannot afford, when they are desperate to secure a decent income in a variety of jobs.

across the country, and some representatives visit their districts regularly. However, it is difficult to identify any attempts to build strong local institutional bases for party politics. While jokes made about certain parties continuing the politics of the former Marxist-Leninist government may be exaggerated, the tendency among politicians to consider themselves capable of pursuing a national agenda without consulting their constituents' views does feel eerily familiar from the socialist period.

What are the prospects for villagers taking a more active role in organizing on their own behalf? The presumably socialist value of collective action has rarely had much purchase in Hungarian village communities, ridiculed as a silly notion impossible to implement when human nature works against it. This is true despite their positive experience in collective farming. Hence it is difficult to imagine villagers rallying to each other's support in mass demonstrations or trade union activities. The complete dismissal of trade unions, because of their history in socialism as tools to be manipulated by managerial elites, has also hampered their potentially positive political role in the postsocialist period. Passivity is a stance wholly in keeping with alienation, and moreover, a safe bet, for whenever Hungarians have taken to the street, to the factory, or to the town hall to rectify years of social injustice, they have been harshly punished for it. This was just as true in 1919, when innocent people were slaughtered by Communists and Radical Rightists alike, as in 1956, when so many thousands lost their lives.

The discomfort many villagers feel with collective efforts is in line with the strong individualist strain of community life. After all, in the late socialist period the most successful means of achieving one's goals was to go one's own way, circumvent the rules, use the back-door approach (the "small gate" or *kis kapu* option). Personal ties were manipulated to one's advantage, and that of one's family. In city and countryside, Hungarians still scurry along to assure their own success, often knowing full well that this may not bode well for others.

The picture I have drawn here is quite grim. But I do harbor some hope. That hope resides in the strength, tenacity, and dignity I have come to know in the people of Sárosd. Reading the first issue of the new newspaper published by the village government, the *Town Crier* (*Kisbíró*, February 1991), I am reminded that much can be done on the local level to improve the quality of life. Newly elected officials hope to win the confidence of their electorate, making promises that things have changed.

It is our goal to inform you, openly and honestly, of our work and our troubles.... The new assembly counts on the ingenuity, knowledge, cooperation, and initiative of the inhabitants. Perhaps one of the greatest mistakes of the past regime was that it didn't trust the people enough. It took control of their fate out of their hands.... Our representative body can only be effective if in the course of time it helps a local self-organizing society to develop, one in which individuals, societies, and associations can work for their own goals. We must recognize that without the functioning of communities, personal desires cannot be realized, nor can true democracy emerge locally, or on higher levels. We would be very satisfied if in the near future we would achieve at least partial results in the creation of democracy.

Kiss Tivadar, mayor

Whether indeed the goodwill expressed in these lines bears fruit, whether a sense of community spirit can flourish, will depend a great deal on the larger political and economic forces in coming years. But I have every confidence in the ability of villagers in Sárosd to weather the storm with humor and integrity.

Consequences

The process of social change during Hungarian socialism, actively promoted by state agencies and party officials, bears similarities to earlier efforts at social change, such as colonialism and missionization. Like their fellow missionaries, party officials envisioned creating a better society, convinced of the righteousness of their cause. Like colonial authorities before them, socialist bureaucrats were fully confident of the modern techniques they employed to transform society. However, the goals set and techniques used were often at odds, generating profound contradictions in the socialist project. These contradictions were patently clear to villagers, who found the socialist experiment an exercise in hypocrisy. While they pined for greater social justice, villagers considered socialism an affront to human nature. Yet the complex social process of confrontation, manipulation, and negotiation now known as "once-existing socialism" substantially altered Hungarian society. Party elites compromised their original goals, devising programs they could live with. Villagers struggled against the state, but were actively engaged in transforming the socialist project. The consequences are clear. "Colonized peoples like the Southern Tswana [or

Hungarian villagers] frequently reject the message of the colonizers, and yet are powerfully and profoundly affected by its media. That is why new hegemonies may silently take root amidst the most acrimonious and agonistic of ideological battles" (Comaroff and Comaroff 1991, 311).

I have argued in this book that socialism honed the culture of capitalism. Categories familiar from the capitalist world—individualism, utilitarianism, economism—all came to characterize the actions and thoughts of Hungarian villagers. I have tried to make clear that these qualities were actively constructed during the socialist period. This has not been an argument about capitalist attitudes and actions, which had developed in the interwar period, being shelved for a time and then reinvigorated in later socialism. Rather, my analysis has demonstrated that the constellation of social practices and attitudes that characterized agrarian life changed in complex and subtle ways over time, initially as the transition to capitalism took hold, and then as socialism replaced it as the dominant political and economic form. The character of work, of attitudes toward work, of family life, of markets, and of money changed in significant ways between World War II and the 1980s. By examining the practices in which nearly all adults participated during the socialist period—complex productive organizations, convoluted relations of private and public intercourse, heated debates over the meaning and morality of community and commerce—one can see that the social practices and attitudes I identified as capitalist in the interwar period differed substantially from those of the late socialist period, attesting to the profound transformation of rural social life. That these changes were not those anticipated tells us much about the complexity of social history.

House of Mirrors

My primary concern from the very beginning of this study has been to construct a theory of society which assumed historical change rather than incorporate history as an afterthought. Society is a historical product; social forms change over time, albeit subtly and in complex ways. The attempt to offer a theory of society as a historical project has vexed and fascinated social scientists from the outset. We have been bequeathed numerous studies, not least the works of the founders of the discipline—notably Marx and Weber—which developed theories concerning the historical development of social forms. Indeed, the question of history and the quandary of society have been implicated in one another since the mid-nineteenth century.

My quest for a historical explanation of contemporary socialism was launched by a curiosity about the dynamics of social action.[8] But an attention to historical explanation was strengthened by my experiences in the field. Villagers possess a strongly historical understanding of their own lives and experience. In the village in particular, and in Hungary generally, history is seen to be a quite active player in everyday affairs. People look to history to explain the present, often seeing their lives fated by a looming and ever present past (Lampland 1990). Villagers thus shared with my theoretical ancestors a keen concern with historical change and social constraints.

My interest in social change was also shared by the practitioners of Marxism-Leninism. Using the tools developed by social analysts and political activists before them, party officials set about to transform society in radical ways. Though I saw committed Marxist-Leninists to be active practitioners and considered myself a mere disinterested observer, we both drew inspiration from Marx. To complicate the picture even further, the history of socialist thought, not least the work of Marx himself, was a product of the long transition to capitalism I wished to understand.

It became clear in the very first stages of this project that theory and ethnography would be intimately intertwined, perhaps more so than in studies of other regions or of different times. Thus, this study has often been a confusing sojourn in a house of mirrors: chronicling the rise of categories that we now take for granted, both as a social phenomenon and as analytic reality. The experiences and explanations for social reality offered by villagers and other Hungarians were the product of the transition to capitalism that I sought to portray. So too, the analytic categories I employ regularly to explain these phenomena—commodities, labor, money, and the quantification of time and space—were artifacts of the social history I was investigating. Thus, the historicity of analysis and social experience has been at the heart of my work from the start.[9]

---

[8] It is interesting to recognize that my question about the *how* of capitalism also reproduces the logic of capitalism itself, that is, the centrality of instrumentalism in capitalist culture and society. For a discussion of the notion of immanent critique central to Marx's political economy, see Benhabib 1986 and Postone 1993.

[9] This reflexivity refers not only to the specific categories of political economy I am chronicling, but also to the attention to the historical process itself. For a discussion of history as a sign of the modern, see Dirks (1990).

One of the reflections in this house of mirrors which has riveted my attention has been the parallel image of commodification and theories of meaning in social analysis. The complex meaning of *dolog* caught my attention in the very first months of fieldwork. My years of city living had taught me that *dolog* meant "thing," but this notion was undermined quickly as I began a serious study of agricultural work. The conflation of action and object in work was quite provocative, suggesting to me an intimate relation between the process of working and the objects produced in that activity. However, it soon became clear that the object most dearly possessed was labor. In other words, labor had come to be seen as an object that could be alienated from one's person and sold to others. This concept of *dolog* came to represent for me a condensed history of commodification. It bespoke the strange phenomenon of commodity fetishism: the curious process through which one could become detached from one's own activity and confront it as a tangible and finite object to be surrendered to others.

As I pursued this analysis further, I was struck by the similarities between the juxtaposition of self and activity in the development of commodification (Barker 1984), and the separation of meaning and action that had characterized social theories since the turn of the century. In a forthcoming book, I argue that in the early phases of the transition to capitalism, labor comes into its own as a category of social value and prestige. However, as the transition continues in the twentieth century, the significance of one's activity as a worker comes to be understood in terms of other qualities or objects, for example, money or time. In other words, labor was no longer seen to be valuable in and of itself, but now had to be considered in light of additional factors. It acquired value only in reference to other qualities of the social landscape. In short, value inhered in the equation, in the relation, in forms of reference.

So too, meaning has been detached from action. It is often portrayed in social analysis as a domain that floats high above everyday social practices; culture is seen by many as external to social practice, rather than understood as constructed and generated in action. In this view, meaning and culture can be constituted only in reference to other qualities or events.[10] Thus, the inherently

---

[10] Elsewhere I have discussed the parallels between the strange manner in which meaning is constituted in capitalism and the development of theories of social action in modern social science, discussing in particular Saussurean theories of linguistic meaning, their anthropological elaboration, and the fetishism of action in capitalism, a process that I refer to as referential (1991). Postone makes similar observations in his

processual quality of meaning has been difficult to capture, though many have devoted significant attention to this question theoretically and ethnographically (Bourdieu 1977; Comaroff 1985; Comaroff and Comaroff 1991; Foucault 1979, 1980a, 1980b; Munn 1986; Taussig 1987; R. Williams 1977).

If indeed the internal split between action and meaning in social analysis may be related to the fetishized displacements of action and object in capitalism, then this may clarify why recent attempts to make social analysis more historical have often foundered. While we may be transfixed by history, we have been puzzled by process. Unable to see meaning as an event, we cannot grasp the processual qualities of social forms. Hence transitions and transformations remain stubbornly elusive in theoretical formulations, and poorly rendered in ethnographic projects. Though the question of referentiality and commodification—and hence the analytical quandary of history and process—has not been explicitly developed in the preceding analysis, it has been a shadowing presence throughout.

---

recent analysis of commodity fetishism: "It seems that objects are accorded significance in capitalism in a different sense than in traditional societies. Their meaning is not so much seen as intrinsic to them, an 'essential' attribute; rather, they are 'thingly' things that *have* meaning—they are like signs in the sense that no necessary relationship exists between the signifier and the signified" (1993, 173, n. 114). The challenge of developing a notion of nonreferential meaning—of meaning as process or action— requires that we struggle against the most central form of constituting objects and value in capitalism.

BIBLIOGRAPHY

Archival Sources

Fehér vármegye és Székesfehérvár szabad királyi város föispánjának bizalmas iratai (FM f.b.i.)
Fejér megye alispánjának iratai (FM a.i.)
Fejér megye alispánjának bizalmas iratai (FM a.b.i.)
Hadtörténéti Levéltar Honvédelmi Miniszterium katonai politikai osztály (HLHM k.p.o.)
Fejér Megyei Levéltár: Gazdasági cselédek és idénymunkások öszeirása, adatgyüjtés (FM g.cs.a.)
Fejér Megyei Levéltár: Jakabszállási cédulagyüjtemény (FM j.c.)
Fejér Megyei Levéltár: Munkástobozás (1914–44) (FM m.t.)
Fejér Megyei Levéltár: Sárosd cédulagyüjtemény (FM s.c.)
Sárbogárdi járás föszolgabirájának iratai (SJ f.i.)
Sárbogárdi járás föszolgabirájának közigazgatási iratai: cselédbérkövetelések, cselédszerzödések (SJ f.i.cs.b.sz.)
Sárosd egyesületek alapszabályai (S e.a.)
Sárosd monográfia
Sárosd Nagyközség iratai (S n.i.)

Newspapers

Fehérvári Népszava, 1946
Fehérvári Napló, 1948, 1953
Fejér Megyei Néplap, 1953, 1956
Fejér Megyei Hírlap, 1960, 1968, 1974, 1982–1984
Kisbiró, 1991

Books and articles

Abel, W. 1980. Agricultural Fluctuations in Europe from the Thirteenth to the Twentieth Centuries. Translated by Olive Ordish. London: Methuen.

Agócs, Peter, and Sándor Agócs. 1994. "The Change Was but an Unfulfilled Promise": Agriculture and the Rural Population in Post-Communist Hungary. East European Politics and Societies 8 (1): 32–57.
Agrarian Theses. 1957. Az Agrárpolitikájának Tézisei, Magyar Szocialista Munkás Párt [Agrarian Theses of the Hungarian Socialist Worker's Party]. Budapest: Kossuth Könyvkiadó.
Alverson, Hoyt. 1978. Mind in the Heart of Darkness. New Haven: Yale University Press.
Antal, László. 1979. Development—with Some Digression (The Hungarian Economic Mechanism in the Seventies). Acta Oeconomica 23 (3–4): 257–273.
―――. 1981. Historical Development of the Hungarian System of Economic Control and Management. Acta Oeconomica 27 (3–4): 251–266.
Ashton, T. H., and C. H. E. Philpin. 1985. The Brenner Debate: Agrarian Class Structure and Economic Development in Preindustrial Europe. Cambridge: Cambridge University Press.
Bajcsy-Zsilinszky, Endre. 1938. Egyetlen Út: A Magyar Paraszt [The Only Road: The Hungarian Peasant]. Budapest: Attila Nyomda R.-T.
Balassa, Iván. 1985. Az Aratómunkások Magyarországon 1848–1944 [Harvest Workers in Hungary, 1848–1944]. Budapest: Akadémiai Kiadó.
Bálint, József. 1983. Társadalmi Rétegzödés és Jövedelmek [Social Stratification and Incomes]. Budapest: Kossuth Könyvkiadó.
Baló, György, and Iván Lipovecz, eds. 1988. Tények Könyve '89 [Fact Book '89]. Budapest: Computer World Informatika K.F.T.
Balogh, Sándor, ed. 1986. Nehéz Esztendök Krónikája, 1949–1953: Dokumentumok [The Chronicle of Difficult Times, 1949–1953: Documents]. Budapest: Gondolat.
Barker, Francis. 1984. The Tremulous Private Body: Essays on Subjection. London: Methuen.
Bauman, Zygmunt. 1974. Officialdom and Class: Bases of Inequality in Socialist Society. In The Social Analysis of Class Structure, edited by Frank Parkin, pp. 129–148. London: Tavistock Publications.
―――. 1990. From Pillars to Post. Marxism Today: 20–25.
Bell, Peter. 1984. Peasants in Socialist Transition: Life in a Collectivized Hungarian Village. Berkeley: University of California Press.

Benhabib, Seyla. 1986. Critique, Norm, and Utopia: A Study of the Foundations of Critical Theory. New York: Columbia University Press.
Benkö, Loránd, ed. 1967. A Magyar Nyelv Történeti-Etimológiai Szótára. 3 Kötet [The Historical-Etymological Dictionary of the Hungarian Language. 3 vols.]. Budapest: Akadémiai Kiadó.
Berend, Iván T. 1983. Gazdasági Útkeresés, 1956–1965: A Szocialista Gazdaság Magyarországi Modelljének Történetéhez [Exploring Economic Solutions, 1956–1965: Toward a History of the Hungarian Model of Socialist Economy]. Budapest: Magvetö Könyvkiadó.

———. 1984. A Megújulás Új Reformhulláma és Megtorpanása, 1965–1976 [The New Wave of Reform of the Regenerated Economy, and Its Recoil]. In Gólyavári Esték: Elöadások a Magyar Történelemröl [Evenings at Gólyavár: Lectures on Hungarian History], edited by Gábor Hanák, pp. 268–277. Budapest: Közgazdasági és Jogi Könyvkiadó.

———. 1990. The Hungarian Economic Reforms, 1953–1988. Cambridge: Cambridge University Press.
Berend, Iván and György Ránki. 1979. Hungary in the Service of the German War Economy during the Second World War. In Underdevelopment and Economic Growth: Studies in Hungarian Social and Economic History, pp. 220–252. Budapest: Akadémiai Kiadó.
Bettelheim, Charles. 1976. Class Struggles in the USSR, First Period: 1917–1923. Translated by Brian Pearce. New York: Monthly Review Press.

———. 1978. Class Struggles in the USSR, Second Period: 1923–1930. Translated by Brian Pearce. New York: Monthly Review Press.
Bibó, István. 1960. Harmadik Út: Politikai és Történeti Tanulmányok [The Third Road: Political and Historical Studies]. London: Magyar Könyves Céh.
Biernacki, Richard. 1995. The Fabrication of Labor: Germany and Britain, 1640–1914. Berkeley: University of California Press.
Blackbourn, David, and Geoff Eley. 1984. The Peculiarities of German History: Bourgeois Society and Politics in Nineteenth-Century Germany. Oxford: Oxford University Press.
Blum, Jerome. 1957. The Rise of Serfdom in Eastern Europe. American Historical Review 62 (4): 807–836.

———. 1978. The End of the Old Order in Rural Europe. Princeton: Princeton University Press.

Bodrogi, Tibor, ed. 1978. Varsány: Tanulmányok egy Észak-Magyarországi Falu Társadalomnéprjazához [Varsány: Studies toward the Social Ethnography of a North Hungarian Village]. Budapest: Akadémiai Kiadó.
Böhm-Bawerk, Eugen von. 1949. Karl Marx and the Close of His System. In Karl Marx and the Close of His System, by Eugen von Böhm-Bawerk and Böhm-Bawerk's Criticism of Marx, by Rudolf Hilferding, edited by Paul Sweezy. New York: Augustus M. Kelley.
Borsai, Ilona. 1968. Summásdalok [Migrant Worker Songs]. In A Parasztdaltól a Munkásdalig [From the Peasant Song to the Worker Song], edited by Imre Katona, János Maróthy, and Antal Szatmári, pp. 95–227. Budapest: Akadémiai Kiadó.
Bourdieu, Pierre. 1977. Outline of a Theory of Practice. Translated by Richard Nice. Cambridge: Cambridge University Press.
Braverman, Harry. 1974. Labor and Monopoly Capital. New York: Monthly Review Press.
Brenner, Robert. 1976. Agrarian Class Structure and Economic Development in Pre-industrial Europe. Past and Present 70: 30–75.
Bukharin, N., and E. Preobraschensky. 1921. A.B.C. of Communism. Translated by P. Lavin. Detroit, MI: Marxian Educational Society.
Burawoy, Michael. 1985. The Politics of Production. London: Verso.
Burawoy, Michael, and János Lukács. 1985. Mythologies of Work: A Comparison of Firms in State Socialism and Advanced Capitalism. American Sociological Review 50: 723–737.
———. 1992. The Radiant Past: Ideology and Reality in Hungary's Road to Capitalism. Chicago: University of Chicago Press.
Buzás, József. 1969. A Székesfehérvári Városi Tanács, Intézöbizottság és Direktórium Szervezete a Tanácsköztarsaság Idején [The Structure of the City Council, Executive Committee and Directorate of Székesfehérvár during the Period of the Soviet Republic]. Fejér Megyei Történeti Évkönyv 2: 255–264.
Central Statistical Office. 1982a. Mezögazdasági Statisztikai Évkönyv 1981 [Statistical Yearbook for Agriculture 1981]. Budapest: Központi Statisztikai Hivatal.
———. 1982b. A Mezögazdasági Kistermelés 1981. I. Kötet [Agricultural Small-Scale Production, vol. 1]. Budapest: Központi Statisztikai Hivatal.
Chatterjee, Partha. 1989. Colonialism, Nationalism, and Colonialized Women: The Contest in India. American Ethnologist 16 (4): 622–633.

Cockcroft, James. 1983. Mexico: Class Formation, Capital Accumulation, and the State. New York: Monthly Review Press.

Comaroff, Jean. 1985. Body of Power: Spirit of Resistance. Chicago: University of Chicago Press.

Comaroff, J., and J. L. Comaroff. 1987. The Madman and the Migrant. American Ethnologist 14 (2): 191–209.

———. 1991. Of Revelation and Revolution: Christianity, Colonialism, and Consciousness in South Africa, vol. 1. Chicago: University of Chicago Press.

Comisso, Ellen, and Paul Marer. 1986. The Economics and Politics of Reform in Hungary. International Organization 40 (2): 421–454.

Cooper, Frederick. 1980. From Slaves to Squatters: Plantation Labor and Agriculture in Zanzibar and Coastal Kenya, 1890–1925. New Haven: Yale University Press.

———. 1992. Colonizing Time: Work Rhythms and Labor Conflict in Colonial Mombasa. In Colonialism and Culture, edited by Nicholas B. Dirks, pp. 209–245. Ann Arbor: University of Michigan Press.

———. 1993. Africa and the World Economy. In Confronting Historical Paradigms: Peasants, Labor, and the Capitalist World System in Africa and Latin America, edited by Prederick Cooper et al., pp. 84–186. Madison: University of Wisconsin Press.

Cooper, Frederick, and Ann L. Stoler. 1989. Tensions of Empire: Colonial Control and Visions of Rule. American Ethnologist 16 (4): 609–621.

Corrigan, Philip, Harvie Ramsay, and Derek Sayer. 1978. Socialist Construction and Marxist Theory: Bolshevism and Its Critique. New York: Monthly Review Press.

Corrigan, Philip, and Derek Sayer. 1985. The Great Arch: English State Formation as Cultural Revolution. Oxford: Basil Blackwell.

Dániel, Arnold. 1931. A Mezőgazdasági Válság és a Magyar Nép Sorsa [The Agricultural Crisis and the Fate of the Hungarian People]. Budapest: Világosság R.-T.

Darvas, József. 1943. Egy Parasztcsalád Története [The History of a Peasant Family]. Budapest: Püski Sándor Dr.

Day, John. 1987. The Medieval Market Economy. Oxford: Basil Blackwell.

Denich, Bette. 1974. Sex and Power in the Balkans. In Woman, Culture, and Society, edited by S. Rosaldo and L. Lamphere, pp. 243–262. Stanford: Stanford University Press.

Dilley, Roy, ed. 1992. Contesting Markets: Analyses of Ideology, Discourse, and Practice. Edinburgh: Edinburgh University Press.
Dirks, Nicholas. 1990. History as a Sign of the Modern. Public Culture 2 (2): 25–32.
Djilas, Milovan. 1957. The New Class: An Analysis of the Communist System. New York: Praeger.
Donáth, Ferenc. 1965. Ki Jogosult a Földre? Szegényparasztok Vitái 1945-ben a Föld Felosztása Körül [Who Is Entitled to Land? The Debates of Poor Peasants in 1945 Surrounding the Distribution of Land]. Agrártörténeti Szemle 7 (1): 60–81.

———. 1969. Földreformellenes Hangulat a Birtokos Parasztságban [Anti-Land Reform Mood among Propertied Peasants]. Agrártörténeti Szemle 11(1–2): 85–99.

———. 1976. Gazdasági Növekedés és Szocialista Mezögazdaság [Economic Development and Socialist Agriculture]. Valóság 9: 18–32.

———. 1977. Reform és Forradalom [Reform and Revolution]. Budapest: Akadémiai Kiadó.

Dús, Agnes, ed. 1984. A Pártélet Kisszótára [The Pocket Dictionary of Party Life]. Budapest: Kossuth Könyvkiadó.
Éber, Ernö. 1932. A Magyar Mezögazdaság Bajforrásainak Kutatása [Research on the Sources of the Trouble with Hungarian Agriculture]. Magyar Szemle 15: 371–378.
Elek, Sándor. 1991. Part-Time Farming in Hungary: An Instrument of Tacit De-Collectivization? Sociologia Ruralis 31 (1): 82–88.
Ellman, Michael. 1980. Against Convergence. Cambridge Journal of Economics 4 (3): 199–210.
Elson, Diane. 1979. The Value Theory of Labour. In Value: The Representation of Labour in Capitalism, edited by Diane Elson, pp. 115–180. Atlantic Highlands, NJ: Humanities Press.
Éltetö, Ödön, and László Vita. 1987. Láthatatlan Jövedelmek és Jövedelem Egyenlötlenség [Invisible Incomes and Income Inequality]. Statisztikai Szemle 65 (8): 750–765.
Erdei, Ferenc. 1941. A Magyar Paraszttársadalom [Hungarian Peasant Society]. Budapest: Franklin Társulat.

———. 1972. Lenin and the Hungarian Co-operative Movement. In Year-Book 1971, edited by Antal Gyenes, pp. 67–78. Budapest: Cooperative Research Institute.

Evans, Grant. 1993. Buddhism and Economic Action in Socialist Laos. In Socialism: Ideals, Ideologies, and Local Practice, edited by C. M. Hann, pp. 132–147. London: Routledge.
Fabian, Johannes. 1983. Time and the Other: How Anthropology Makes Its Object. New York: Columbia University Press.

Farkas, Gábor. 1980. Politikai Viszonyok Fejér Megyében 1919-1945 [Political Affairs in Fejér County, 1919-1945]. Budapest: Akadémiai Kiadó.
Farkas, Zoltán. 1983. Munkások, Érdek- és Érdekeltségi Viszonyai [Relations and Levels of Interest among Workers]. Szociológia 1-2: 27-52.
Fazekas, Károly. 1984. Wage and Performance Bargaining on the Internal Labor Market. In Wage Bargaining in Hungarian Firms: Studies of the Institute of Economics, pp. 29-88. Budapest: Hungarian Academy of Sciences.
Fehér, Ferenc, Ágnes Heller, and György Márkus. 1983. Dictatorship over Needs. New York: St. Martin's Press.
Fehér, Lajos. 1972. Current Problems of Our Co-operative Policy. In Year-Book 1971, edited by Antal Gyenes, pp. 9-48. Budapest: Cooperative Research Institute.
Fehér, Zoltán. 1975. Bátya Néphite [The Folk Beliefs of Bátya]. Folklór Archivum 3: 9-207.
Fehérváry, Krisztina. 1995. Building the New Socialist City: Modernism and Modernization in Stalinist Hungary. M.A. thesis, University of Chicago.
Fél, Edit, and Tamás Hofer. 1969. Proper Peasants: Traditional Life in a Hungarian Village. Viking Fund Publications in Anthropology no. 46. Budapest: Corvina Press.
Fencsik, László, ed. 1980. Politikai Kisszótár [Political Pocket Dictionary]. Budapest: Kossuth Könyvkiadó.
Ferge, Zsuzsa. 1986. Fejezetek a Magyar Szegénypolitika Történetéből [Chapters from the History of Hungarian Policies toward the Poor]. Budapest: Magvető Kiadó.
Foucault, Michel. 1979. Discipline and Punishment. Translated by A. Sheridan. New York: Vintage Books.
———. 1980a. Power/Knowledge: Selected Interviews and Other Writings, 1972-1977. Edited and translated by C. Gordon. New York: Pantheon Books.
———. 1980b. The History of Sexuality. Translated by R. Hurley. New York: Vintage Books.
Fredrickson, George. 1981. White Supremacy: A Comparative Study in American and South African History. New York: Oxford University Press.
Friedman, Milton, and Rose Friedman. 1979. Free to Choose. New York: Harcourt Brace Jovanovich.
Gábor, R. István, and Péter Galasi. 1978. A "Másodlagos Gazdaság": A Szocializmusbeli Magánszféra Néhány Gazdaságszociológiai Kérdése [The "Secondary Economy": Several Economic

Sociological Questions on the Private Sphere Characteristic of Socialism]. Szociológia 3: 329–343.
Galasi, Péter, ed. 1982. A Munkaeröpiac Szerkezete és Müködése Magyarországon [The Structure and Function of the Labor Market in Hungary]. Budapest: Közgazdasági és Jogi Könyvkiadó.
Garon, Sheldon. 1987. The State and Labor in Modern Japan. Berkeley: University of California Press.
Gordon, Andrew. 1985. The Evolution of Labor Relations in Japan: Heavy Industry, 1853–1955. Cambridge: Harvard University Press.
Goven, Joanna. 1993. The Gendered Foundations of Hungarian Socialism: State, Society, and the Anti-Politics of Anti-Feminism, 1948–1990. Ph.D. diss., University of California, Berkeley.
Gross, Jan. 1988. Revolution from Abroad: The Soviet Conquest of Poland's Western Ukraine and Western Byelorussia. Princeton: Princeton University Press.
Grossman, Gregory. 1977. The "Second Economy" of the USSR. Problems of Communism 26 (5): 25–40.
Guha, Ranajit. 1983. Elementary Aspects of Peasant Insurgency in Colonial India. Delhi: Oxford University Press.
Guha, Ranajit, and Gayatri Chakravorty Spivak, eds. 1988. Selected Subaltern Studies. New York: Oxford University Press.
Gunst, Péter. 1975. A Magyar Mezögazdaság Technikai Fejlödése és annak Akadályai (A XVIII. Század Végétöl 1945-ig) [The Technical Development of Hungarian Agriculture and Its Barriers (From the End of the Eighteenth Century until 1945)]. Agrártörténeti Szemle 17: 42–53.
Hacking, Ian. 1990. The Taming of Chance. Cambridge: Cambridge University Press.
Hankiss, Elemér. 1979. Társadalmi Csapdák [Social Snares]. Budapest: Magvetö Kiadó.
———. 1984. "Második Társadalom"? Kiserlet egy Fogalom Meghatározására és egy Valóságtartomány Leirására ["Second Society"? An Attempt to Define a Concept and Describe the Character of Reality]. Valóság 11: 25–44.
Hann, C. M. 1980. Tázlár: A Village in Hungary. Cambridge: Cambridge University Press.
———. 1985. A Village without Solidarity: Polish Peasants in Years of Crisis. New Haven: Yale University Press.
Haraszti, Miklós. 1978. A Worker in a Worker's State. Translated by Michael Wright. New York: Universe Books.
Hare, P. G., H. K. Radice, and N. Swain. 1981. Hungary: A Decade of Economic Reform. London: George Allen and Unwin.

Hart, Keith. 1982. The Political Economy of West African Agriculture. Cambridge: Cambridge University Press.
Haskell, Thomas L., and Richard F. Teichgraeber III, eds. 1993. The Culture of the Market: Historical Essays. Cambridge: Cambridge University Press.
Havel, Vaclav. 1988. Anti-Political Politics. *In* Civil Society and the State, edited by John Keane, pp. 381–398. London: Verso.
Heller, András. 1937. Cselédsor: A Mezögazdasági Cselédek Helyzete 1935-ben, Különös Tekintettel a Székesfehérvári Járásra [Servanthood: The Situation of Agricultural Servants in 1935, with Special Consideration of the District of Székesfehévár]. Budapest: Szent István Társulat.
Héthy, Lajos, and Csaba Makó. 1974. Workperformance, Interests, Powers, and Environment: The Case of Cyclical Slowdowns in a Hungarian Factory. European Economic Review 5: 141–157.
―――. 1978. Munkások, Érdekek, Érdekegyeztetés [Workers, Interests, Bargaining]. Budapest: Gondolat.
Hidas, Peter I. 1983. A Mezögazdasági Termelés és a Parasztság Magyarországon az Elsö Világháború Idején [Agricultural Production and the Peasantry in Hungary during the First World War]. Agrártörténeti Szemle 25: 14–29.
Hollos, Marida, and Bela C. Maday, eds. 1983. New Hungarian Peasants: An East Central European Experience with Collectivization. New York: Columbia University Press.
Hobsbawm, E. J. 1964. Labouring Men: Studies in the History of Labour. London: Weidenfeld and Nicolson.
Humphrey, Caroline. 1983. Karl Marx Collective: Economy, Society, and Religion in a Siberian Collective Farm. Cambridge: Cambridge University Press.
Huseby-Darvas, Éva. 1984. Community Cohesion and Identity Maintenance in Rural Hungary: Adaptations to Directed Social Change. Ph.D. diss., University of Michigan.
Ihrig, Károly. 1935. A Mezögazdaság Irányítása [The Supervision of Agriculture]. Magyar Szemle 23: 123–131.
Illyés, Gyula. 1967. People of the Puszta. Translated by G. F. Cushing. Budapest: Corvina Press. Originally published 1936.
Janos, Andrew C. 1982. The Politics of Backwardness in Hungary, 1825–1945. Princeton: Princeton University Press.
Jávor, Kata. 1978. Kontinuitás és Változás a Társadalmi és Tudati Viszonyokban [Continuity and Change in Social and Cognitive Relations]. *In* Varsány: Tanulmányok egy Észak-Magyarországi Falu Társadalomnéprjazához [Varsány: Studies toward the So-

cial Ethnography of a North Hungarian Village], edited by Tibor Bodrogi, pp. 295–374. Budapest: Akadémiai Kiadó.
Jenei, Károly. 1969. Az Agrárkérdés Fejér Megyében 1918–1919-ben [The Agrarian Question in Fejér County in 1918–1919]. Fejér Megyei Történeti Évkönyv 2: 87–106.
Jenkins, Robert. N.d. Gender Differentials in Earnings Determination in the Contemporary Hungarian Labor Force. Paper presented at the annual meeting of the Population Association of America.
Johnson, Chalmers, ed. 1970. Change in Communist Systems. Stanford: Stanford University Press.
Jordanova, Ludmilla. 1989. Sexual Visions: Images of Gender in Science and Medicine between the Eighteenth and Twentieth Centuries. Madison: University of Wisconsin Press.
Jowitt, Ken. 1983. Soviet Neotraditionalism: The Political Corruption of a Leninist Regime. Soviet Studies 35 (3): 275–297.
Juhász, Pál. 1973. A Mezőgazdaság Fejlödésében Megjelenö Tehetetlenségröl [Inertia Appearing in the Development of Agriculture]. Közlemények 93. Budapest: Szövetkezeti Kutató Intézet.
———. 1976. A Mezőgazdasági Szövetkezetek Dolgozóinak Rétegzödése Munkajellegcsoportok, Származás és Életút Szerint [The Stratification of Workers at Agricultural Cooperatives according to Work Grouping, Background, and Life History]. Szövetkezeti Kutató Intézet Évkönyv 1975: 241–277.
———. 1979. Adatok és Hipotézisek a Mezőgazdasági Szövetkezetek Állandó Dolgozóinak Rétegzödéséröl [Data and Hypotheses Concerning the Stratification of the Permanent Workers of Agricultural Cooperatives]. Társadalom-tudományi Közlemények 2: 62–81.
———. 1983. Agrárpiac, Kisüzem, Nagyüzem [Agrarian Market, Small and Large-Scale Enterprises]. Medvetánc 2: 117–139.
———. 1984. The Transformation of Management, Work Organization, and Worker Endeavours in Hungarian Cooperative Farms. Budapest: Cooperative Research Institute.
———. N.d. Szövetkezet és Iparosodás [Cooperative and Industrialization]. Manuscript.
Juhász, Pál, and Bálint Magyar. 1984. Néhány Megjegyzés a Lengyel és a Magyar Mezőgazdasági Kistermelö Helyzetéröl a Hetvenes Években [Several Comments on the Situation of Polish and Hungarian Petty Commodity Producers in the 1970s]. Medvetánc 2–3: 181–208.
Kalecki, M. 1964. Theory of Economic Dynamics. London: Allen and Unwin.

Kára, Judit. 1973. Nagydobronyi Szokások és Hiedelmek [Customs and Beliefs from Nagydobrony]. Folklór Archivum 1: 40–63.
Károly, János. 1904. Fejér Vármegye Története. 5 Kötet [The History of Fejér County. 5 vols.]. Székesfehérvár: Csitári Kö- és Könyvnyomdája.
Károlyi, Mihály. 1931. Tiétek a Föld! Üzenet a Magyar Földmivesszegénységnek [The Land Is Yours! A Message to the Agrarian Poor of Hungary]. Paris.
Katona, Imre. 1958. A Vándormunkások Toborzása a Kapitalista Magyarországon [The Recruitment of Itinerant Workers in Capitalist Hungary]. Ethnographia 69: 29–52.
———. 1968. A Mezőgazdasági Munkásság Dalai [The Songs of the Agricultural Working Class]. In A Parasztdaltól a Munkásdalig [From the Peasant Song to the Worker Song], edited by Imre Katona, János Maróthy, and Antal Szatmári, pp. 21–93. Budapest: Akadémiai Kiadó.
Katona, Imre, Zoltán Simon, and Imre Varga. 1955. A Rigmusköltészet: Tanulmányok. [Rhymed Poetry: Studies]. Budapest: Müvelt Nép Tudományos és Ismeretterjesztö Kiadó.
Katznelson, Ira. 1981. City Trenches: Urban Politics and the Patterning of Class in the United States. New York: Pantheon Books.
Kay, Geoffrey. 1979. Why Labour Is the Starting Point of Capital. In Value: The Representation of Labour in Capitalism, edited by Diane Elson, pp. 46–66. Atlantic Highlands, NJ: Humanities Press.
Kemény, István. 1982. The Unregistered Economy in Hungary. Soviet Studies 34 (3): 349–366.
Kenedi, János. 1981. Do It Yourself: Hungary's Hidden Economy. London: Pluto Press.
———. 1986. Why Is the Gypsy the Scapegoat and Not the Jew? East European Reporter 2 (1): 11–14.
Kerbolt, László. 1934. A Beteg Falu [The Sick Village]. Pécs: Dunántúl Pécsi Egyetemi Könyvkiadó és Nyomda R.-T.
Kerék, Mihály. 1934. Földbirtokpolitika [Policy on Landed Property]. Budapest: A Magyar Szemle Társaság.
———. 1939a. Kollektivizmus vagy Földreform? [Collectivism or Land Reform?] Magyar Szemle 37: 231–234.
———. 1939b. A Magyar Földkérdés [The Hungarian Land Question]. Budapest: Mefhosz Könyvkiadó.
Kertesi, Gábor, and György Sziráczki. 1985. Worker Behavior in the Labor Market. In Labour Market and Second Economy in

Hungary, edited by Péter Galasi and György Sziráczki, pp. 216–245. Frankfurt: Campus.

Kideckel, David. 1976. The Social Organization of Production on a Romanian Cooperative Farm. Dialectical Anthropology 1: 267–276.

———. 1977. The Dialectic of Rural Development: Cooperative Farm Goals and Family Strategies in a Romanian Commune. Journal of Rural Cooperation 5 (1): 43–61.

———. 1982. The Socialist Transformation of Agriculture in a Romanian Commune, 1945–1962. American Ethnologist 9 (2): 320–340.

———. 1993. The Solitude of Collectivism: Romanian Villagers to the Revolution and Beyond. Ithaca: Cornell University Press.

Kligman, Gail. 1988. The Wedding of the Dead: Ritual, Poetics, and Popular Culture in Transylvania. Berkeley: University of California Press.

Kogutowicz, Károly. 1930. Dunántúl és Kisalföld Irásban és Képben [Transdanubia and the Small Alföld in Words and Pictures]. Szeged: M. Kir. Ferenc József Tudományegyetem Földrajzi Intézete.

Konrád, George. 1984. Antipolitics. Translated by Richard E. Allen. San Diego: Harcourt Brace Jovanovich.

Konrád, György, and Iván Szelényi. 1976. Social Conflicts of Underurbanization: The Hungarian Case. In Social Consequences of Modernization in Communist Societies, edited by Mark Field, pp. 162–178. Baltimore: Johns Hopkins University Press.

———. 1979. The Intellectuals on the Road to Class Power. Translated by Andrew Arato and Richard F. Allen. New York: Harcourt, Brace and Jovanovich.

Kornai, János. 1959. Overcentralization in Economic Administration: A Critical Analysis Based on Experience in Hungarian Light Industry. Translated by John Knapp. Oxford: Oxford University Press.

———. 1980a. A Hiány [Shortage]. Budapest: Közgazdasági és Jogi Könyvkiadó.

———. 1980b. "Hard" and "Soft" Budget Constraint. Acta Oeconomica 25 (3–4): 231–246.

———. 1992. The Socialist System: The Political Economy of Communism. Princeton: Princeton University Press.

Kornfeld, Móric. 1935. A Földbirtokpolitika körül [About Policy toward Landed Property]. Magyar Szemle 23: 211–218.

Kovács, Imre. 1937. Néma Forradalom [Silent Revolution]. Budapest: Cserépfalvi Könyvkiadó.

Kovácsné Orolin, Zsuzsa. 1976. A Mezőgazdaságban Dolgozó Nők Élet- és Munkakörülményei [The Life and Work Conditions of Women Working in Agriculture]. In Nők—Gazdaság—Társadalom [Women, Economy, Society], edited by Egon Szabady, pp. 205-224. Budapest: Kossuth Könyvkiadó.

Kriedte, Peter, Hans Medick, and Jürgen Schlumbohm. 1981. Industrialization before Industrialization: Rural Industry in the Genesis of Capitalism. Cambridge: Cambridge University Press.

Kula, Witold. 1976. An Economic Theory of the Feudal System: Towards a Model of the Polish Economy, 1500-1800. Translated by Lawrence Garner. London: New Left Books.

———. 1986. Measures and Men. Translated by R. Szreter. Princeton: Princeton University Press.

Kulcsár, László, and András Szijjártó. 1980. Iparosodás és Társadalmi Változások a Mezőgazdaságban [Industrialization and Social Changes in Agriculture]. Budapest: Közgazdasági és Jogi Könyvkiadó.

Kulcsár, Rózsa. 1985. The Socioeconomic Conditions of Women in Hungary. In Women, State, and Party in Eastern Europe, edited by Sharon Wolchik and Alfred Meyer, pp. 195-213. Durham: Duke University Press.

Laki, Mihály. 1993. The Changes for the Acceleration of Transition: The Case of Hungarian Privatization. East European Politics and Societies 7 (3): 440-451.

Lampland, Martha. 1989. Biographies of Liberation: Testimonials to Labor in Socialist Hungary. In Promissory Notes: Women in the Transition to Socialism, edited by Sonia Kruks, Rayna Rapp, and Marilyn B. Young, pp. 306-322. New York: Monthly Review Press.

———. 1990. The Politics of History: Historical Consciousness of 1847-1849. Hungarian Studies 6 (2): 185-194.

———. 1991. Pigs, Party Secretaries, and Private Lives. American Ethnologist 18 (3): 459-479.

———. 1992. Corvée, Maps, and Contracts: Agricultural Policy and the Rise of the Modern State in Hungary during the Nineteenth Century. Working Paper 3.2, Center for German and European Studies. Berkeley: University of California.

———. 1994. Family Portraits: Gendered Images of the Nation in Nineteenth Century Hungary. Eastern European Politics and Societies 8 (2): 287-316.

———. N.d. Transforming Objects: The Transition from Feudalism to Capitalism in Nineteenth-Century Hungary. Manuscript.

Laqueur, Thomas. 1990. Making Sex: Body and Gender from the Greeks to Freud. Cambridge: Harvard University Press.

Le Goff, Jacques. 1980. Labor Time in the "Crisis" of the Fourteenth Century: From Medieval Time to Modern Time. In Time, Work, and Culture in the Middle Ages, translated by Arthur Goldhammer, pp. 43–52. Chicago: University of Chicago Press.

Lencsés, Ferenc. 1982. Mezögazdasági Idénymunkások a Negyvenes Években [Agricultural Seasonal Workers in the 1940s]. Budapest: Akadémiai Kiadó.

Le Roy Ladurie, Emmanuel, and Joseph Goy. 1982. Tithe and Agrarian History from the Fourteenth to the Nineteenth Centuries: An Essay in Comparative History. Translated by Susan Burke. Cambridge: Cambridge University Press.

Lewin, Moshe. 1974. Political Undercurrents in Soviet Economic Debates: From Bukharin to the Modern Reformers. Princeton: Princeton University Press.

Ludden, David. 1992. India's Development Regime. In Colonialism and Culture, edited by Nicholas B. Dirks, pp. 247–287. Ann Arbor: University of Michigan Press.

Lukács, Georg. 1971. History and Class Consciousness: Studies in Marxist Dialectics. Translated by Rodney Livingstone. Cambridge, MA: MIT Press.

Lukács, László. 1983. Vándoralakok, Vándormunkások és a Területi Munkamegosztás Kelet-Dunántúlon [Vagrants, Itinerant Workers, and the Territorial Division of Labor in Eastern Transdanubia]. Alba Regia, Annales Musie Stephani Regis 20: 185–199.

Macartney, C. A. 1956. October Fifteenth: A History of Modern Hungary, 1929–1945. Edinburgh: Edinburgh University Press.

Magyar, Bálint. 1986. Dunaapáti, 1944–1958: Dokumentumszociográfia. 3 Kötet [Dunaapáti, 1944–1958: Documentary Sociography. 3 vols.]. Budapest: Müvelödéskutató Intézet és Szövetkezeti Kutató Intézet.

———. 1988. 1956 és a Magyar Falu [Nineteen Fifty-six and the Hungarian Village]. Medvetánc 2–3: 207–212.

Makó, Csaba. 1985. A Társadalmi Viszonyok Erötere: A Munkafolyamat [The Labor Process: An Arena of Social Struggle]. Budapest: Közgazdasági és Jogi Könyvkiadó.

Mallon, Florencia. 1983. The Defense of Community in Peru's Central Highlands: Peasant Struggle and Capitalist Transition, 1860–1940. Princeton: Princeton University Press.

Manchin, Robert, and Iván Szelényi. 1987. Social Policy under State Socialism. In Stagnation and Renewal in Social Policy: The Rise and Fall of Policy Regimes, edited by G. Esping-

Anderson, L. Rainwater, and M. Reins, pp. 102–139. White Plains: M. E. Sharpe.
Mandel, Ernest. 1978. From Stalinism to Eurocommunism: The Bitter Fruits of "Socialism in One Country." Translated by Jon Rothschild. London: New Left Books.
Mandel, Ernest, and Alan Freeman, eds. 1984. Ricardo, Marx, Sraffa: The Langston Memorial Volume. London: Verso.
Marrese, Michael. 1985. Hungarian Agriculture: In the Right Direction. Working Paper, Northwestern University.
Marks, Shula, and Richard Rathbone, eds. 1982. Industrialization and Social Change in South Africa. London: Longman.
Márkus, István. 1973. Az Utóparasztság Arcképéhez [Toward a Portrait of the Post-Peasantry]. Szociológia 1: 56–67.
Marx, Karl. 1967. Capital. Translated by Samuel Moore and Edward Aveling. Edited by Frederick Engels. New York: International Publishers.
———. 1973. Grundrisse: Foundations of the Critique of Political Economy. Translated by Martin Nicolaus. New York: Vintage Books.
Mátyus, Aliz, and Katalin Tausz. 1984. Maga-ura Parasztok és Uradalmi Cselédek [Peasants Who Are Masters of Themselves and Manorial Servants]. Budapest: Magvetö Kiadó.
McCloskey, Donald. 1985. The Rhetoric of Economics. Madison: University of Wisconsin Press.
Meek, Ronald. 1956. Studies in the Labour Theory of Value. New York: Monthly Review Press.
———. 1976. Is There an "Historical Transformation Problem"? A Comment. Economic Journal 86: 342–347.
Merey, Klára. 1956. A Mezögazdasági Munkásság Mozgalmai a Dunántúlon 1905–1907-ben [The Movements of the Agricultural Working Class in Transdanubia in 1905–1907]. Budapest: Szikra.
Michnik, Adam. 1985. Letters from Prison and Other Essays. Translated by Maya Latynski. Berkeley: University of California Press.
Móra, Magdolna. 1969. A Szociális Helyzet Fejér Megyében a Tanácsköztársaság Idején [The Social Situation in Fejér County at the Time of the Soviet Republic]. Fejér Megyei Történeti Évkönyv 2: 73–86.
Móricz, Zsigmond. 1982. Úri Muri [A Lordly Spree]. Budapest: Móra Könyvkiadó. Originally published 1927.
———. 1977. Rokonok [Relatives]. Budapest: Móra Könyvkiadó. Originally published 1932.

Morishima, M. 1974. The Fundamental Marxian Theorem: A Reply to Samuelson. Journal of Economic Literature 12 (1): 71–74.

Morishima, M., and G. Catephores. 1976. Is There an "Historical Transformation Problem"? A Reply. Economic Journal 86: 348–352.

Moskoff, W. 1978. Sex Discrimination, Commuting, and the Role of Women in Rumanian Development. Slavic Review 37 (3): 440–456.

———. 1982. The Problem of the "Double Burden" in Romania. International Journal of Comparative Sociology 23 (1–2): 79–88.

Mosse, George L. 1985. Nationalism and Sexuality: Middle-Class Morality and Sexual Norms in Modern Europe. Madison: University of Wisconsin Press.

Munkaegységkönyv [Work Unit Book]. 1960. Budapest: Mezögazdasági Kiadó.

Munn, Nancy. 1986. The Fame of Gawa: A Symbolic Study of Value Transformation in a Massim (Papua New Guinea) Society. Cambridge: Cambridge University Press.

Nagengast, Carole. 1991. Reluctant Socialists, Rural Entrepreneurs: Class, Culture, and the Polish State. Boulder: Westview Press.

Nagy, Imre. 1946. Magyar Földreform-Tervek [Plans for a Hungarian Land Reform]. In Agrárproblémák: Tanulmányok—Birálatok, 1938–1940 [Agrarian Problems: Studies, Criticism, 1938–1940], pp. 101–113. Budapest: Szikra Kiadás.

Nee, Victor. 1989. Peasant Entrepreneurship and Politics of Regulation in China. In Remaking the Economic Institutions of Socialism: China and Eastern Europe, edited by Victor Nee and David Stark, pp. 169–207. Stanford: Stanford University Press.

———. 1991a. Peasant Entrepreneurs in China's Second Economy: An Institutional Analysis. Economic Development and Cultural Change 39 (2): 293–310.

———. 1991b. Social Inequalities in Reforming State Socialism: Between Redistribution and Markets in China. American Sociological Review 56 (3): 267–282.

Nee, Victor, and David Stark, eds. 1989. Remaking the Economic Institutions of Socialism: China and Eastern Europe. Stanford: Stanford University Press.

Némethy, Béla. 1938. A Magyar Földbirtokpolitika Feladatai [The Tasks of Hungarian Policies toward Landed Property]. Magyar Szemle 34: 127–133.

Némethy, Endre. 1938. Kemenesaljai Hiedelmek a Földmüvelés és Állattenyesztés Köréböl [Beliefs Concerning Agriculture and Animal Husbandry from Kemenesalja]. Ethnographia 49: 230–231.

Nötel, R. 1986. International Finance and Monetary Reforms. In The Economic History of Eastern Europe, 1919-1975, vol. 2, Interwar Policy and the War and Reconstruction, edited by M. C. Kaser and E. A. Radice, pp. 520-563. Oxford: Clarendon Press.
Nove, Alec. 1983. The Economics of Feasible Socialism. London: George Allen and Unwin.
O. Nagy, Gábor. 1,966. Magyar Szólások és Közmondások [Hungarian Sayings and Proverbs]. Budapest: Gondolat Kiadó.
Orbán, Sándor. 1972. Két Agrárforradalom Magyarországon: Demokratikus és Szocialista Agrárátalakulás, 1945-1961 [Two Agrarian Revolutions in Hungary: Democratic and Socialist Agrarian Transformation, 1945-1961]. Budapest: Akadémiai Kiadó.
Pach, Zsigmond Pál. 1961. A Magyarországi Agrárfejlödés Elkanyarodása a Nyugat-Európaitól (a Feudalizmusból a Kapitalizmusba való Átmenet Magyarországi Sajátosságainak Kérdéséhez) [The Turning Away of Hungarian Agrarian Development from That of Western Europe (Toward the Question of the Hungarian Particularities of the Transition from Feudalism to Capitalism)]. Agrártörténeti Szemle 111: 1-7.
Packard, Randall M. 1989. The "Healthy Reserve" and the "Dressed Native": Discourses on Black Health and the Language of Legitimation in South Africa. American Ethnologist 16 (4): 686-703.
Pakulski, Jan. 1985. Bureaucracy and the Soviet System. Manuscript.
Palmer, Robin, and Neil Parsons, eds. 1977. The Roots of Rural Poverty in Central and Southern Africa. Berkeley: University of California Press.
Pamlényi, Ervin, ed. 1973. A History of Hungary. London: Collet's.
Pelczynski, Z. A. 1988. Solidarity and "The Rebirth of Civil Society." In Civil Society and the State, edited by John Keane, pp. 361-380. London: Verso.
Petánovics, Katalin. 1987. Vállus: Egy Summásfalu Néprajza [Vállus: An Ethnography of a Migrant Village]. Budapest: Akadémiai Kiadó.
Pető, Iván, and Sándor Szakács. 1983. A Gazdasági Intézményrendszer Átalakítása 1948-49-ben [The Transformation of the Economic Institutional System in 1948-1949]. Medvetánc 4: 205-229.
———. 1985. A Hazai Gazdaság Négy Évtizedének Története, 1945-1985. I. Kötet: Az Újjáépítés és a Tervutasításos Irányítás Idöszaka. 1945-1968 [The History of Four Decades of Domestic Economy, 1945-1985, vol. 1, The Period of Rebuilding and Com-

mand Economy Planning, 1945–1968]. Budapest: Közgazdasági és Jogi Könyvkiadó.
Pine, Francis. 1993. "The Cows and Pigs are His, the Eggs are Mine": Women's Domestic Economy and Entrepreneurial Activity in Rural Poland. *In* Socialism: Ideals, Ideologies, and Local Practice, edited by C. M. Hann, pp. 225–242. London: Routledge.
Porter, Theodore. 1986. The Rise of Statistical Thinking, 1820–1900. Princeton: Princeton University Press.
Postan, M. M. 1973. Medieval Trade and Finance. Cambridge: Cambridge University Press.
Postone, Moishe. 1993. Time, Labor, and Social Domination: A Reinterpretation of Marx's Critical Theory. Cambridge: Cambridge University Press.
Répássy, Helga. 1987. Társadalmi-Gazdasági Változások a Nöi Munkaeröpiacra a Mezögazdaságban [Social, Economic Changes in the Women's Labor Market in Agriculture]. Társadalomkutatás 4: 121–131.
Rév, István. 1987. The Advantages of Being Atomized: How Hungarian Peasants Coped with Collectivization. Dissent: 335–350.
Rofel, Lisa. 1992. Rethinking Modernity: Space and Factory Discipline in China. Cultural Anthropology 7 (1): 93–114.
Róna-Tas, Ákos. 1989. Everyday Power and the Second Economy in Hungary: Large Consequences of Small Power. Comparative Study of Social Transformations Paper no. 28. Ann Arbor: Comparative Study of Social Transformations.
———. 1990. The Second Economy in Hungary: The Social Origins of the End of State Socialism. Ph.D. diss., University of Michigan.
———. 1994. The First Shall Be Last? Entrepreneurship and Communist Cadres in the Transition from Socialism. American Journal of Sociology 100 (1): 40–69.
———. In press. The Second Economy as a Subversive Force: The Erosion of Party Power in Hungary. *In* The Waning of the Communist State: Economic Origins of Political Decline in China and Hungary, edited by Andrew G. Walder. Berkeley: University of California Press.
———. N.d. The Small Transformation: The Social History of the Private Sector in Hungary. Manuscript.
Roseberry, William. 1983. Coffee and Capitalism in the Venezuelan Andes. Austin: University of Texas Press.
Rupp, Kalman. 1983. Entrepreneurs in Red: Structure and Organizational Innovation in the Centrally Planned Economy. Albany: State University of New York Press.

Sabel, Charles, and David Stark. 1982. Planning, Politics, and Shop-Floor Power: Hidden Forms of Bargaining in Soviet-Imposed State-Socialist Societies. Politics and Society 11 (4): 439–475.
Sampson, Steve. 1984. Rumors in Socialist Romania. Survey 28: 142–164.
———. 1985–1986. The Informal Sector in Eastern Europe. Telos 66: 44–66.
Samuelson, P. A. 1971. Understanding the Marxian Notion of Exploitation: A Summary of the So-Called Transformation Problem between Marxian Values and Competitive Process. Journal of Economic Literature 9 (2): 399–431.
Sanderson, Susan R. Walsh. 1984. Land Reform in Mexico, 1916–1980. New York: Academic Press.
Sándor, Pál. 1955. A XIX. Századvégi Agrárválság Magyarországon [The Agrarian Crisis in Hungary at the End of the Nineteenth Century]. A Magyar Tudományos Akadémia Társadalmi-Történeti Tudományok Osztályának Közleményei: 31–81.
Sándor, Pál, György Szabad, and Antal Vörös. 1952. Parasztmozgalmak a Habsburg Önkényuralom Idején, 1849–1867 [Peasant Movements during the Period of Habsburg Absolutism, 1849–1867]. Budapest: Müvelt Nép Könyvkiadó.
Sárkány, Mihály. 1978. A Gazdaság Átalakulása [The Transformation of the Economy]. *In* Varsány: Tanulmányok egy Észak-Magyarországi Falu Társadalomnéprjazához [Varsány: Studies toward the Social Ethnography of a North Hungarian Village], edited by Tibor Bodrogi, pp. 63–150. Budapest: Akadémiai Kiadó.
Schneider, M., and V. Juhász, eds. 1937. Magyar Városok és Vármegyék Monográfiája: Fejér Vármegye [Monograph of Hungarian Cities and Counties: Fejér County]. Budapest: Cserépfalvi Könyvkiadó.
Sewell, William Jr. 1980. Work and Revolution in France: The Language of Labor from the Old Regime to 1848. Cambridge: Cambridge University Press.
Sider, Gerald. 1986. Culture and Class in Anthropology and History: A Newfoundland Illustration. Cambridge: Cambridge University Press.
Simonyi, Ágnes. 1983. A Teljesítménynövelés Feltételei a Munkaszervezetben [The Conditions for Increasing Output in the Organization of Work]. Budapest: Munkaügyi Kutató Intézet.
Solt, Katalin, and Endre Szigeti. 1990. A Magyar Szociáldemokraták Közgazdasági Nézetei (1890–1945) [The Economic Views

of Hungarian Social Democrats (1890–1945)]. Budapest: Akadémiai Kiadó.
Somogyi, Imre. 1942. Kertmagyarország Felé [Toward a Garden Hungary]. Budapest: Magyar Élet Kiadása.
Sozan, Michael. 1985. A Határ Két Oldalán [On Both Sides of the Border]. Paris: Irodalmi Ujság Sorozata.
———. 1986. The Jews of Aba. East European Quarterly 20 (4): 179–197.
Sraffa, P. 1960. The Production of Commodities by Means of Commodities. Cambridge: Cambridge University Press.
Staniszkis, Jadwiga. 1991. "Political Capitalism" in Poland. East European Politics and Societies 5 (1): 127–141.
Stansell, Christine. 1982. City of Women: Sex and Class in New York, 1789–1860. Urbana: University of Illinois Press.
Stark, David. 1980. Class Struggle and the Transformation of the Labor Process: A Relational Approach. Theory and Society 9 (1): 89–130.
———. 1986. Rethinking Internal Labor Markets: New Insights from a Comparative Perspective. American Sociological Review 51: 492–504.
———. 1989. Coexisting Organizational Forms in Hungary's Emerging Mixed Economy. In Remaking the Economic Institutions of Socialism: China and Eastern Europe, edited by Victor Nee and David Stark, pp. 137–168. Stanford: Stanford University Press.
———. 1990. Privatization in Hungary: From Plan to Market or from Plan to Clan. East European Politics and Societies 4 (3): 351–392.
Stark, David, and Victor Nee. 1989. Toward an Institutional Analysis of State Socialism. In Remaking the Economic Institutions of Socialism: China and Eastern Europe, edited by Victor Nee and David Stark, pp. 1–31. Stanford: Stanford University Press.
Stewart, Michael. 1987. "Igaz Beszéd"—avagy Miért Énekelnek az Oláh Cigányok? ["True Speech," or Why Do Gypsies Sing?]. Valóság 1: 49–64.
———. 1990. Gypsies, Work, and Civil Society. In Market Economy and Civil Society in Hungary, edited by Chris Hann, pp. 140–162. Portland, OR: Frank Cass.
———. 1993. Gypsies, the Work Ethic, and Hungarian Socialism. In Socialism: Ideals, Ideologies, and Local Practice, edited by C. M. Hann, pp. 187–203. London: Routledge.
Stoler, Ann. 1985. Capitalism and Confrontation in Sumatra's Plantation Belt, 1870–1979. New Haven: Yale University Press.

———. 1989. Making Empire Respectable: The Politics of Race and Sexual Morality in Twentieth Century Colonial Cultures. American Ethnologist 16 (4): 634–660.
Swain, Nigel. 1985. Collective Farms Which Work? Cambridge: Cambridge University Press.
———. 1992. Hungary: The Rise and Fall of Feasible Socialism. London: Verso.
Sweezy, Paul, and Charles Bettelheim. 1971. On the Transition to Socialism. New York: Monthly Review Press.
Szabó, Károly, and László Virágh. 1984. A Begyüjtés "Klasszikus" Formája Magyarországon (1950–1953) [The "Classical" Form of Procurement in Hungary (1950–1953)]. Medvetánc 2–3: 159–179.
Szabó, László. 1968. Munkaszervezet és Termelékenység a Magyar Parasztságnál a XIX–XX. Században [Work Organization and Productivity among the Hungarian Peasantry in the Nineteenth-Twentieth Centuries]. Szolnok: A Damjanich János Múzeum Közleményei.
Szabó, Zoltán. 1937. A Tardi Helyzet [The Situation in Tard]. Budapest: Cserépfalvi Könyvkiadó.
Széchenyi István. 1830. Hitel [Credit]. Pest: Petrózai Trattner J. M. és Károlyi István Könyvnyomtató-Intézetében. Facsimile edition: 1991, Budapest: Közgazdasági és Jogi Könyvkiadó.
Szekeres, Endre. 1936. A Földhaszonbérlet Reformja [Reform of Land Leases]. Magyar Szemle 28: 15–23.
Szelényi, Iván. 1982. The Intelligentsia in the Class Structure of State Socialist Societies. In Marxist Inquiries, edited by M. Burawoy and T. Skocpol, pp. 287–326. Chicago: University of Chicago Press.
Szelényi, Iván, in collaboration with Robert Manchin, Pál Juhász, Bálint Magyar, and Bill Martin. 1988. Socialist Entrepreneurs: Embourgeoisement in Rural Hungary. Madison: University of Wisconsin Press.
Szelényi, Iván, and Szonja Szelényi. 1990. Az Elit Cirkulációja? [The Circulation of the Elite?]. Kritika 9: 8–10.
Szuhay, Miklós. 1961. Termelésfejlesztés és Vetésterület-Korlátozás mint az 1929–1933-as Válságból való Kilábalás Eszközei [Development of Production and Restriction on Area Sown as Devices for Recovering from the 1929–1933 Crisis]. Agrártörténeti Szemle 3: 413–425.
———. 1962. Az Állami Beavatkozási Politika Szerepe a Mezögazdasági Terményértékesítésben az 1929–33-az Gazdasági Válság Idöszakában [The Political Role of State Intervention in

the Sale of Agricultural Produce during the Period of the Economic Crisis of 1929–1933]. Agrártörténeti Szemle 4: 157–178.
Taussig, Michael. 1987. Shamanism, Colonialism, and the Wild Man. Chicago: University of Chicago Press.
Thompson, E. P. 1963. The Making of the English Working Class. New York: Vintage Books.
———. 1967. Time, Work-Discipline, and Industrial Capitalism. Past and Present 38: 56–97.
Tilly, Louise A., and Joan W. Scott. 1978. Women, Work, and Family. New York: Holt, Rinehart and Winston.
Tinbergen, J. 1961. Do Communist and Free Economies Show a Converging Pattern? Soviet Studies 12 (4): 333–341.
Tribe, Keith. 1978. Land, Labour, and Economic Discourse. London: Routledge and Kegan Paul.
Trouillot, Michel-Rolph. 1989. Discourses of Rule and the Acknowledgment of the Peasantry in Dominica, W.I., 1838–1928. American Ethnologist 16 (4): 704–718.
Trotsky, Leon. 1937. The Revolution Betrayed: What Is the Soviet Union and Where Is It Going? Translated by Max Eastman. London: Faber and Faber.
Vági, Gábor. 1981. A Közelmúlt és Értékelése egy Magyar Faluban [The Recent Past and Its Appraisal in a Hungarian Village]. *In* Folyamatos Jelen: Fiatal Szociográfusok Antológiája [Continuous Present: An Anthology of the Works of Young Sociographs], edited by György Berkovits and István Lázár, pp. 197–221. Budapest: Szépirodalmi Könyvkiadó.
Vajda, Mihály. 1983. Lukács and Husserl. *In* Lukács Reappraised, edited by Ágnes Heller, pp. 107–124. New York: Columbia University Press.
———. 1988. East-Central European Perspectives. *In* Civil Society and the State, edited by John Keane, pp. 333–360. London: Verso.
van Onselen, Charles. 1976. Chibaro: African Mine Labour in Southern Rhodesia 1900–1933. London: Pluto Press.
———. 1982. Studies in the Social and Economic History of Witwatersrand, 1886–1914. New York: Longman.
Vasary, Ildiko. 1987. Beyond the Plan: Social Change in a Hungarian Village. Boulder: Westview Press.
Verdery, Katherine. 1983. Transylvanian Villagers: Three Centuries of Political, Economic, and Ethnic Change. Los Angeles: University of California Press.

———. 1991. National Ideology under Socialism: Identity and Cultural Politics in Ceausescu's Romania. Berkeley: University of California Press.
Virág, István. 1947. Új Honfoglalása [New Conquest]. Székesfehérvár.
Volgyes, Ivan. 1985. Blue-Collar Working Women and Poverty in Hungary. In Women, State, and Party in Eastern Europe, edited by Sharon Wolchik and Alfred Meyer, pp. 221–233. Durham: Duke University Press.
Volgyes, Ivan, and Nancy Volgyes. 1977. The Liberated Female: Life, Work, and Sex in Socialist Hungary. Boulder, CO: Westview Press.
Vörös, Antal. 1966. A Paraszti Termelö Munka és Életforma Jellegének Változásai a Dunántúlon, 1850–1914 [Changes in the Character of Peasant Productive Labor and Life-Style in Transdanubia, 1850–1914]. Történelmi Szemle 9 (2): 162–186.
Wädekin, Karl-Eugen. 1982. Agrarian Policies in Communist Europe: A Critical Introduction. The Hague: Allanheld, Osmun.
Walder, Andrew. 1986. Communist Neo-Traditionalism: Work and Authority in Chinese Industry. Berkeley: University of California Press.
Williams, Raymond. 1977. Marxism and Literature. Oxford: Oxford University Press.
Wolchik, Sharon, and Alfred Meyer, eds. 1985. Women, State, and Party in Eastern Europe. Durham: Duke University Press.
Wood, Ellen Meiksins. 1991. The Pristine Culture of Capitalism: A Historical Essay on Old Regimes and Modern States. London: Verso.
Zavada, Pál. 1984. Teljes Erövel: Agrárpolitika, 1949–1953 [With Full Force: Agrarian Policies, 1949–1953]. Medvetánc 2–3: 137–158.
———. 1986. Kulákprés: Dokumentumok és Kommentárok egy Parasztgazdaság Történetéhez [Kulak Press: Documents and Commentaries toward a History of Peasant Economy]. Budapest: Müvelödéskutató Intézet.

INDEX

A.B.C. of Communism, 235–6
Agrarian Theses (Az MSZMP agrárpolitikájának tézisei), 168–76; emphasis of on local conditions, 173–4; and imposition of market restrictions, 172–3; and role of agrarian specialists, 171–2
agricultural production, 2, 6, 13–4, 17; and agrarian specialists, 140–1, 171–2, 202–3, 207, 248; alienation of peasants from, 159–60, 167, 198; changed by Hungarian socialism, 22–3, 167, 171, 226, 231, 331–3, 337; changes in character of, 226; and changes in labor force, 201, 204–7, 277, 338; correlated with religious calendar, 323–4; and debates over form and rate of collectivization, 174–6; and debates over land reform, 115–8; distinction between manual and professional workers in, 172, 194, 205–7, 224, 229, 248; food shortages, 216–7; and household-based production, 174, 214–6, 217, 305, 308; impact of antikulak policies on, 152; importance of, 125–6, 129, 134, 175, 198; involvement of tradesmen in, 56, 57; and issues of local control, 114, 124, 125–6, 134–5, 173, 174, 193, 194, 213, 217, 224–5, 251–4; mechanization of, 200–7, 231, 232, 277, 338, 350; and new landowners, 109, 130; in nineteenth-century Hungary, 104–7; and persistence of private production, 170, 173, 175, 176, 179, 182, 183, 184; policy contradictions in, 153–4, 177, 277–8, 310; and problems with labor in cooperatives, 195–7, 198–9; and programs to improve skills and knowledge in, 52, 172, 202–3, 225; quotas in, 109–10, 134, 141, 149, 152, 153, 154, 158, 168, 172; role of state in, 90, 109–10, 129, 131–3, 134, 135–6, 154, 157–8, 168–76, 200–1, 217, 231; scholarly studies of, 3–4, 24, 25; sharecropping to boost levels of, 198–200; use of terrorist tactics to promote, 154–5; war damage to, 113–4. See also collectivization; cooperatives; second economy; sharecropping
arató (harvesters), 58. See also harvesters

Bartha, Andor: renter of Jakabszállás, 29, 96; treatment of manorial workers, 96–8
Bibó, István, 331
Biernacki, Richard, 9n. 6, 10
Bukharin, N., 235, 236, 237

capitalism: contrasted with socialism, 17, 234; distinctive attributes of, 1, 2, 10, 11, 14, 274; forms of domination in, 12; historical analyses of, 15–16; rise of in Hungary, 103–8; as sociohistorical process, 15, 357
centeredness. See movement
China: influence of on economic policy, 176
class: in agrarian communities, 27, 139, 206, 231–2, 294; relations shaped by labor exchange, 51
collectivization, 1, 2, 109, 231; class differences in impact of, 25, 26–7, 124, 136; consumer cooperatives as transitional phase in, 173–5; debates over form and rate of, 174–6; and focus on male head of household, 180–1, 182–3, 199; impact of on character of work, 6, 25, 173, 183,

201–2, 222, 225; as industrialization policy, 160, 212; and mechanization of production, 200–7, 231, 232, 277–8; need for reassessment of privations of final phase of, 191–2; peasants' resistance to, 138, 159, 162, 164, 168, 177–8, 179–80, 180–1, 189, 191–2, 195, 197, 231, 338; as political weapon, 152–3; and professionalization of managers, 172, 206–7, 224, 248; propaganda promoting, 179–87; and relation between property and production, 109, 110, 196, 206; and scientific socialism, 186–7, 219–20; and state's distrust of landed peasantry, 136–7, 175; as unique to socialism, 4; use of family imagery in, 180–2, 185. *See also* agricultural production; Communist party; cooperatives; land reform; socialism

commodification, 7–16, 167, 219, 359; culture of, 33–4; defined, 11; gender effects in, 327; as historical process, 13–4, 332–3; in interwar period, 15; promoted by socialism, 1, 5, 7, 274; and role of markets, 6; and shift in character of second economy, 312–3, 342; and social domination, 12; in village life, 16, 170, 329, 342. *See also* capitalism; commodity fetishism; markets; second economy; socialism; transition from socialism

commodity fetishism, 5, 11, 12; changes in over time, 13–14, 15; Marx's theory of, 7–10; and wage labor, 9–10. *See also* capitalism; commodification; markets; socialism

Communist party, 161; agrarian policies of, 114–5, 118–9, 136–8, 141–3, 152–3, 154, 157–8, 168–76, 207–17; Central Committee of, 174, 213, 216; cultural hegemony of, 282–3, 289n. 13; efforts of to separate property and production, 137, 138, 152–3, 159, 165, 173, 196, 226–7, 337; fears concerning in Sárosd, 65, 94–5; ideological commitment of to supremacy of industry and industrial worker, 139, 180, 204, 214, 216–7, 292; investment policies of, 169, 198, 200–1, 218; and overlap between language and life, 237; policy of toward agrarian specialists, 140–1, 171–2, 223–4; position of on land reform, 116, 117, 118–9, 124; reversal in position of regarding private production, 311, 345; state takeover by, 135–6.

See also planning; socialism; transition from socialism

consciousness: changes in to accomplish agricultural reform, 169–70, 197; transformation of through handbooks and dictionaries of socialist principles, 237

consumer cooperatives. *See* collectivization

control, ideology of, 35–6, 199, 223; and censure of movement, 36; and definition of humanity, 84, 101; gender bias in, 43, 199; importance of in postwar years, 130–1; moral implications of, 73; and negative characterization of migrant laborers, 71–2; new interpretation of, 137; reasserted in revolution of 1956, 162, 164; role in determining social position, 40–2, 130; and sharecropping, 53, 199; and social identity, 60–1. *See also* diligence; labor; "possessing activity"; social identity

cooperatives, 116, 117n. 8; appeal of, 144–5, 164, 176–7, 201–2, 350–1; changes in state policy regarding, 169–76, 203–4, 213–4, 231–2, 250–1; changes in workforce of, 201, 204–7, 223, 224, 229, 231, 277, 278; changes in workweek of, 322–3; decision-making processes of, 193–4, 213, 229, 230, 247–8, 249–50, 251, 253–4; "defeminization" of labor in, 184–5, 282n; diversification of into nonagricultural activities, 213–4, 251; equipment shortages on, 200; hierarchies in, 177–9, 180–1, 193–4, 205–6, 225–6, 229, 230, 248; and household-based production, 174, 196, 198, 213, 214–5, 232; impact of mechanization on value of work in, 205–7, 277; likened to large family, 180–2, 185, 187, 195, 226n; members' wages linked to position in organizational hierarchy of, 205, 219, 221, 223, 251, 258–9; membership criteria, 146, 170; and option to return to private farming, 145–6, 157–8, 188, 350–1; planning in, 193, 247–70; prejudices against, 150, 152, 170, 179; pressures to join, 179–80, 181–2, 189–90; as refuge for new landowners, 148–9, 158–9; and rent payments to members, 177, 178n; repudiated by landowning peasants, 164, 170, 177–8, 188, 191–2; role of managers and planners in, 172, 179, 223–5, 229, 230, 232, 248–50, 251, 253–4,

258–9; in Sárosd, 144, 188–98, 275–6; seasonal workers employed by, 199; sharecropping contracts in, 198–99, 231; shift to monetary payment system in, 218–9, 221–2, 222–3, 232; social relations of production in, 202–4, 206–7, 223, 224–5, 229, 230, 232, 277, 278; subordination of women in, 183–5, 221; as test sites for industrial reforms, 212–3; transitional role of, 169, 201, 207, 209, 212; variety in organizational forms of, 176, 188, 212, 213–4, 225; work ethic in, 2, 178, 188, 195, 198–9, 205n, 216, 222, 261, 303–4, 315–6; work units as form of remuneration in, 197–8, 219, 221–2, 336; *See also* agricultural production; collectivization; Dózsa Farm; Harmony Farm; Kalocsai cooperative; Progress Farm; Rákóczi Farm; Sárosd

craftsmen: *See* tradesmen

cseléd (manorial workers), 73, 74–5. *See also* manorial workers

Csillag, 37, 73, 90, 97; absorbed into state farm, 146

day laborers, 40; category defined, 62–3; and control over labor among, 43; familial division of labor among, 63; forms of rebellion among, 70; and membership of in Rákóczi Farm, 188; and movement in work process of, 63; party affiliation of, 63–4; and "people fairs," 63n; use of cash among, 63

dictionaries: role of in socialist state, 235, 236, 237–8

diligence: class differences in, 100, 109, 131, 149, 150, 177, 294; cyclical quality of, 323; among harvesters, 60; in household-based production, 215, 291, 314, 325–6, 327, 340; importance of, 53, 130–1, 166, 197, 218, 239, 266, 268, 288, 292, 313n, 314, 316, 321, 340; among migrant laborers, 70; not associated with cooperative production, 195, 196–7, 205, 263, 303–4, 315–6; role of in socialism, 197, 218; and sharecropping, 53, 198–9; and social identity, 36, 101–2, 130–1, 150, 314, 316; and social value, 108. *See also* control, ideology of; labor; social identity

division of labor: in cooperatives, 172, 183–5; among day laborers, 63; among landowners, 47–8; among manorial workers, 82–3, 84; role of gender in, 183–5, 299, 325, 339–40

Djilas, Milovan, 16, 22

dolog (work), 359; etymology of term, 9, 44, 320; relation to work-unit payment systems, 222

Donáth, Ferenc, 118, 142, 162, 174–5, 176, 193, 195, 196

Dózsa Farm, 188; retention of private lands in, 188–9

Éber, Ernö, 116

economism, 2, 5, 6, 21, 246, 270, 336; associated with planning, 270, 271, 339; effects of in socialism, 20

Elek, István, 29, 90, 97, 111

elites: advantages and privileges accruing to as managers and planners, 213, 223, 233–4, 251, 256, 258, 259, 262–5; composition of in Sárosd, 40, 53–6; as disproportionately represented in cooperative management, 194–5; ebbing of control of cooperative production, 206–7; as managers in cooperatives, 205, 206–7, 248, 279; postwar control of equipment and livestock, 129, 130; role of in acceptance of cooperatives, 179–81, 189, 194; role of in socialism, 22, 244–5, 336; social clubs among, in Sárosd, 57–8; social ties among, 265–6; technocratic, 167, 207, 223, 229, 230, 231–2, 248, 283. *See also* Communist party; inequality; socialism; technocracy

enterprise democracy, 280, 281–2, 283

Erdei, Ferenc, 44–5, 70, 85, 102

Eszterházy, Count, 29, 40, 48, 54, 63, 80, 81, 92, 95; dominance of in village life, 275–6

exchange, 44, 320; among landowners, 49, 50–1; postwar conditions of, 130

exploitation, 22, 53, 219

falusi (villager), 40

Fehér, Ferenc, 22, 237

Fejér County, 37, 80, 94, 254; described, 28; differences in culture of villages in, 345; effects of land reform in, 124; migrant laborers in, 67; postwar agricultural production in, 126; and Soviet Republic, 89–90

feudalism, 3, 10, 13; source of value in, contrasted with capitalism, 107–8; and

transition to capitalism in Hungary, 103–8
food shortages, 216–7

gazda, 43, 226; as applied to landowners, 54; as applied to overseers, 79, 80, 86
Gazdakör: *See* Peasant Circle
Gerö, Ernö, 160
Goven, Joanna, 185
Great Plain, 28, 54, 133, 143; and land reform, 119, 123, 125, 128
Gypsies. *See* Roma

handbooks: as source of guidance in socialism, 235–6
Harmony Farm, 192, 200, 230, 251, 255, 350; changes in character of work at, 279–80, 283–4; committees composing elected directorate of, 280–2; decision-making processes at, 280, 281–2; distribution of lands for private production, 281; and diversification into nonagricultural activities, 282; and economies of scale, 277–8, 286; efforts to prevent abuses by workers, 216; isolation of members by social group at, 192–3, 278; leadership positions filled by village elites, 194–5; membership, 276; operative management of, 279; organizational structure of, 193, 194, 225, 230, 278–9, 286; physical layout described, 275; planning goals of, 257; spatial representation of expanding managerial power at, 282–4
harvesters, 40; anguish over poverty among, 61–2; category defined, 58–9; and control over labor among, 59–60; diligence of, 60; as employees of landowners, 48–9; familial division of labor among, 59; forms of rebellion among, 70–1; membership of in Rákóczi Farm, 188
Heller, Ágnes, 22, 237
Heller, András, 79, 80, 82, 83
hónapos (migrant laborers), 66, 67. *See also* migrant laborers
household-based production (háztáji), 174, 188, 196, 198, 272, 336; and centrality of kitchen in domestic space, 290, 298–9, 302, 303; as challenge to legitimacy of socialist state, 216–7; enlarged role of, 214–6; and presence of animals, garden plots, and outbuildings, 293–4, 296, 298–9, 301; term defined, 214–5. *See also* agricultural production; cooperatives; markets; second economy; sharecropping; transition from socialism
Hungarian Communist Party (Magyar Kommunista Párt), 161. *See also* Communist party
Hungarian Land Credit Institute (Magyar Földhitelintézet), 92
Hungarian Workers' Party (Magyar Dolgozók Pártja), 135n

Ihrig, Károly, 132
Illyés, Gyula, 27, 77n. 32, 84, 85
Independence Front (Függetlenségi Front), 135
Independent Smallholders (Független Kisgazda Párt), 135n
individualism: as capitalist trait, 1, 2; promoted by socialism, 5, 20, 234, 246
industrial workers: improved conditions of, 168; as "instructors of the people," 139, 179; return to agriculture of, 204; as superior in Communist ideology, 139, 182, 292
inequality: between agrarian and industrial workers, 139, 182, 204–5; in capitalism, 12; generated by unequal access to state monies, 218, 251–2; between men and women, 184–6, 221–2, 251, 299, 325; between workers and managers, 179, 180–1, 195, 223, 229, 248, 258–9
intelligentsia: emergence of as new class, 244–5; in Sárosd, 54; as village outsiders, 57–8. *See also* elites

Jakabszállás, 29, 37, 67, 73, 92, 127; as a center of production for Harmony Farm, 192, 275, 278; direct involvement of renter in, 81; living conditions of migrant laborers at, 66–7; not affected by Soviet Republic, 90–1; physical layout described, 276; strike at, 70–1; transformed into cooperative, 144, 146, 188; treatment of workers at, 96–7. *See also* Harmony Farm; Progress Farm
Jews: prewar treatment of, 106; property of expropriated, 115, 116; in Sárosd, 57, 58, 111, 113; treatment of in war years, 111, 113n;
Jowitt, Ken, 18, 20
Juhász, Pál, 195

Jurcsek, Béla, 130

Kalocsai cooperative, 188
Kemény, István, 267
Kenedi, János, 259, 266, 348
Kerbolt, László, 33, 37n. 1, 47, 64, 65, 79, 87–8, 98
kinti (outsider), 40
Konrád, György, 21
Kornai, János, 239, 243, 245, 255
Kovács, Imre, 71, 83
kulaks, 124, 136, 155; abuse in designation as, 142–3, 144; impact on agricultural production of policies against, 152, 157; special quotas for, 141, 142; as state enemies, 136–7, 138, 139, 170–1. *See also* landowners

labor: age differences in attitudes toward, 318–9, 321–3, 332; agricultural, in Hungary, 13–14, 24, 25, 173; as category unique to capitalism, 11; cyclical quality of, 323; differences in motivation for, 328–9; differences in terms used to describe, 319–20; differential value of in cooperatives, 205–7, 248, 282; distinction made between physical and nonphysical, 224, 283, 286, 315, 321, 349; and divergent career choice patterns, 205–6; gender differences in attitudes toward work, 324–7, 328; importance of control over, 35–6, 40–2, 43, 45, 59–60, 84, 100–3, 147, 162, 177, 182, 223, 316, 318–9, 330, 336, 340; meaning of in socialism, 110, 173, 197, 203–4, 218, 223, 226, 314, 349; mobility of restricted by state, 147; as object, 8–9, 329, 330, 342; postwar identification with time and/or money, 1, 2, 14–5, 167, 173, 218, 222, 223, 232, 261, 307, 312, 313, 325, 328, 342, 349; and separation of family members, 59, 60–1, 197–8, 299–300, 324–5; social value of, 3, 9, 12, 14, 166, 288, 294, 314, 323, 329–30, 359; temporal aspects of workday, 167, 322, 323; value of education and skill in, 223–4, 229, 315n, 319, 338. *See also* commodification; commodity fetishism; control, ideology of; diligence; movement; "possessing activity"; social identity; wage labor

labor recruitment, 133; role of state in, 133–4
labor theory of value, 8n. 4, 166
labor unrest: in prewar Hungary, 107; in Sárosd, in interwar period, 96–8
land: and class differences in meaning of ownership of, 61n, 89, 109, 115, 149–50; importance of ownership of for social identity, 40, 41, 42–43, 177, 182, 226, 231, 294, 357
landowners, 40, 226; attitudes of toward new landowners, 130, 131, 149–50; category defined, 46–7; as employers of harvesters, 48–9; familial division of labor among, 47–8; multiple terms designating, 54; persistence of commitment to private production among, 170, 177, 189, 190, 191–2; as renters and sharecroppers, 52–3; as "sober calculators," 186, 187; types of institutional support for, 52, 54; use of cash among, 51–2; use of machinery among, 49; as village elites, 53–4. *See also* kulaks
land reform, 25; administrative aspects of, 124–5; and debates over property rights vs. economic rationality, 115–7; and ethnic and regional differences in property distribution, 122–4; expropriation of nonland materials during, 121; as favoring agrarian poor, 121–2; and importance of agricultural production, 125–6, 129; lack of formal debate over, 119; and Land Claims Committee in Sárosd, 127–8; and obstacles facing new landowners, 129–30; politics of in interwar period, 91–2; postwar struggles over at local level, 114, 124, 125–6; reactions of middle peasantry to, 117–8; role of cultural and racial politics in, 116; in Sárosd during Soviet Republic, 89–90, after WWII, 126–7. *See also* collectivization; Nagyatádi Land Reform; new landowners
leisure, 283, 337; and aging, 318; space for in villagers' homes, 291, 295, 298, 299, 303; villagers' attitudes toward, 288, 300, 313, 314, 316–8, 330
Lenin, V., 180, 241; and interest in Taylorism, 241
levente (compulsory military training), 48; paramilitary groups, 5
Lukács, György, 167

machinery: absence of in agriculture, 200–1; changes in status of occupations associated with, 205–7; expropriation of in land reform, 121, 129; use of in Sárosd, 49, 64, 311–2; villagers' resistance to use of, 201–2. *See also* agricultural production; collectivization

Magyar, Bálint, 162

manorial estates: community life on, 78; disciplinary practices on, 84–5; holdings of in Sárosd, 38–9; and legal invulnerability of owners and renters, 97–8; and legal recourse of owners and renters against workers, 98–9; migrant laborers on, 66; occupational hierarchy on, 77; organizational hierarchy on, 80–1; as outside village life, 73; social categories of workers on, 40; social hierarchy on, 76–7; structural similarities with early collectives, 25–6; studies of life on, 27; women's work on, 82–3. *See also* Bartha, Andor; Csillag; Elek, István; Eszterházy, Count; Fejér County; Jakabszállás; manorial workers; Pusztasárosd; Sárosd

manorial workers, 40; attitude toward theft among, 86; as beneficiaries of interwar land reform, 93; category defined, 73–6; as comparison group with subsistence peasants, 26, 28; control over labor among, 84, 85–6; daily work routines of, 81–3, 84; dehumanization of, 60, 84–5, 101, 102; diligence of, 83; living conditions of, 78–80; market isolation of, 83–4; marriage patterns among, 77–8; membership of in Kalocsai cooperative, 188; movement as characteristic of work process of, 73; salaries of, 7, 83; and sharecropping, 82; social identity of, 84, 85–6; state's efforts to encourage loyalty among, 87–8; strategic work slowdowns among, 85–6; theft among, 98–9; "uncenteredness" of, 73–4. *See also* control, ideology of; manorial estates; movement; "possessing activity"

markets, 5, 6, 264; class differences in exposure to, 102; and commodity fetishism, 8–9, 315; isolation of manorial workers from, 83–4; meaning of value in, 348; regulation of in agriculture, 172–3, 209, 257n; role in defining villagers' social positions, 36, 300; state's efforts to eliminate, 110, 146, 187; villagers' resistance to, 36, 102, 108, 315, 337, 342n. *See also* capitalism; commodification; commodity fetishism; feudalism; socialism; traditionalism; transition from socialism

Márkus, György, 22, 237

Marx, Karl, 7, 180, 235, 358; and commodity fetishism, 7–12

Marxism-Leninism, 233, 238, 358; and dialectical and historical materialism, 236–7; scientific principles of, 233, 235, 236–7, 255, 338–9

meaning, 359, 360; theories of, in anthropology, 15n. 14

melléküzemág (subsidiary industrial units of cooperatives), 213

merchants, 54; as village outsiders, 40, 56–8

middle peasantry: *See* landowners

migrant laborers, 40; category defined, 66; and coercive character of work, 72; control over labor among, 71–2; dehumanization of, 71; forms of rebellion among, 70–1; living conditions of, 66–7; movement as characteristic of work process of, 66; religious fervor among, 71; temporal dimensions of labor contracts of, 323n; as village outsiders, 66. *See also* manorial estates

Ministry of Agricultural Deliveries, 152

Ministry of the Interior: and state intervention in rural life, 4–5

morality: attitudes toward, 2–3, 5–6; defined by work, 14; identified with stability, 87, 337; replaced by scientific rationalism, 235; as sociohistorical product, 16; state supervision of, 4–5. *See also* control, ideology of; diligence; labor; movement; "possessing activity"

movement: as antithesis of centeredness, 73; and control over labor, 102, 337; extent of differences in by social group, 60, 61, 63, 66; gender effects in, 102; as index of morality, 87; and social identity, 36; and social position, 73–4, 102. *See also* control, ideology of; "possessing activity"; social identity; and specific social groups [e.g., day laborers]

munkás (worker), 44n. 7

Nagy, Imre, 110, 160, 165, 167; and changes in agricultural policy, 157–8

Index

Nagyatádi Land Reform, 40, 91; and division of land in Sárosd, 92–3; and political agitation in Sárosd, 94–8
National Council Supervising Land Reform (Országos Földbirtokrendezö Tanács), 124
National Economic Work Commission Office (Országos Gazdasági Munkaközvetitö Iroda), 133
National Land Office (Országos Földhivatal), 125
National Peasant Party (Nemzeti Parasztpárt), 135n; position of on land reform, 117, 118, 124
National Planning Office, 169
Nee, Victor, 18
Néppárt (People's Party), 56
Népszabadság, 175
new landowners: alliance of state with, 136n; difficulties facing, 129, 130, 131, 171; movement of to cooperatives, 131, 148–9; prejudices against, 131, 149–50; support for private property among, 138
nincstelen ("have nothings"), 37, 63

Oláh, György, 91
Orbán, Sándor, 122

paraszt (landowners), 46, 54. *See also* landowner
Peasant Circle, 52, 54, 57
planning: and avoidance of consideration of political consequences in, 244, 245, 271; "bargaining" in, 255, 256; and creation of new vocabulary, 257, 258; and economism, 270, 271, 339; as essence of socialism, 234, 243, 245, 254–5, 339; and falsification of plan figures, 242–3, 249, 255; and incremental approach to economic reform, 239–40; influence of noneconomic factors in, 252, 253–4, 256–7, 259, 262–6, 339; and necessity of fulfilling goals, 242–3, 249, 256, 338; political character of in socialist state, 233–4, 256–7; as process of quantification, 219, 240–1, 242, 243–4, 245, 254, 255; as reification of political practice, 243–4, 245, 254–5; role of county and state agencies in, 250–4, 255; role of second economy in, 309; as scientific activity, 187, 238, 254–5, 271, 338. *See also* socialism; transition from socialism

politics: defined, 233
Popular Democratic Republic, 90
"possessing activity": age and gender differences in, 318, 329–30; and alienation of manorial workers, 86; challenged by new importance of money, 313, 318–9, 323; and class differences, 100, 101, 108, 288; gender differences in, 100–1; ideology of strengthened in postwar years, 130–1, 337; and movement, 73; as principle of action, 44–5; relation of to work-unit remuneration, 222, 223; and social identity, 84, 149–50, 318; term defined, 44; threatened by socialist policies, 110. *See also* control, ideology of; diligence; labor; movement
Preobraschensky, E., 235, 236, 237
priests, 54, 263; as enemies of the state, 137; as part of village elites, 54–6
private/public dichotomy, 2, 21, 232, 246–7, 270, 333; age and gender effects in, 274, 339, 340; and character of domestic activity, 287–303; defined by capitalism, 6; political implications of, 247, 331; as reflecting broader oppositions in socialist society, 270, 273–4, 339; rejected by wealthy villagers, 61; 232; role in production, 273–4, 303; and social identity, 328, 333, 339
production: social relations of, 35, 100, 202–4, 275; and workers' consciousness, 19. *See also* agricultural production; capitalism; collectivization; cooperatives; markets; socialism
Progress Farm (Haladás), 158; abandoned, 162, 188; described, 144; founded by Communist party members, 145; membership, 147–8, 153
protection (*protekció*), 262, 265, 266, 286, 355. *See also* elites; socialism
puszta, 28, 37, 57, 88
Puszta/Kispuszta, 29, 297; as a center of production for Harmony Farm, 192, 275, 278, 279
Pusztasárosd, 37, 73

Rákosi, Mátyás, 155, 157, 160, 165
Rákóczi Farm, 188
reform: and decentralized decision-making, 209, 213, 217–8, 224, 250, 252, 257, 264, 266; pattern of, in Hungary, 25

Reform of the Economic Mechanism, 207–17, 239, 257
rents: paid to cooperative members, 177, 178n
revolution of 1956, 110, 338, 355; effects of on agrarian policy, 167–8; and repudiation of cooperatives, 162, 164, 168; rural participation in, 160, 162–5; supporters of designated as class enemies, 170–1n
Roma, 46, 260; effects of state policy toward, 165; excluded from land reform, 122–3; and ideology of control, 131; and movement, 102–3; as state collectors during Stalinist period, 156n; treatment of during war years, 111n; as village outsiders, 58
Roman Catholic Youth League, 46; dual institutional role of, 48, 52; membership, 52, 57
Róna-Tas, Ákos, 180

Sárosd: absence of strong middle class in, 143; actions of Land Claims Committee in, 127–8; archival material concerning, 32, 33; corruption among officials in, 261–8; demographics of villagers, 37–8; differences in domestic life in, described, 287–303; distribution of lands in, 29, 37–9, 127–8; efforts of Communist party to promote private production in, 311; fieldwork in, described, 30–2; interwar capitalist character of, 36; interwar unemployment in, 64, 65, 96; interwar strikes in, 96, 99; location of, 29; and outsiders, 40; and pressures on villagers to join cooperatives, 189–92; quota abuses in, 155–6; revolution of 1956 in, 161–2, 163; social groups in, described, 29, 40, 46–88; terrorism in during Stalinist period, 156; villagers' anti-intellectual bias, 248; villagers' fears of decollectivization, 351, 352; villagers' lack of interest in political matters, 245, 269–70, 354–5; villagers' reactions to interwar political agitation, 95–6; villagers' reactions to land reform, 126–7, 128; war damage to, 111–4
second economy, 2, 19, 21, 24, 204, 273, 304, 338; blurring of boundaries in, 304–5; differences between agricultural second economy and private enterprise, 304–5, 341; differences in motivation to participate in, 318–9, 333–4; growth of petty commodity production in, 305–6, 312–3, 331–2; increased official acceptance of, 170, 216, 217, 232, 260, 309, 310; and monetary wages, 222n; pigs as icon for, 313–4; as realm of individual control, 330, 331, 333; role of household-based production in, 215–6, 217, 273, 288; role of in supplementing income, 216, 260, 306–7; role of women in, 185n, 186, 215, 216, 326, 327, 339–40; success of linked to divergent ideologies of labor and identity, 328, 332, 340, 343; term defined, 304; value of labor in, 229, 261, 339, 340
self-interest. See elites; utilitarianism
sharecropping, 51, 52–3, 231, 320–1; as solution to problems with cooperative production, 198–200
Smallholders' Party (Kisgazda Párt): position of on land reform, 117–8, 119, 124, 351
social change: attributes of, 16, 357–8; role of state in, 4, 5, 336, 356
Social Democratic Party, 94, 135; position of on land reform, 116
social identity: basis for in socialism, 110, 165; and consumption, 300; defined by control over labor, 3, 41–2, 53, 101, 110, 318, 325, 336–7; and diligence, 36, 177, 178, 294; among harvesters, 59–60; among manorial workers, 84, 85–6; reproduced in cooperatives, 178–9, 180–1; threatened in cooperatives, 181–3; of women, linked to domestic labor, 325–7. See also control, ideology of; diligence; labor; "possessing activity"
social scientism. See socialism; Marxism-Leninism
socialism: abuse of power and/or privilege in, 224–5, 233–4, 251, 258, 261–7; capitalist attributes of, 5, 17, 20–1, 247, 304, 333, 335, 340, 357; centrality of "one nation, one factory" model in, 180; contrasted with capitalism, 21–2; corruption in, 260–4, 266–8, 304; denial of divergent interests in, 180, 195, 256–7; and dismissal of importance of ideology and politics in, 2, 245, 260–1, 270, 312, 335; and forms of domination, 12–3, 237; and human nature, 13, 209, 218, 234, 246, 259, 260–1, 271, 335, 355, 356; impact of on village life, 2–5, 7, 282–3, 286, 336,

356; and increased importance of rationality and profitability, 257; limitations in analyses of, 16, 335–6; and modernization theories, 17–8; and neotraditionalist theories, 18, 19; and the new institutionalism, 18–9; planned economy in, 5, 19–20, 21, 132, 146, 153–4, 180, 182, 209, 217–8, 234, 238; and privilege, 262–7; private/public dichotomy in, 2, 21, 24, 246, 270, 273–4, 303, 330, 331, 340; as product of contradictions of capitalism, 15–6; as promoting commodity fetishism, 5, 312, 333; rejection of public institutions of, 274–5; and restricted labor mobility, 146, 147; rhetoric of, 181–2, 268–70; role of quantification in, 240–1, 242, 243–4, 245, 249; scientific rationality in, 186–7, 219, 221, 233–8, 241, 254–5; secondary status of peasantry in, 139, 180, 182–3; and social change, 1, 336, 356, 357; state control of culture in, 282–3, 324n; and state employment, 6–7; and state promotion of private economic activity, 308–9, 310; support for goals of, 2, 202, 245, 269, 315, 353; and treatment of women, 183–6; withdrawal of state from domestic sphere, 185–6; and workers' consciousness, 13; unique effects of, in Hungary, 22–3, 356. *See also* capitalism; commodity fetishism; commodification; Communist party; planning; transition from socialism

Socialist Worker's Party of Hungary (Magyarországi Szocialista Munkáspárt), 94

Soviet Republic, 89–91, 126, 161. *See also* land reform; Sárosd

Soviet Union, 157, 160, 312, 347; and crushing of Hungarian revolution, 161–2, 167; and influence of on agricultural policy, 132, 135, 165, 176, 221; presence of Russian troops, 112, 113, 119, 126, 134

specialization: attitudes toward, 223–4. *See also* technocracy

Stark, David, 18

state. *See* Communist party; socialism

state farms, 132, 145, 146, 149, 169; distinguished from cooperatives, 207, 209n; machinery on, 200; prejudices against, 150

summás (migrant laborers), 66, 67. *See also* migrant laborers

surplus, 51, 315
Szabó, Zoltán, 64n. 27, 67–8, 91
Szelényi, Iván, 20–1, 305, 309, 331–2, 342
Székesfehérvár, 28, 45, 46, 280, 287, 289, 295, 331, 332; as industrial center, 292, 293, 296, 349; shepherds' guild feast held in, 324

Tausz, Katalin, 85
technocracy, 248; rise of, 223–31, 283; and skills acquisition, 202–3, 213, 225, 229, 248; use of power in, 224, 225
tradesmen, 54; as village outsiders, 40, 56–8
trade unions, 7, 147, 195n. 11, 355
traditionalism: as hallmark of rural communities, 3–4, 36, 103, 311
Transdanubia, 29, 114, 123, 124
transition from socialism, 341; class bias in literature concerning, 341; defined by reintroduction of market economy, 341–53; defined by shift to multiparty system, 341, 353–5; and impact of price fluctuations in commodities and labor, 342–3, 344, 347; and increasing anomie, 353; and role of elderly in sustaining second economy, 343–4; and villagers' attitudes toward moneymaking, 349; and villagers' belief in objective value, 348; and villagers' conception of state's role in managing economy, 348–9; and villagers' dependence on cooperative farming, 350–2; and villagers' inexperience with investment, 346–7; and villagers' inexperience with risk-taking, 342, 344–6, 349. *See also* second economy; socialism

unemployment: resolved by land reform, 115; in Sárosd, 64, 65, 352
utilitarianism, 20, 209, 265, 336, 339; age effects in, 329–30; as capitalist trait, 1, 2; as motivation for private production, 173; as part of human nature, 173, 260–1, 266, 268, 271, 339; promoted by socialism, 5, 173, 234, 246, 260; self-interest of elites, 264–5; and theft, 260

virilisták (list of taxpayers), 53, 111

wage labor, 52, 63, 165, 219, 299; attitudes toward, 184, 222–3; and commodity fetishism, 9–10; in cooperatives, 203–4, 218–9, 221–2, 232; and inequality in remuneration, 221–2, 223, 251; on manorial estates, 7, 9n. 7; state intervention in, 133; temporal aspects of workday, 322–3. *See also* capitalism; commodification; commodity fetishism; markets; women

Walder, Andrew, 18

Weisz, Widow (Mrs. Berthold Weisz), 29, 81, 89n. 43, 92, 276

women: and class differences in attitudes toward husbands' drinking, 292–3n; and control over labor, 43, 100–1, 199, 299; as fieldworkers, 59; managerial functions of, 48, 184–5, 232, 290; and movement, 102; as property, 190n. 10; responsibilities of on manorial estates, 82–3, 84; restrictions on full participation of in managerial, technical, and professional functions, 185–6; and sanctions against engaging in leisure activities, 317, 326; as scapegoats in patriarchal households, 47n. 9; as sharecroppers, 199, 229n, 320–1; and socialist iconography, 182; as state collectors during Stalinist period, 156n; and stress, 293n. 16, 322, 326; and subservient position in cooperatives, 183–5, 253n, 325; temporal aspects of labor among, 324–5; and wage labor, 165, 184–5, 221–2, 251, 325; workplace safeguards for at Harmony Farm, 280. *See also* division of labor; second economy; socialism; and specific social groups [e.g., day laborers]

work. *See* labor

work ethic. *See* control, ideology of; diligence; labor

work units: and scientific management, 219, 221; use of as remuneration for labor in cooperatives, 219

worker-peasant solidarity, 136, 139, 142–3, 165

workers' councils: and parallels with cooperatives, 164

Year of Transition (late 1947 to late 1948), 135–8

Made in the USA
Middletown, DE
22 November 2015